W9-CAA-017

NEW DIRECTIONS IN ATTRIBUTION RESEARCH

Volume 1

NEW DIRECTIONS IN ATTRIBUTION RESEARCH

Volume 1

EDITED BY

JOHN H. HARVEY
VANDERBILT UNIVERSITY

WILLIAM JOHN ICKES
UNIVERSITY OF WISCONSIN

ROBERT F. KIDD
UNIVERSITY OF WISCONSIN

LAWRENCE ERLBAUM ASSOCIATES, PUBLISHERS
1976 Hillsdale, New Jersey

DISTRIBUTED BY THE HALSTED PRESS DIVISION OF

JOHN WILEY & SONS

New York Toronto London Sydney

Lawrence Erlbaum Associates, Inc., Publishers
62 Maria Drive
Hillsdale, New Jersey 07642

Distributed solely by Halsted Press Division
John Wiley & Sons, Inc., New York

Library of Congress Cataloging in Publication Data
Main entry under title:

New directions in attribution research.

Includes indexes.
1. Attribution (Social psychology)–Address essays, lectures. I. Harvey, John H., 1943–
II. Ickes, William John. III. Kidd, Robert
HM132.N475 153.7'34 76-26028
ISBN 0-470-98910-6

Printed in the United States of America

This book is dedicated to Fritz Heider, whose work serves as an enduring reminder that we are all attribution theorists.

Contents

Preface

Until recently, attribution was one of the few areas in social psychology where systematic theorizing outstripped empirical efforts. As is often the case, however, provocative theorizing provided the impetus as well as the context for the emergence of empirical investigations. Since the publication of an earlier collection of papers on the topic (Jones, Kanouse, Kelley, Nisbett, Valins, & Weiner, 1972), attribution has become one of the most heavily researched areas in social psychology. The present volume was undertaken because of the burgeoning amount of work being done in the mid-1970s on attributional phenomena and the subsequent need for an updated volume dealing with these phenomena.

Definitions of the term "attribution" have varied considerably and are therefore not easily dealt with in a cursory fashion. Attributional approaches, which originally grew out of work on person perception, generally refer to the conditions associated with the individual's attempt to find structure in his own behavior and the behavior of others. In essence, attributional approaches see the "man on the street" as operating like a scientist, obtaining information from his social surroundings and trying to discern the causes and consequences of ongoing behavioral and environmental events. In no small way, these approaches reflect the rich cognitive emphasis in the field of contemporary psychology as a whole.

In this collection, we have attempted to bring together works which provide a broad sample of current attributional research. Additionally, we have included some articles that are designed to balance these more data oriented chapters by providing broader theoretical and historical perspectives. The major themes of this book are the personal and interpersonal consequences of attributional processes. In each chapter authors deal with the application of an attributional framework as a means of answering particular psychological questions. For example, under what conditions does a person attribute freedom to himself when undertaking a given behavior? How do his causal ascriptions influence his feelings of control over life events or his motivation to perform a task that is intrinsically interesting? How will his attributions about the cause of a belief or an underlying emotional state affect the stability of the belief itself? When will he be reluctant to give help to a dependent other or reciprocate another's favor? These and other questions are discussed in the following chapters.

Our organizational scheme is to some degree arbitrary. As a way of introducing and placing in a historical perspective the sections on current theoretical and empirical work, we begin with a conversation with the eminent pioneer in attribution theory, Fritz Heider. The majority of the subsequent chapters, which are concerned with contemporary theoretical and empirical issues, have been organized into two main sections dealing with attributional phenomena at the personal and interpersonal levels of analysis. The volume concludes with a section devoted to theoretical analysis and integration.

An increasing number of investigators are presently doing quality research on attributional processes. Because of this continued strong interest, the editors and publisher are now in the process of preparing a second volume that will follow the same general format as the present one.

A number of people have contributed valuable assistance in the planning and execution of this project. During the volume's inception, Bernard Weiner, Larry Wrightsman, and several of the contributors to the book provided helpful advice. Larry Erlbaum's counsel and encouragement were of great value during the preparation and organization of the manuscripts. Yvette Harvey provided a major contribution in the form of technical assistance. We would also like to express our gratitude to the Departments of Psychology at Vanderbilt, Ohio State, and Wisconsin and to our colleagues at these universities for various forms of assistance during the project. In particular, we are grateful to Vanderbilt University for funding a large portion of the copying and mailing expenses involved in the editorial work. Finally, this project was carried out with the assistance of an NIMH grant (MH 25736) and a Vanderbilt Research Council grant to John Harvey and an NIMH grant (MH 26646) to William Ickes and Robert Kidd.

NEW DIRECTIONS IN ATTRIBUTION RESEARCH

Volume 1

Part I

HISTORICAL PERSPECTIVE

Fritz Heider received his doctorate degree in philosophy and psychology from the University of Graz, Austria, in 1920 after promising to repay his father for his university education by raising pigs. While not being able to trace completely the origins of his work on attribution and person perception, Heider acknowledges that these topics have intrigued him since his graduate days. In reaction to traditional psychology's emphasis on such topics as psychophysics and stimulus–response learning, he began in the 1920s what became a lifelong career of trying to direct psychologists' attention to what he feels are the more important problems of interpersonal relations.

Despite its early beginning, Heider's work received only token attention for almost four decades. Psychologists' lack of response to ideas that seemed "so obvious" as to preclude scientific investigation distressed Heider at various points from 1921, when he gave his first talk on person perception, until 1958, when his book *The Psychology of Interpersonal Relations* was published. With this work, social psychologists at last began to take seriously Heider's contention that "scientific" psychology could learn much from the conceptual explication of "common-sense" psychology. Today it is impossible to deny the impact of Heider's thinking in the areas of person perception and attribution, as attested to by the numerous references to his work in this volume.

Since retiring in 1968 as Distinguished Professor at the University of Kansas, Fritz Heider has continued to devote much of his time to thinking and writing about psychology. These pursuits have taken the tangible form of some 16 large notebooks of ideas on the shelf of his study. He is almost invariably found with a pencil and pieces of paper in his shirt pocket, which he uses to jot down ideas as they occur to him.

This conversation is a condensed version of a longer discussion between the editors and Heider recorded at the Heiders' home in Lawrence, Kansas, in June 1975. At the request of the editors, Grace Heider, Fritz's wife and professional colleague for the last 42 years, joined in the discussion. Grace also holds a doctorate degree in psychology and is a former professor of psychology at Kansas.

1

A Conversation with Fritz Heider

Editors: Professor Heider, most social psychologists view you as the founder of the attributional approach. We would like to know more about the major influences that directed your thinking about psychology in general, and about attribution and person perception in particular.

Fritz Heider: In general, my thinking about psychology was influenced by two important circumstances. One was that during the teens and up to 1920, I wanted to become a painter. I was very involved with drawing and painting and always had an easel in my study. That was really a source of my interest in perception.

I still think that anyone who studies perception should draw because this is the best way to get the basic knowledge and experience. As long as one's knowledge of perception rests only on experimental results, I think his knowledge is very limited. Experiments are very good for the purpose of testing an idea, but you usually cannot get new ideas from them. But you *can* get new ideas from being involved with perception, drawing and painting especially. Things such as constancy phenomena, for example. You have to get them in your blood in order to draw.

Editors: Are you saying that drawing provides a way of directly experiencing and interacting with the phenomena you are talking about?

Fritz Heider: That is right.

Editors: And that offers some perspective and a source of ideas?

Fritz Heider: Yes. I remember a statement by [Sigmund] Koch, to whom the APA [American Psychological Association] assigned the task of editing the six volumes of *Psychology: The Study of a Science.* At the end of the task, Koch became very discouraged with the whole state of psychology. He said that the one thing most psychologists lack is the know-how, the direct experiential knowledge.

Editors: How did your own experience as a painter bear upon your subsequent theorizing as a psychologist? What did you learn from painting that later came out in the form of psychological theory?

Fritz Heider: Well, a great deal. For example, I used the term depth to refer to the difference between the proximal stimulus and the distal stimulus. Experience teaches us that we are involved with relevant life things, things which are mostly in the distance area. For example, I am involved with whether I can swim in that river or whether this board on which I am working will break. These things are in the distance, but we get information about them through something that immediately touches us. And this bridge between the immediate influence on us and the objects that are distant—that is attribution. I see something (or something is given in experience) and then I make a bridge to what is most relevant and important—namely, the dispositional characters, the more stable characters of the environment, the abilities and intentions of other persons which are relatively stable. This process is very important. We get the information about these distant objects and events through something that is more proximal. Our construction of what "things" are there in the environment, the causal relations among them, what happens to them, how they interact, and so on—all these are products of the attribution process.

My first attempt to discuss these issues was in my 1927 paper. "Thing and Medium." The medium is that which allows us to get information about things. The medium is that which touches us, but which we do not change. By means of the perceptual apparatus, we then try to correctly "read" this information. For people like James Gibson, the whole interpretation is already in the stimulus proper, which can be questioned. We get the information through the stimulus proper, but we have to have some apparatus to read the information. It is as if you give a person a book in a foreign language; the information is all there, but he cannot read it. So there is something to the idea that the information has to be read through organizing processes and so on.

Editors: We are beginning to understand how drawing and painting had an important influence on your thinking about psychology, but a minute ago you said that there was another experience which also influenced you greatly. What was that?

Fritz Heider: The second important circumstance was a bit later. Especially after World War I, in 1918–1919, it was a very hard time in Austria. There was half-starvation, and people never knew whether they would be able to get food. We did not have electric lights; we used a little glass filled with oil with a wick on top instead. When I wrote my thesis, I worked in my room about 2 hours each morning with no heat at all. I had to wear thick gloves and write with them on.

We lived on the edge of the city, and in the mornings at about 11:00, I would go into the woods, chop down a little tree, and start a fire.

Anyway, it was a very difficult time when people got very tense and were easily angry with each other. It was often the case that two people whom I liked very much would quarrel with each other. Usually each one came to me and complained about the other. I found that when one came and presented his position, he was very plausible; and when the other came, he was plausible also. But I wondered why these two people did not see things in the same way. The situation was completely different for them. Well, that led me into the problems of fitting together these points of view and reconciling them to some "objective" situation. And that was one major source of my interest in interpersonal relations.

Editors: Did you find much encouragement to pursue your interests in psychology?

Fritz Heider: That is another story. At first, when I finished my studies at the gymnasium, I wanted to become a painter, but my father said that I really should have an ordinary job, since painting, writing, music, etc. should only be hobbies (he had many hobbies). He was an architect and encouraged me to become one, too, and so I started to study architecture in Graz where we lived. But I did not like it very much–I had to spend a lot of time copying details from Greek temples–and I gave it up; I did not do much for the rest of the year. Then the next year my father said, "Well, you can get into the government–study law, and then you can earn your living and paint in your spare time." So the next fall, I began to study law. In Austria, legal training begins with the old Roman law, and there is lots of memorizing at which I was never very good, so I gave that up too.

My father kept reminding me that I had to do *something.* I said that I would like to go the University and audit courses. I said that I did not want to take an exam or get a PhD; I just wanted to learn. And my father said, okay, that he would allow me 4 years for that, but that afterward I would have to find something useful to do. Since the family had some land in the country, he suggested that I raise pigs. I thought that 4 years of education sounded wonderful, and so I certainly agreed to that. And then I started to take all kinds of courses–art, literature, premed–and gradually I got more and more into philosophy and psychology.

Editors: How did you manage to escape your obligation to raise pigs for a living?

Fritz Heider: Before I answer that I want to tell you more about my studying at Graz. There was a great philosopher there who was also a psychologist. His name was Meinong, and like Husserl, he had been a student of Brentano. At that time he had a very great influence on the English philosophers: Russell, G. E.

Moore, and so on. They all wrote long papers on Meinong's system, and today there are still people who study him.

Meinong was really a very imposing personality, and now I realize that some of the ideas that I developed about common-sense psychology originally came from him. But when I studied with him, I did not like his theorizing at all because it was too logical and dry. I had begun my studies with no intention of seeking a degree, but after my third year, as I saw that so many of the people who had started with me had gotten their PhDs, I thought I might try too. I went to Meinong and told him I would like to try to write a thesis, and he said it was okay and gave me a topic on the subjectivity of sense qualities, something half-way between philosophy and psychology. I worked on it, and at Christmas time in 1919 I went to tell him what I was doing. He said, "Yes, it's all right," and in March, 1920, I handed it in. It was accepted, and so I got a PhD.

Editors: What was your reaction to psychology as it was developing then?

Fritz Heider: When I went to the university and started psychology, I thought that interpersonal relations was the most important thing psychologists should talk about. But they were busy with verbal learning, sensations, and those other topics that are still around in so-called experimental psychology—a term which, by the way, is very much misused because people in other areas like personality and social psychology are also experimentalists.

Editors: How did your early contemporaries react to your own interests? For instance, how were your ideas about interpersonal relations received?

Fritz Heider: Well, I started talking about these things very early, in 1920 or 1921. There was sort of a popular science group in Graz who had an invited speaker every month or so. One month they invited me to speak, and so I gave a talk on person perception, on how we interpret what we see about other persons. I gave some examples from everyday life, but they must have expected something very different. It all seemed very obvious to them; they just laughed at my examples. And they left without a response. But there was one professor of philosophy (not Meinong) who was very supportive. He was a wise person, and he, anyway, was impressed with what I said. I went with him and his wife to their home, and we talked about it. They were both very pleased with the talk.

Editor: But, generally speaking, there was not much interest in the kind of psychology you wanted to do.

Fritz Heider: That is right, and for quite a long time, I did not want to get into a university position. I got my PhD from Meinong in 1920, and during the next year I worked in Graz, doing vocational guidance testing with school children and that job, by the way, saved me from starting in the pig business. But in 1921, I became restless and thought I should get out of Graz. I had an uncle who lived in Berlin, and I decided to go there. I had no idea what was going on in Berlin, but right away I went to the Psychological Institute and began to audit Köhler's and Wertheimer's courses and very soon became acquainted with [Kurt] Lewin. I had many, many, long discussions with him. From that time on, I was in continuous contact with Lewin until he died in 1947.

Editors: Those must have been marvelous conversations. Here were you with a painter's background and a concern for real human relationships, and here was Lewin who wanted to translate psychology into topological, almost physical science terminology. Did you have a common basis of understanding when you sat down to talk?

Fritz Heider: Oh, sure . . . Let's see. I went to Berlin in 1921, just as Lewin was doing the work that led up to his topological psychology and all the rest. I was there for several years. Then I was away, wandering in Italy, Czechoslovakia, etc., until 1926.

When I came back to Berlin in the winter of 1926, Lewin told me about topological psychology. I remember standing with him one day waiting for a bus. The sidewalks were covered with a light, powdery snow, and Lewin was drawing topological diagrams in the snow with the point of his umbrella.

Editors: That is quite a striking image. Were there other people watching you?

Fritz Heider: No, but I remember that the diagrams were becoming more substantial and elaborate as he continued to draw them. That picture has always remained with me.

Editors: What were you doing during this time? You obviously had been doing more than just wandering.

Fritz Heider: I did some painting and spent a lot of time reading plays and novels. I found that literature had a lot to say about psychology and expressed some very great insights about human relations. And that is what I thought psychologists should be worrying about.

I also took little jobs from time to time, but I always tried to have some leisure for my real work. It was like a disease with me; I just thought all the time about psychology. When I had no job, no way of making money by writing about psychology, I did it anyway.

But the young people in psychology are so chased around today. You have to get jobs and apply for grants. You cannot do what I did–taking hikes for weeks, going down to Florence to stay for the winter–you cannot do that. My cousin used to subsidize my travels by giving me a small monthly allowance while I was hiking around. It was very meager, about $25 a month, but one could live on it in Europe in those days.

Grace Heider: During this time Fritz's father wrote his niece and said that he did not know who was supporting Fritz, but that whoever it was, he was not grateful to him.

Editors: So he was still after you to settle down. When did you finally do it?

Fritz Heider: When I was 30 I decided to go job hunting; it was a very definite decision at the time.

Grace Heider: Then he was suddenly offered jobs in Vienna, Graz, and Hamburg, all at once. J. Allesch, whom he had come to know in Berlin, told him that if he went to Vienna, he would spend all his time in coffee houses and dancing with girls. He advised him to go to Hamburg—he would get some work done there.

Fritz Heider: Yes, Hamburg had fewer distractions than Vienna.

Grace Heider: He took Allesch's advice and went to Hamburg. William Stern was a professor there and a well-known psychologist. He also got to know Cassirer, a philosopher who had influenced Lewin's thinking; and he saw a good deal of Heinz Werner, who later came to this country.

In the meantime Koffka had been given a 5-year research appointment at Smith College, something quite unheard of then, before the days of grants. I was a member of Koffka's laboratory and also connected with a well-known school for the deaf at Northampton. The administrators of the school wanted to build a research department and asked Koffka to oversee its development. Koffka was willing to do it, if he could find a more advanced person to work under him. I was only a recent MA at that time. He tried various people in this country (this was before the depression) without finding anybody who seemed to be the right person and then wrote to Stern at Hamburg. Stern approached Fritz and asked him if he would like to go to America. Fritz thought that it would be fun for a year, so he took a leave of absence and came. Then we met and were married. At first we had no real plans as to where we would settle, but then the Hitler period came, and we were very glad to be here instead of in Germany.

Editors: And during this time you were still being influenced by Lewin as well?

Fritz Heider: That is right. Lewin's analysis was concerned with actions and means–end processes, and when I was first exposed to it I tried right away to apply it to interpersonal relations. But I could not. It got too complicated. I told Lewin about my difficulties and showed him what I had done, and he said that he could not find a direct way to use it to represent interpersonal relations either. But I learned from Lewin that one must have a system of terms to represent that which one investigates. Lewin's source for that idea was Cassirer. Lewin looked around for mathematical systems that could be used to represent psychological processes and found topology. For a while he thought that this would be the psychology of the future.

Editors: You said "for a while." Do you mean that he changed his mind?

Fritz Heider: Yes, I think so. And after World War II, when he started the Center for Group Dynamics at MIT [Massachusett's Institute of Technology], which later moved to the University of Michigan, he probably leaned on it less and less.

Editors: It sounds like at that point he was more interested in the kind of phenomena which you had been interested in all along—the dynamics of interpersonal relations.

Fritz Heider: Not quite. At Iowa, Lippit came along and helped steer him, so to speak, toward a study of group processes. The people working with him there were studying topics like frustration and level of aspiration but did much less with dyadic interpersonal relations.

Editors: Why do you think there was an emphasis upon the group as opposed to the dyad?

Fritz Heider: I believe that was Lippitt's influence, and it probably fitted in with what Lewin himself was ready for at that moment. Lippitt originally came from Springfield College in Massachusetts, where there was a teacher (I do not know his name or much about him) who was interested in athletic groups and other sorts of groups as well. He was the "seed," so to speak, and through his influence on Lippitt and Lippitt's influence on Lewin, this interest in groups eventually developed into the sensitivity training groups at Bethel, Maine, and the whole encounter group movement.

Editors: So all of this began with a single nameless teacher from Springfield, Massachusetts, who was interested in athletic teams? [Editor's note: It was later learned that this teacher was Harold Seashore.]

Fritz Heider: That is what was underway at the MIT Center for Group Dynamics. Lewin and his group began by studying simple role-playing situations, a good deal like what Moreno was doing with his psychodrama. For instance, once when I was visiting they were playing out an encounter in a factory between a foreman and a worker.

Editors: Did you still maintain contact with Lewin during this period?

Fritz Heider: We often spent some time together in the summers. And when he was at MIT, living near Boston, and we were in Northampton, we saw each other fairly often. But there was a time during this period when he was so intensely involved with his own work that we took walks together and that sort of thing but did not have any real discussions. The last time, in the fall of 1946,

I remember that he came to visit us, and we had a good talk. He was beginning to show a renewed interest in theoretical problems. But that was the last time I saw him. He died the next year.

Editors: It sounds like your friendship with Lewin was a long and rewarding one. He was obviously a very stimulating and influential man. In addition to Lewin, there were many other figures in the history of psychology who were contemporaneous with you. Brunswik, for example—did he have much impact on your thinking about perception and attribution?

Fritz Heider: No, Brunswik did not have much direct impact on my thinking. You see, I had published "Thing and Medium" in 1927 and a second paper in 1930. Brunswik's work came out later. He often said that he was very much influenced by *my* papers. And what he developed later, about probability, I did not accept. But we were very good friends; I was very close to him and Else [Frenkel-Brunswik]. As long as I was in Europe, I looked them up whenever I went to Vienna. But he really did not influence my writing. Of course, Lewin had a strong influence, and Wertheimer, too.

Grace Heider: Wertheimer influenced his thinking about balance, especially unit formation.

Fritz Heider: Yes, he did—and Köhler, too. Tolman was another influence. He was in Vienna for a half-year or so, and he was close to the Brunswiks. He knew about their thinking, and his own ideas made use of the concept of cognitive maps. His cognitive approach was fairly close to mine, and he encouraged me very much.

Tolman was a delightful person. The last time I saw him he said, "Senility is catching up with maturation."

Editors: As an American psychologist, Tolman was a bit of a renegade, especially during that period in psychology's history. And speaking of American psychologists, when you made the move from Europe and encountered people like Watson and Skinner in the U.S., what did you think about their ideas and their approach to the study of behavior?

Fritz Heider: Well, I think Skinner has said some very interesting things. But for me, his approach is restricted by throwing out the most interesting part of psychology. Reward and punishment are indeed very important. But I prefer the way Lewin treats these topics in his book *Dynamic Psychology*, which was translated by Adams and Zener. In Lewin's paper on reward and punishment, the issue is stated in an entirely different way—in a cognitive way—which to me is much more sensible.

I always use a simple example to illustrate the difference between conditioning and cognitive learning. Some years ago the psychology department at Kansas was in another building, and there was a door in that building with the knob on the left side. As a result of some remodeling, the door was changed and the knob put on the right side. For years, I had been "conditioned" to open it from the left. After the change, if I was thinking about it—was alert, so to speak—I opened the door correctly right away. There was no long period of extinction and the training of a new response. But if I was thinking of something else, very often I fell back to attempting to reach to the left. So this "stupid" kind of learning does work; we are partly very stupid and at times act just like Skinner's rats. But not always, for when we are alert our cognitions can overrule these habits.

Nevertheless, in many cases we still behave on the level of rote learning and conditioning. I suspect that especially for personality development that is important—phobias, for instance, where we know there is nothing to fear, but the habit persists. These phenomena are very important, but since the Skinnerians have neglected their cognitive aspects, they are treated in what seems to me a superficial, simplistic way. I think that on the whole where the Skinnerians have the best success is with animal training, with retarded persons, or with very young children.

Editors: Well, your own approach to the study of human behavior was certainly a far cry from Skinner's. How did American psychologists react to your attempt to redefine the field?

Fritz Heider: Well, certainly there were times when I expected to have more response than I got, for instance, in the 1940s. A year before Lewin died, in 1946, he arranged for a meeting at Harvard on social psychology and social perception. I gave an hour-long talk on social perception, and I thought that at last there would be some interesting discussion. Jerome Bruner and Gordon Allport were there, among others, but after the talk nobody said anything; they just quietly trooped out. I was very disappointed. Roger Barker was there too; I think he was the only one who said he was impressed. But, otherwise, no response. And I thought that these were things that were so interesting; certainly there would be some discussion. But, no, nobody said anything.

Grace Heider: Maybe the reason they walked out was that they had to get home for dinner. . . . Lewin thought you had something.

Fritz Heider: Yes, he did. I met Allport and Bruner again and again, and they never referred to that talk. But they were always very friendly.

Grace Heider: They probably thought that it sounded too easy—to obvious.

Editors: Were the concepts of the "naive psychology" that you used in *The Psychology of Interpersonal Relations* the same concepts that you used in the talk at Harvard?

Fritz Heider: Yes, but possibly the time for proposing that we pay attention to the common-sense psychology was not yet ripe. Psychologists were still trying to show that everyday people could not understand psychology without a formal education. And, of course, at that time *phenomenology* was a bad word. Some people thought that this was phenomenology. I do not think that. But there are still some people who say I am a phenomenologist.

Editors: But regardless of how people tried to label your ideas, from your talk to the science group at Graz to your talk at Harvard 25 years later, there was no real change in attitude. The response was the same in both cases. The real irony is that if you look closely at the hardest scientific data available in social psychology today, the interpretations all tend to be based on intuitive psychology.

Fritz Heider: Yes. Usually there is an independent variable and a dependent variable, and what is investigated is the connection between the two. But this is where there are special difficulties in creating experiments in social psychology. The independent variable can never be directly imposed. Instead, there is a certain manipulation by which the independent variable is supposed to be varied. Also, the dependent variable is not directly observed, but there is some kind of external symptom of this variable which is supposed to vary. Now the basis for connecting the independent and dependent variables to the particular operations used for manipulation and measurement is usually some form of "naive psychology" which is not investigated, which is taken for granted! I think that is why in so many experiments, there are so many explanations possible–because this naive psychology has not been thought through clearly.

Editors: You have been talking about what you call "naive or common-sense" psychology as opposed to what is traditionally thought of as "scientific psychology." Obviously, your ideas about common-sense psychology were developing a long time before you published them in *The Psychology of Interpersonal Relations* in 1958.

Grace Heider: Actually, the writing of the book began nearly 15 years earlier, in 1943.

Fritz Heider: Yes, but it is difficult to identify a particular time when the ideas themselves developed. I can point to a particular time when the idea of balance originated, but I cannot point to a particulat time when the idea of attribution originated. There was already something of that in my doctoral thesis. Attribution, of course, is just a close relative of perception.

The ideas for the book developed over quite a long period during which I collected notes and read and analyzed short stories, novels, plays, and cartoon strips. The symbolic language in the back of by book is an attempt to provide a system which is a little like topology, namely, a system of terms with which one can represent interpersonal relations. I translated many examples into these terms. But I did not use the symbolic language in the book itself because it is too hard to read.

Sometimes I analyzed very simple things, Aesop's fables, for instance. That Aesop's fables are not analyzed or mentioned more in psychological literature, I think is a disgrace! Think about the "Fox and the Grapes," for instance. There are so many things to be analyzed in that fable. A good example of attribution can also be found in the expression, "Hunger is the best cook." That represents an attributional link between a state of the person and some dish which he is eating. There is also a wonderful little French saying, a children's verse actually, which says, "This animal is very vicious; if one attacks it, it defends itself." If you think about this saying, there is a very interesting attribution process going on here.

Editors: It seems that your work took place on two levels. On one level you were trying to describe symbolically what happens in interpersonal relations (using your symbolic language system) and on another level you were attempting to explain behavior–not merely to describe it.

Fritz Heider: You know, this symbolic language is not quite just description. In order to develop it, I had to discover the basic concepts of common-sense psychology, and that is what I tried to explain in the beginning of the book. I had to analyze and think through simple expressions to bring out the essential meaning–as people have done, for instance, with probability. Our naive or common-sense notions of probability have also been philosophically analyzed; it is exactly the same thing. I just analyzed the concepts which are already there. I do not know if one should say this is theory, but it is not just description. If you try to translate some ordinary events into these symbols, you will see that you have to think a lot about it; there is already a lot of theory there.

Editors: When you were translating plays, short stories, etc. into your symbolic language, did you ever get flashes of insight and begin to see that the connections that were developing had implications for new relationships you had not really thought out or written about?

Fritz Heider: Yes. One example is from Ibsen's play, "A Doll's House," which I analyzed. From there I got the idea of "latent imbalance," an imbalance which is not yet effective. It does not have any consequences because the characters do not yet know about it. It is the essence of many dramatic situations because the audience knows that the imbalance exists and that if the characters find about it, there will be a tremendous tragedy.

The main characters in "A Doll's House" are Helmer and his wife Nora, who love each other. Nora had done something 10 years ago which Helmer, as a banker, would think is terrible if he knew about it. She had forged her father's signature and really sacrificied herself, so to speak, for Helmer, because the forged signature gave them the means to go to Italy. (The doctor had told Helmer that he was quite ill and that if he did not get to warmer climate, he would soon die.) So Nora forged the signature and thought that if Helmer found out about it, he would appreciate her sacrifice and be very grateful. In balance terms, it is very simple: there is Helmer, Nora, and the forging of the signature. There is a plus relationship between Nora and Helmer; a plus relationship between Nora and the forging; but a minus between Helmer and the forging. However, the last relationship, since Helmer does not know about it yet, is still latent. It is like a stick of dynamite lying there; everybody knows it could explode at any moment. That is dramatic.

Editors: We agree with you, that seems to be an important structural base of tragedy.

Fritz Heider: Yes, it is one kind of a very dramatic situation.

Editors: This seems to be one aspect of your ideas which could easily be elaborated theoretically and tested. Since the publication of *The Psychology of Interpersonal Relations,* people have expanded on a number of your ideas. Your ideas about person perception have been investigated to some degree, but what aspects of your attribution theory do you think have been well developed or neglected since you wrote about them?

Fritz Heider: Well, keeping up with all the literature on attribution is difficult, but I think that the work is most developed in the area of attributions of success and failure, especially [Bernard] Weiner's studies. And then there is [H. H.] Kelley's work on causal schemata. But there are many other things that have not yet been adequately treated in attributional terms. For instance, the attribution of motivation–answers to the question, "Why did he do it?" not "Why did he succeed?" And the whole thing which I am not quite clear about–which I call "target attribution."

Laura Huxley, the wife of Aldous Huxley, wrote a book called *You Are Not The Target,* in which she suggests that everyone thinks he is the most important person. This perception becomes one of the ingredients of causal attribution. If something in the environment is seen as very important, people will probably attribute to it. But in the case of target attribution, when someone else does something, my tendency will be to think that it was because of me. "He did it either to harm me or to benefit me; I was causally important." This is especially true of the person who falls in love. If it is a man, he is inclined to think that everything the girl does is intended to repulse or attract him.

Editors: Why does he give himself such an exaggerated weight as a causal variable in her behavior? Why does he view his own behavior toward her as so influential? Is it egocentrism?

Fritz Heider: We are all a little egocentric.

Editors: That is true. Most of us would like to see our own behavior as quite influential in affecting the behavior of others. But is it true that some people chronically, and most people occasionally, like to think that they are *not* the cause of another person's behavior, and to minimize their role?

Fritz Heider: Well, of course, if it does not flatter them to see themselves as causal . . . But I think there are many cases–in children especially–where a person may prefer to get hostile attention than no attention at all.

Editors: What you have just said indicates that a possible motivation for this bias of weighting oneself very heavily in accounting for the behavior of another person is the need to be taken into account by that person–to receive attention; to be validated as a person yourself; to be considered important by others; or, possibly, to consider yourself as important.

Fritz Heider: Yes, but of course the most important person does not have to be the self. It can be also another person.

Editors: If that were the case–if a person considered his wife, say, to be the most important person in his life–might he then shift his focus and give his own behavior less weight in causing particular effects and put a greater emphasis on her behavior?

Fritz Heider: Yes, it seems quite possible.

Editors: We seem to have been crossing the border of other-perception into the area of self-perception. Over the years, have you given much thought to self-perception, and how it may or may not differ from other-perception?

Fritz Heider: Yes, in many cases, self-perception is similar to other-perception. Nietzsche said that "the thou is older than the I." That is, the image of the other is older than the image of the self.

I remember that a friend of mine in Vienna did some early studies of self-perception. He simply went around asking people about their experience of the self. One woman he interviewed was very prominent in management. I asked him, "What kind of self-image does she have?" He said, "That is an interesting case; she does not have any self-image. She does not perceive herself at all; she just acts on other people."

Editors: Her attention was constantly directed outward and never inward?

Fritz Heider: That is right. But, offhand, one would think that she had a tremendous self-image and saw herself as very important. We thought that if other people rated her egocentrism, she would be rated very high.

Editors: In your book, the idea of egocentric attribution was emphasized. Is it still something that you believe in?

Fritz Heider: Yes. These are essentially balance tendencies. One is inclined to attribute to oneself good things, but one suffers when one has to attribute to oneself something that is not so good . . .

Editors: This is in the area of intersection between your balance and attribution conceptions–is that right?

Fritz Heider: Yes, the two are very close together; I can hardly separate them. Because attribution, after all, is making a connection or a relation between some event and a source–a positive relation. And balance is concerned with the fitting or nonfitting of relations.

Editors: Would it be accurate to say, then, that attribution is a basic unit of analysis in a balance situation?

Fritz Heider: It may be, but remember that there can be balance or imbalance in situations where attribution is not one of the cognitive elements. If you dislike a person and you have to work with him, this is an imbalanced situation, but questions about attribution are not directly involved.

In Romeo and Juliet, to take another example, there is a strong plus relationship between them, and then there is a group membership: Juliet is a Capulet, and Romeo is a Montague. And there is a minus relationship between the two families. Imbalance exists here without the need to represent causal attribution as an element.

Editors: How is the recognition of balance or imbalance in a situation like this experienced? And how does the affect associated with that recognition vary? For example, in what kind of imbalanced situations might Juliet examine her cognitions and feel very rational and unemotional about them, saying "Yes, I recognize that an imbalance exists, but I do not feel particularly upset about it." What do you think separates these situations in which a person acknowledges the imbalance and is not greatly affected by it from those in which the person is deeply affected?

Fritz Heider: Well, I have recently thought a lot about units and an old distinction within units which is not used much any more. There are some units like similarity which are in a way not very strong. Similarity is something that the person can recognize, but, in itself, it is not real. You may think of it in contrast to a unit of cohesiveness, in which two things are "tied together." This unit of cohesiveness is very different from similarity; right away, it has effects. In a cohesive relationship, if one element starts to move, then the other has to move; there is a real interaction between them. The same difference comes out a little bit when sociologists talk about class terms (all the 10-year-olds, for instance). This "class" is not a real unit compared to an interaction unit–a club, for instance, where people really interact. There seem to be two kinds of units, and the affective consequences of imbalance may be different for each.

Editors: It seems that imbalance and the negative affect associated with it would be strongest when it has a real meaning for the person in terms of what he does, in terms of his behavior, his existing commitments . . .

Fritz Heider: That is right. But I would be caution about making generalizations.

Editors: It does not seem that attribution theorists, and social psychologists in general, have recognized the overlap between your balance conception and your attribution conception, and that seems to be something we are at least hinting at in this conversation.

Fritz Heider: That is right. There are many relations between balance and attribution.

Editors: One thing we have not talked much about is the functional value of attribution and balance tendencies. A number of theorists have suggested that we make attributions in order to aid our prediction and control, and that the need for control is the fundamental motivation in attributional behavior. Do you believe that?

Fritz Heider: I think that there is probably a good deal of truth in that, even if it is control in the sense that we want to avoid something. But we want to get the information that is given to us by the environment in order to adjust to its important features. Yes, I think that this is true, although maybe at times it is just curiosity . . . curiosity about something that we do not want to or have no reason to control. I think that is also possible.

Editors: So curiosity is a motive, but one which has no instrumental value?

Fritz Heider: That is right. After all, there are some people who study psychology for the instrumental value, but there are others who are just curious about it.

Editors: A number of attribution theorists have been making the point that by attributing personality traits (stable dispositions) to others, we are able to achieve an illusion of control by believing that we have rendered their behavior more predictable.

Fritz Heider: Well, that, of course, gets into individual differences. I can imagine that there are quite a few people for whom everything is in terms of control over the environment. But there are other people who do not want to control; they are just curious. They are concerned about people not because of a need for control but because they like them.

Editors: You have indicated that like all of us, you have not kept up with everything in attribution theory. But what is your general impression about the status of the theory, and where do you think it could profitably go in the future?

Fritz Heider: It is all-pervasive and important. I think it can go anywhere. I recently talked to [Philip] Zimbardo about it, and he expressed it very clearly. He realizes, first of all, that he always used some kind of attribution without formulating it, and that it is everywhere. Attribution is part of our cognition of the environment. Whenever you cognize your environment you will find attribution occurring.

Part II

ATTRIBUTION AT THE PERSONAL LEVEL

Attributional approaches assume that people are motivated to seek meaning in their own behavior as well as in the world about them. For this reason, attributional processes are important in the individual's attempts to understand and interpret the possible causes for his own actions, feelings, and attitudes. These inferences about one's own behavior and internal states are called "self-attributions" and form the basis for the present section. In each of the papers in this section, the authors examine the attribution process at the personal level.

Attributions often are motivated by a person's desire to maintain control in an uncertain and unpredictable world. In the first paper in this section, "Attributions of Personal Control," Camille Wortman explores some theoretical and empirical work predicated on the assumption that causal attributions often are made to enhance feelings of control over one's environment. Some of these considerations are applied to the individual's attempts to comprehend everyday accidents and even such "uncontrollable" life events as natural disasters, unemployment, or the death of a loved one.

Mark Snyder's paper, "Attribution and Behavior: Social Perception and Social Causation," provides a more ideographic view of perceived control. According to his analysis, people vary in the degree to which they see their own and others' behavior as situationally versus dispositionally controlled. Although there is theory and evidence suggesting that actors tend to see their behavior as primarily controlled by environmental forces, Snyder presents some data which indicate that actors may nonetheless vary in a general disposition to see their behavior as situationally or dispositionally caused.

Notions of control and responsibility for outcomes are closely related to feelings of personal freedom in the decision-making process. People often assess and make attributions about the extent to which they are free to behave in a choice situation, and the fact that these attributions may have real implications for their behavior is a strong counterpoint to B. F. Skinner's supposition that freedom is illusory. John Harvey, in his paper, "Attribution of Freedom," reviews the rapidly accumulating work on the determinants and consequences of self-attributions of freedom.

While relegating freedom to the status of a superstitious belief, Skinner bolsters his claim for the omnipotence of schedules of reward and punishment in controlling human behavior. In "Schedules of Reinforcement: An Attributional Analysis," Stanley Rest contrasts the traditional stimulus–response perspective with a cognitive–attributional interpretation of intermittent reinforcement schedules in human learning paradigms. Human subjects tested in a type of modified Skinner setting are assumed to analyze the pattern of responses and reinforcements and make judgments about the causal factors that control the degree to which behavior and reward covary. Such judgments appear to influence the person's expectations about obtaining future rewards as well as his subsequent responding.

In a related exploration of traditional reinforcement notions, "The Self-Perception of Intrinsic Motivation," Michael Ross presents an account of the mediating role of attributions in human motivation. He discusses the consequences of offering extrinsic rewards for performing tasks that were initially intrinsically motivated. Under certain conditions, rewards have been found to reduce interest in an activity that was once performed for its own sake. Ross' work is concerned with a person's attempts to understand his own motivation based on the perceived reward structure of the task. His findings may have important implications for the design of reward structures in token economies and classroom situations.

While attributions at the personal level are useful in maintaining control and discerning some stability in the patterns of external contingencies, they are also used by people to assess their own internal states, such as feelings and attitudes. Focusing on a person's perceptions of his own feelings, Michael Storms and Kevin McCaul analyze the role of self-attributions in the exacerbation of emotional behavior. In "Attributional Processes and Emotional Exacerbation of Dysfunctional Behavior," they present a model describing the effects of causal perceptions on the perpetuation of unwanted and anxiety provoking emotional states and behaviors such as insomnia, neurosis, and stuttering. The model represents the convergence of several different theoretical perspectives and is applied to the detailed analysis of stuttering behavior.

In "Extensions of Objective Self-Awareness Theory: The Focus of Attention–Causal Attribution Hypothesis," Shelley Duval and Virginia Hensley propose a theory of conscious attention which posits two states of human consciousness, one in which a person's attention is "focused," and one in which attention is "unfocused." The authors suggest that a person's perception of the cause of an event is influenced by the degree to which the event and its possible causes are similar on the focalization dimension. For example, it is hypothesized that a person will attribute greater responsibility for an event to himself when his attention is focused on himself than when it is focused elsewhere.

In the realm of attitude change, dissonance and attribution approaches have often been viewed as "warring" factions. Mark Zanna and Joel Cooper, in "Dissonance and the Attribution Process," present a series of studies which illustrate how these two perspectives can peacefully coexist. Dissonance, created by engaging in attitude-discrepant acts with minimal justification, has been postulated to induce attitude change because of its aversive arousal properties. To date these properties have been either assumed to exist or, at best, only weakly evidenced following the performance of counterattitudinal behaviors. Employing an attributional approach, Zanna and Cooper present some intriguing data on the nature and quality of the feelings aroused in the typical forced compliance situation.

2

Causal Attributions and Personal Control

Camille B. Wortman

Northwestern University

In this chapter, I will examine the view that people make causal attributions in order to enhance their feelings of control over their environment. First, some theoretical statements that are based on this assumption will be reviewed. Empirical evidence for the notion will then be examined. The chapter will conclude with a discussion of some of the functions that might be served by making such attributions.

PREVIOUS THEORETICAL STATEMENTS CONCERNED WITH CONTROL

The idea that people strive for mastery and control of their environments is certainly not new and has played a central role in many theoretical statements (see, for example, de Charms, 1968; Hendrick, 1943; Kelly, 1955; White, 1959; Woodworth, 1958). The importance of a controllable, predictable world is evident in many of the theories that social psychologists have advanced in recent years. For example, cognitive-consistency models of behavior (see Abelson, Aronson, McGuire, Newcomb, Rosenberg, & Tannenbaum, 1968) are based on the assumption that people are motivated to believe that events follow from one another in a predictable and orderly fashion. In his theory of psychological reactance, Brehm (1966) postulated that people are motivated to believe that they are free to do as they please. According to Brehm, people will become motivationally aroused when they are exposed to attempts to control their behavior. This arousal, called reactance, leads individuals to try to restore their freedom to behave as they choose. Kelley (1971) has argued that people's motivation to maintain control has a pervasive influence on their attributions of causality. According to Kelley, "The purpose of causal analysis—the function it serves for the species and the individual—is effective control. . . . Controllable factors will have a high salience as candidates for causal explanation. In cases of ambiguity or doubt, the causal analysis will be biased in its outcome toward controllable factors [pp. 22–23]."

The assumption that people are motivated to maintain control is also evident in two theoretical statements concerning reactions to victimization (Lerner, 1970, 1971, 1975; Walster, 1966).[1] Since these models make explicit predictions concerning the relationship between desire for control and causal attributions, they will be discussed in some detail. In a paper concerned with how people assign responsibility for an accident, Walster (1966) asserted that:

> As the magnitude of the misfortune increases . . . it becomes more and more unpleasant to acknowledge that "this is the kind of a thing that could happen to anyone." . . . If a serious accident is seen as the consequence of an unpredictable set of circumstances, beyond *anyone's* control or anticipation, a person is forced to concede the catastrophe could happen to him. If, however, he decides that the event was a predictable, controllable one, if he decides that *someone* was responsible for the unpleasant event, he should feel somewhat more able to avert such a disaster. [p. 74].

This analysis suggests that a person's assignment of blame for an accident will be affected by his[2] desire to avoid the frightening thought of his own victimization. One prediction that can be derived from this view is that people will be more likely to assign blame as the negative consequences of an accident become more serious. In order to test this reasoning, Walster conducted an experiment in which subjects were exposed to a tape-recorded account of a car accident. The car owner's behavior was held constant in the various conditions, but subjects were led to believe that the consequences of the accident were either mild or severe. The results were complex but, in general, the car owner was viewed as more responsible for the accident when the consequences were severe.

Subsequent experimenters employing a similar methodology have failed to replicate this finding, however (see, for example, Shaver, 1970; Shaw & Skolnick, 1971; Walster, 1967). Unfortunately, many of these studies are characterized by methodological problems (see Chaikin & Darley, 1973, and Lerner, 1975, for a more detailed discussion); these problems have probably contributed to the confusing and contradictory pattern of results. For example, subjects in most of these experiments were simply asked to read and comment on a written account of an accident. As Lerner (1975) has pointed out, subjects in such studies are probably not very motivated or involved. According to Lerner, the motivation to believe in a predictable, controllable world is likely to

[1] Another theoretical model that applies to this topic is Shaver's (1969) "defensive attribution" hypothesis. Although this model has received some support (see Chaikin & Darley, 1973), it is not directly relevant to perceived control considerations and will not be discussed here.

[2] For convenience, his or he will be used to signify his/her or he/she.

affect a person's attributions of causality only when that person is confronted with an outcome that is personally threatening. Chaikin and Darley (1973) examined attributions of responsibility for an accident in a more involving setting and found some evidence for Walster's (1966) hypothesis. In this study, subjects witnessed a videotape of a "real" accident in the laboratory. The more severe the consequences of the accident, the less willing subjects were to assign responsibility for the accident to chance.

Walster's analysis implies that the specific target(s) we select for blame will be influenced by our desire to control (that is, prevent) the recurrence of the accident. Some support for this view was attained in an early sociological investigation by Bucher (1957). Bucher conducted a series of interviews with residents of Elizabeth, New Jersey, in 1951–1952. During that time, there were three airplane crashes in the vicinity within a 3-month period. On the basis of these data, Bucher (1957) reached the following conclusions:

> Responsibility was laid where people thought the power resided to alleviate the conditions underlying the crashes. It was not instrumentality in causing the crashes which determined responsibility but ability to do something to prevent their recurrence. . . . A considerable number of respondents mentioned figures who indirectly carried out actions leading to the crashes, such as pilots, mechanics, those manning the control tower, etc. But in no case were these figures held responsible. . . . Even when these lesser figures were accused of incompetence, the responsibility was shifted upward onto higher authorities. . . . People traced the line of responsibility upward in the hierarchy of authority and placed it where, in their eyes, the power to remedy the situation lay [pp. 471–472].

Walster (1966) suggested that a person's desire to avoid victimization affects not only his assignment of blame, but also his tendency to punish those who cause accidents: "He can protect himself by putting people like the ones responsible away—isolating them so that they cannot cause calamities, or reforming them so they will not cause them [p. 74]." Sociologists studying assignment of blame for disasters (for example, Bucher, 1957; Veltford & Lee, 1943) have also found a desire to punish potentially guilty parties. However, as Veltford and Lee (1943) have noted, this does not necessarily represent a desire for control. Such a result may also be viewed as a form of "scapegoating," in which people work off the frustrations, anger, shock, and horror brought on by the disaster. In analyzing reactions to the tragic Cocoanut Grove nightclub fire in Boston which killed 498 people, Veltford and Lee asserted that the public's fixing of blame and call for punishment helped to relieve blocked emotional reactions or frustrations about what had occurred. This "scapegoating" hypothesis would also predict a relationship between severity of consequences and attribution of blame, since people are likely to feel more upset if the consequences are severe than if they are mild. It may be possible to discriminate between these two views by examining the conditions under which people are most willing to punish transgressors. For example, if punishment stems from a desire to prevent or control future recurrences of the accident, people should be more willing to punish an offender if the punishment is to be made public and act as a deterrent to others or if there is evidence that the punishment will prevent the individual from committing the offense in the future. Along these lines, Kelley (1971) has suggested that our motivation to control negative outcomes may make us more likely to blame or punish others when a negative outcome stems from shortcomings that are modifiable: "In responding to another person's negative actions, there is no corrective gain to come from blaming him if his physical or mental incompetence was involved, but there is considerable possibility for effective control if it was a matter of his intention, attitude, or motivation [p. 24]."

Drabeck and Quarantelli (1967) have pointed out, however, that regardless of why this process occurs, blaming and punishing a person possibly responsible for an accident is unlikely to be effective in preventing future recurrences of the event. They argue that attributing blame to individuals draws attention away from more fundamental systemic causes. According to these investigators, such attributions are naturally counterproductive because they create the illusion that corrective action of some sort is being taken. Of course, people's attempts to gain control through their causal attributions are interesting whether or not such attributions are effective in this regard. And such attributions may serve important psychological needs for the individual, even if they are "counterproductive" in producing societal changes.

In short, Walster's (1966) model has received some support from both laboratory experiments and sociological investigations, although there are competing explanations for some of these findings that have yet to be ruled out. A second model that makes specific predictions about the relationship between a desire to control one's outcomes and assignments of causality is Lerner's (1970, 1971, 1975) just-world theory. He has argued that individuals are motivated to believe in a "just world," where people get what they deserve and deserve what they get. According to Lerner, the issue of deserving begins to gain meaning in childhood. During development, the child learns to forego immediate gratification and instead to endure deprivation, effort, and pain so that he may achieve desirable outcomes in the long run. Lerner (1971) has maintained that people want to believe that such sacrifices will pay off: "If the person becomes aware that someone else—who lives in and is 'vulnerable' to the same environment—has received undeserved suffering or failed to get what he deserved, the issue must arise as to whether the person himself can trust his environment [p. 8]."

Since evidence of an unjust world threatens a person's sense of control over his potential outcomes, he should attempt to convince himself that there is some correspondence between people's behavior and their outcomes. When confronted with a victim of misfortune, a person will be motivated to believe that the victim earned his suffering. He will either blame the person for the outcome or (if the victim is clearly innocent) derogate the personal characteristics of the victim. Consistent with this reasoning, it has been found that subjects derogate the personal qualities of a fellow subject who is assigned to receive painful electric shocks, even if it is made clear that the outcome occurred by chance (see Lerner & Matthews, 1967).

This theoretical orientation has generated an enormous amount of research over the past few years; since these studies have been summarized elsewhere (see Lerner, Miller, & Holmes, 1976), they will not be reviewed here. The similarities and differences between Lerner's and Walster's models might be noted, however. It can be predicted from both models that people will be motivated to punish those who are clearly guilty of perpetrating negative outcomes. Walster has suggested that punishment is motivated by a desire to prevent the recurrence of the accident; according to Lerner, people wish to punish the guilty because it helps maintain the belief that there is a relationship between one's behavior and outcomes. Both theories also predict that blame or derogation will increase as the magnitude of negative consequences increase, since serious negative consequences are more threatening than mild ones. However, the theories differ somewhat in their predictions of who will be blamed or derogated. Walster has suggested that we prefer to blame someone for an accident in order to avoid the thought of our own victimization. This person need not be the victim, however. Of course, if the person who perpetrated the accident is also victimized by it, Walster suggests that we would prefer to blame the victim than to conclude that chance was responsible. Her view implies, however, that if the victim is clearly innocent, and if there is someone else who might be plausibly blamed for the accident, the victim will not necessarily be blamed or derogated. In contrast, Lerner suggests that it is the victim–the person who has suffered the negative consequences–who will be blamed or derogated. If the victim is clearly innocent and blame is impossible, then the personal characteristics of the victim will be derogated in an effort to see him as deserving of his fate. In fact, the clearer it is that the victim is truly innocent, the more negatively his personal qualities should be rated, since an innocent victim should threaten a subject's view of the just world more than a victim who is behaviorally responsible for his fate (see Jones & Aronson, 1973).

These theories also make different predictions concerning reactions to accidents with positive consequences. Since Walster's (1966) model implies that we make attributions to feel protected from catastrophes, it would seem to be of little relevance to such cases. Another interpretation of Walster's hypothesis (Shaw & Skolnick, 1971) is that the more positive the consequences become, the more motivated we are to blame chance. By doing so, we are able to believe that the consequences may happen to us. Lerner's model makes a different prediction: The more desirable the consequences, the more we should come to believe that the recipient's behavior was especially meritorious or that his personal characteristics make him worthy of the consequences. Both Lerner (1965) and Apsler and Friedman (1975) have found support for the latter view.

Although the theoretical statements of Walster (1966) and Lerner (1970, 1971, 1975) differ in focus, both are consistent with the notion that people assign responsibility so as to maintain control over their environment. But as noted previously, there are alternative explanations for many of the experimental findings. For this reason, the research generated by these theories provides indirect evidence at best for control motivation. However, several experiments have been specifically designed to study people's perceptions of causality and control. In the remainder of this chapter, I will review a representative sample of these experiments and discuss some possible consequences of exaggerating one's personal control.

SOME EVIDENCE CONCERNING ATTRIBUTIONS AND CONTROL

Evidence that people's attributions are affected by their desire to control their environment can be divided into three major categories. First, many studies have shown that subjects erroneously perceive a causal connection between outcomes that occur together by chance. Second, several investigators have found results which imply that people exaggerate their ability to influence uncontrollable outcomes, such as games of chance and "uncontrollable" life events. Finally, there is evidence to suggest that people underestimate the extent to which their behavior is controlled by situational or external forces. Each of these kinds of evidence will be considered in turn.

The Illusion of Contingency

Several experiments provide suggestive evidence that subjects are reluctant to entertain the hypothesis that events occur together by chance. In many studies, subjects who are presented with events that occur in succession tend to assume that the two are related or that one predicts the other, even though this is not the case.

In an interesting set of experiments, Chapman and Chapman (1967) have found that subjects have difficulty assessing the contingency between various events. These investigators were interested in understanding why clinical psychologists persist in the use of diagnostic tests that have been shown to lack validity, that is, why people persist in maintaining beliefs in the face of disconfirming evidence. According to Chapman and Chapman, clinicians who use the Draw a Person test often indicate that they have observed a tendency for patients who are paranoid to show elaboration of the eye in their drawings. This belief occurs despite the fact that in four separate studies such a relationship was not substantiated. The authors posed the following question: Do these results occur because clinicians are rigid in their unwillingness to accept research findings, or is there something about the diagnostic task that produces a spurious belief in a relationship between particular symptoms and diagnoses? In order to shed light on this question, Chapman and Chapman conducted a study in which clinically naive undergraduates were asked to examine "actual" clinical protocols. In these protocols, various types of drawings (for example, elaborated eyes, big head) were randomly paired with various personality disorders. Although the pairings were random, the undergraduates "rediscovered" the same relationships between the test responses and the types of problems that clinicians report observing. The authors argued that these results occurred because both clinicians and laymen have a priori hypotheses about the relationship between certain types of drawings and certain problems (for example, drawing a large head is indicative of insecurity about one's intelligence, drawing male genitals is indicative of homosexuality for males). They then misperceive the data as confirming these hypotheses. Subjects seem to ignore instances when the hypothesis is not confirmed while focusing on cases when the two symptoms are paired as evidence in support of the hypothesis. This tendency for prior beliefs about causation to affect the intake and processing of information occurred even if subjects were given repeated exposure to the experimental materials and even if they were offered a substantial financial reward ($20) for making accurate observations. Using different stimulus materials, Starr and Katkin (1969) were able to replicate this illusory correlation effect. In their study, sentences from the Rotter Incomplete Sentence Blank were randomly paired with statements supposedly reflecting the major problem of a patient who had completed the sentence. Again, illusory correlations emerged, suggesting that the phenomenon is relatively robust.

Subjects' failure to process correlational information correctly has also been reported in experiments by Smedslund (1963) and by Ward and Jenkins (1965). In the latter study, subjects were asked to examine data from a number of different cloud-seeding experiments that had supposedly been conducted in different states throughout the country. Each "experiment" was presumably carried out over a 50-day period when clouds were present. The authors experimentally manipulated how the information was presented to the subjects. For some subjects, the events (seeding or no seeding followed by rain or no rain) were presented serially. A second group received the information in an organized numerical summary. A third group received the information both ways, with the serial information preceding the summary. The actual amount of covariation varied from "experiment" to "experiment." After examining the data for a given "experiment," subjects were asked to indicate the amount of control exerted by cloud-seeding over rainfall. Although this paradigm can be criticized as possibly noninvolving from the subject's point of view, some interesting findings emerged. Only the group that received the summary without the serial information was able to make reasonably accurate judgments about the relationship between seeding and rainfall. When the information was presented serially, as it normally is in real life, subjects were unable to judge the correct relationship between the two variables. The group that received serial information followed by a summary were also unable to make accurate judgments. Apparently, these subjects developed hypotheses about the relationship between the two variables on the basis of the serial information, and were not able to process the summary information objectively.

All of the aforementioned experiments indicate that subjects make errors when asked to judge the contingency between two outcomes or events. Jenkins and Ward (1965) investigated some factors that might affect subjects' judgments of contingency between their own behavioral responses and the outcomes that follow them. These investigators conducted a series of experiments to study subjects' judgments of contingency in a two-response, two-outcome situation. On each trial of a given problem, the subject was required to press one of two buttons, and one of two possible outcomes was programmed to appear.

The investigators speculated that a feeling of illusory contingency might occur if one of the two outcomes was more desirable than the other and was programmed to appear with great frequency independent of the subject's behavior. They reasoned that if the subject were to press a particular button and receive a desirable outcome, the subject would be likely to continue pressing the same button. The investigators felt that the predominance of one response together with one outcome might produce a spurious belief in control. They speculated that a change in the character of the subject's task, which encouraged him to alternate the two responses, might reduce this spurious belief in control.

Each subject was asked to work on five problems, in which the actual contingency between his responses and the outcomes, as well as the number of times the desired outcome appeared, were experimentally varied. On each of 60 trials, subjects were instructed to press one of two response buttons and then press a "test" button. Immediately after the test button was pressed, one of two outcomes was illuminated on a display panel. For half of the subjects one of the outcomes was designed to be more desirable than the other. The two outcomes were "score" and "no score"; subjects were instructed to attempt to produce the score outcome as frequently as possible. For the remaining subjects the two outcomes were neutral symbols, and they were instructed to learn how to produce each of them at will on any given trial. These subjects were told to indicate which outcome they were trying to produce before each trial began. This second set of instructions was expected to produce a more balanced use of the two responses and, as a result, a more accurate judgment of contingency.

Subjects were told that for each problem, the relationship between their button pressing and the outcomes could range from no control (no relationship between the two) to complete control (the response button and the outcomes are perfectly correlated). After the 60 trials for a given problem, they were asked to indicate on a numerical scale the degree of control they had over the outcomes. The possibility that the correct judgment for a given problem might be one of zero control was explicitly stated. In order to determine whether active involvement with the task would enhance the subject's feelings of personal control, Jenkins and Ward also included observer–subjects. An observer–subject was paired with each actor–subject and was instructed to watch the actor's behavior and judge his degree of control over the outcomes.

The results revealed that regardless of the instructions that the subjects received or whether they performed or observed the task,[3] the judged amount of contingency was entirely unrelated to the actual degree of contingency. As expected, the judgment of control was positively correlated with the frequency of successes (the frequency with which subjects produced the score response or attained the response they were trying to produce). The authors speculated that subjects' failure to understand how much control they had might simply be a matter of miscommunication. Perhaps subjects were using "control" to mean "getting what you want," rather than the ability to influence their outcomes. In later studies, the investigators added other dependent measures, such as asking the subjects on what percentage of trials they could produce each of the outcomes at will. However, employing these measures did not improve subjects' accuracy in judging contingency.

This research suggests that if a person attempts to produce a certain outcome, and that outcome occurs, he is very likely to ignore the possibility that the outcome occurred by chance. The contingency between the subject's intention and an outcome is apparently regarded as very persuasive evidence of personal control. Interestingly, subjects seemed to conclude that producing an intended outcome was evidence of their control even when this happened rarely. Failure to produce the desired outcome did not seem to be taken as evidence against personal control. For example, on one of the problems, subjects received only 8 successes in 60 trials, and there was no actual contingency between their behavior and the appearance of the desired outcome. Nevertheless, almost half of the subjects felt that they had some control.

As Miller and Ross (1975) have noted, these studies on illusory correlation suggest a plausible alternative explanation for the finding that subjects are more likely to attribute success to their own abilities and skills, while attributing failure to factors in the environment and situation. Some investigators have interpreted such results as evidence of subjects' motivation to view themselves positively (see, for example, Wortman, Costanzo, & Witt, 1973). However, in most performance settings, people probably intend to succeed. Perhaps it is the contingency between their intent and the positive outcome that leads subjects to attribute success to themselves. When they intend to succeed and then fail, there is no such contingency. Given the lack of "fit" between their intentions and the

[3] In a more recent study, however, Langer and Roth (1976) have found evidence that involved subjects believed they had more control over a chance task (predicting coin tosses) than observers thought they had.

outcome, a situational or environmental attribution may seem appropriate. It would be interesting to employ both intentions and outcomes as independent variables in the same experiment and examine the effects on self-attribution. Perhaps subjects who intend to fail will make a situational attribution for success.

With all of the recent interest in attribution processes, it seems surprising that there have not been more experiments concerning people's assessment of contingency between their responses and their outcomes. Although there is little experimental evidence on this topic, this issue has generated a great deal of theoretical discussion (see Kelley, 1967, 1971, 1972). For example, Kelley (1967, 1971) has suggested that in order to assess causality, people look for evidence of covariation between their behavior and various outcomes. According to Kelley, if people vary or change their behavior and notice a corresponding change in the environment, they are likely to attribute causality to themselves. If they vary their behavior, but notice no change, an external attribution is likely. Several interpersonal influence studies have indicated that subjects attribute causality to themselves if they attempt to instruct or help another and the other improves. But if their attempts are met by repeated poor performance on the other's part, they attribute this failure to the other (see Miller & Ross, 1975, for a review). As Kelley (1967, 1971) has repeatedly stressed, such data do not necessarily reflect motivational processes on the part of subjects to view themselves positively.

> . . . Insofar as the teacher tried harder or varied her instructional method after B's poor initial performance, the information pattern available to her would show a strong positive covariation between her own behavior and that of the improving student, but negative or noncovariance between her own behavior and that of the consistently poor performer. By the covariation principle, the former would warrant a self-attribution, and the latter, not. Much needed in this regard are experimental procedures in which the degree of covariance of this sort is varied independently of the trend in the other person's behavior [1971, pp. 19–20].

Throughout this discussion it has been implied that a person's judgments of personal control are affected by the temporal relationship between responses and outcomes. Michotte (1963) has found the temporal sequence of events to be very important in people's perceptions of physical causality. In one study, subjects were shown a movie of a square coming into contact with a second square, which then began to move. Almost all subjects perceived the movement of the second square to be caused by the first one. However, if a brief delay (one-fifth of a second) was imposed between impact and the movement of the second square, the impression of causality was eliminated. Surprisingly, subjects' perceptions of social events occurring in various temporal sequences, or of temporal variations between their responses and outcomes, have received almost no investigation. In the aforementioned research by Jenkins and Ward (1965), for example, the temporal delay between the subjects' responses and outcomes was held constant. We might well imagine situations in which the temporal sequence of events might delude people into thinking that they are producing a certain outcome when, in fact, their behavior is irrelevant. Both Jones and Gerard (1967) and Kelley (1971) have discussed the possible ramifications of such errors in dyadic interaction situations.

In addition to understanding subjects' reactions to various temporal sequences between their behavior and outcomes, it may be interesting to study the conditions under which subjects are willing to covary their behavior to obtain information about causality. We might speculate that, at least in the short run, learning about causality is incompatible with affecting the environment. When a person is trying to produce a certain outcome, he is likely to try several responses, perhaps simultaneously. A person who has noticed that the leaves are rapidly dropping from his schefflera may move it into the sunlight, increase the water, provide fertilizer, and perhaps even talk to it more often. Thus, it will not be clear precisely which cure, or subset of cures (if any), is responsible should the plant return to health. Trying one option at a time, and waiting to see if it works, is obviously impractical in many settings. In fact, the more urgent or serious the problem, the more likely a person may be to try all conceivable remedies. Thus, it may be those situations in which we most need accurate causal information that we are least likely to get it.

The Illusion of Control

Chance outcomes. According to investigators cited previously, subjects often conclude that there is a causal relationship between events that occur in succession by chance. There are also experiments which suggest that under certain circumstances, people believe that they can influence chance events. In an interesting participant-observation study of crapshooting behavior, Henslin (1967) has argued that the players had developed several hypotheses concerning control of the dice. For example, the players uniformly believed that a hard throw would produce a high number and a soft throw a low number. Players also believed that control over the dice would be maximized if the shooter displayed concentration and effort. The first time he played, Henslin was continually admonished, "Take your time! Don't throw 'em out so fast! Take your time and work on it! [p. 319]."

According to Henslin, the most important requirement for correct shooting is confidence. Crap shooters believe that confidence imparts control to the shooter. For this reason, shooters sometimes express supreme confidence in the outcome of their throw. For example, a player will frequently say, "There's a (desired point)" while throwing the dice in the air. Similarly, players feel that accidentally dropping the dice negatively affects the outcome of the next throw. Presumably, such an accident indicates that the dice are out of the dropper's control.

In short, Henslin's research suggests that there are several techniques used by crapshooters to enhance their perceived control over the game. Shooters seem unwilling to acknowledge that the outcome of the game is influenced by chance. According to Henslin (1967) "Failure does not represent 'the absence of control' but, rather, that someone's or something's control over the dice was greater than that of the shooter's. It is never that it was merely chance [p. 325]."

Keeping in mind the findings of Chapman and Chapman (1967) concerning the reluctance to relinquish a hypothesis in the face of negative evidence, it is not hard to understand how these beliefs might be maintained among crap shooters. Chapman and Chapman's research suggests that when a certain action (for example, throwing the dice softly) and the expected outcome (getting a low number) occur in succession, that fact will be taken as evidence in support of the rule. But negative instances (a soft throw and a high number or a hard throw and a low number) are unlikely to diminish belief in the rule.

Some of the ideas inherent in Henslin's investigation have been followed up in an intriguing series of studies by Langer (1975). Langer's experiments were designed to investigate what she refers to as an illusion of control. According to Langer, an illusion of control is an expectancy of a personal success probability inappropriately higher than the objective probability would warrant. She hypothesized that if aspects of a skill situation such as competition, choice, familiarity with the stimulus or response, or involvement are introduced into a chance situation, they lead the subject to believe that he can exert some control over the chance event.

Langer's first study supported Henslin's observation concerning the importance of confidence when gambling. In Langer's experiment, subjects were induced to play a chance-determined card game with an experimental confederate. The game involved drawing for the high card. The independent variable was whether the confederate acted confident or nervous. In the "dapper" condition, the confederate gave the appearance of being confident and outgoing and was dressed in stylish clothes. In the "schnook" condition, the confederate appeared to be shy and nervous, behaved awkwardly, and was dressed in ill-fitting clothes. As predicted, the subjects bet significantly less money when competing against the dapper confederate than when competing against the schnook. According to Langer, such a difference should not occur if the subjects regard the task as governed completely by chance. Perhaps the subjects felt more confident when their opponent appeared to be a schnook. Or perhaps they inferred that their opponent would perform better when he appeared to be confident than when he appeared to be nervous and unsure of himself. The results can also be viewed in terms of the just-world hypothesis discussed earlier. In order to check on the manipulation, Langer included a rating scale which indicated that subjects viewed the dapper confederate as more interpersonally competent than the schnook. They may have felt that the dapper person "deserved" to win a great deal, but that the schnook deserved little, and made their bets accordingly.

In order to investigate the effects of choice, familiarity, and involvement on expectations of control in a chance setting, Langer (1975) conducted three separate experiments in which subjects were participants in lotteries. For each study, the lottery was conducted at their place of employment. Tickets cost $1 each, and the winning ticket was worth approximately $50 in the first lottery and $25 in the remaining lotteries. In the first experiment, Langer manipulated whether subjects could choose their lottery ticket. Half of the subjects were permitted to select their ticket from a box of tickets; the remaining subjects were given a ticket. Near the end of the lottery, subjects were asked to indicate the amount for which they were willing to sell their ticket. This measure was considered to be indicative of the value of the ticket for the individual involved. The results were as predicted: Subjects in the choice condition required an average of $8.67 for their ticket, while no-choice subjects required only $1.96.

Langer's second lottery study was designed to replicate and extend the previous one. Both familiarity of the stimulus and choice were manipulated. For half of the subjects, the lottery tickets were familiar (letters of the alphabet); for the remaining subjects, they were unfamiliar (line drawings of novel symbols). Half of the subjects were asked to select a ticket; the remaining subjects were given a ticket. As predicted, subjects regarded their ticket as more valuable if they had chosen it and if it had been familiar.

Langer investigated the effects of passive involvement, or thinking about the desired outcome, in her third lottery study. In this experiment, each lottery ticket consisted of a three-digit number. Half of the subjects received one digit each day for 3 days and thus were required to think about the lottery on three separate occasions. The remaining subjects received their whole number at one time. The results indicated that subjects regarded their ticket as more valuable in the involvement than in the noninvolvement condition.

In another experiment conducted outside of a "lottery" setting, Langer investigated the effects of familiarization, or practice, as well as active involvement, on subjects' estimates of the likelihood of success on a chance task. Subjects were shown an apparatus containing three paths and were told that the machine was preprogrammed so that on a random basis one of the paths would sound off a buzzer when traveled with a stylus. It was explained that the object of the game was to select the path that would set off the buzzer. Familiarization was manipulated by allowing half of the subjects to practice with the task for a few minutes. The remaining subjects had no opportunity for practice. Subjects in the high-involvement condition were instructed to operate the stylus themselves; for subjects in the low-involvement condition, the experimenter manipulated the stylus on the route that the subject had selected. Before beginning the task, subjects were asked to rate how confident they were that they would select the correct path. As predicted, confidence increased with familiarization and involvement.

Some factors that lead individuals to exaggerate their control over chance outcomes were also examined in a pair of experiments by Wortman (1975). Some earlier studies had suggested the possibility that initiating or causing a chance outcome (such as actively throwing the dice, pulling out the slip, etc.) might induce a spurious belief in control. For example, Lerner and Matthews (1967) conducted a study in which pairs of subjects were led to believe that one of them would be assigned to a shock condition, while the other would be in a more desirable condition. The decision as to which subject was assigned to each condition was to be determined strictly by chance–by which of two slips of paper each subject selected from a bowl. Interestingly, subjects tended to view the person who picked out the first slip, and thus "caused" both outcomes, as more responsible for both subjects' fates than the experimenter or the other subject, even though the outcomes were obviously determined by chance. Since this experiment provided suggestive evidence that merely initiating an outcome might produce a spurious belief in control, Wortman (1975) decided to examine this variable further.

In addition, Wortman examined a second variable: whether the subject had foreknowledge of what he wanted to attain. In an earlier experiment, Strickland, Lewicke, and Katz (1966) provided suggestive evidence that this factor might affect perception of control. Subjects were told that they would each make a bet and throw a pair of dice 30 times. It was explained that subjects could earn money if they won their bets. Strickland et al. were primarily interested in the level of conservatism or riskiness that would be exhibited by their subjects. They predicted that subjects who were asked to bet after they had thrown the dice but did not know the outcome would bet more conservatively than subjects who were asked to bet before throwing the dice. The results supported this hypothesis. Wortman (1975) hypothesized that these findings may have occurred because subjects who were asked to bet before throwing the dice felt that they could influence the outcome. Since subjects in both conditions initiated or "caused" the outcome by throwing the dice, these results implied that causality alone may not be sufficient to result in feelings of control. On the basis of this reasoning, Wortman conducted two experiments to test the following hypothesis: An individual is likely to feel that he can control a chance outcome only if he initiates the outcome and has knowledge of what he hopes to obtain before doing so.

In the first study, subjects were shown two consumer items and told that they would get to win one by a chance drawing. Two marbles of different colors were placed in a can and mixed up. One-third of the subjects were told that the experimenter would pick a marble to determine their prize and were told beforehand which marble stood for which prize. Another third were told to select a marble to determine their prize and were told beforehand which marble stood for which prize. The remaining subjects were told to select a marble to determine their prize but were not told until after they had picked their marble which marble stood for which prize. Subjects then received a marble which led them to win either the item they preferred or the item they did not prefer. Dependent measures included subjective perceptions of control, choice, and responsibility. The results strongly supported the hypothesis: Subjects who "caused" their own outcome and knew beforehand what they hoped to attain perceived themselves to have more control over the outcome, more choice about which outcome they received, and more responsibility for their outcome than subjects in the remaining conditions. These results were replicated in a second experiment.

Why did the foreknowledge variable have such a strong effect on subjects' feelings of control over a chance outcome? And, in general, what processes mediated the subjects' perceptions of control in the results reported by Langer (1975) and Wortman (1975)? On tasks such as throwing dice (Strickland et al., 1966) or choosing the correct marble (Wortman, 1975), only if people know what they hope to attain (a certain number or color) can they attempt to exert control. Subjects in these situations may have made an effort to get a certain number or marble–by throwing or reaching in a certain way, for example–and this attempt at exercising control may have led to actual feelings of control. It is possible that many of Langer's (1975) results were also mediated by subjects' differential attempts to exert control. In contrast to subjects who merely received a ticket, subjects who could choose their ticket probably made some effort to select the "correct" one. Similarly, subjects may have put more effort into selecting from among "familiar" than "unfamiliar" tickets. And subjects who practiced or who expected to be actively involved in manipulating a task probably tried harder (or expected to try harder) to exert control than subjects who did not.

Henslin's (1967) participation–observation studies have implied that people who believe they can control chance outcomes expend a great deal of effort to do so. One interpretation of Wortman's (1975) and Langer's (1975) results is that the reverse may also be true: The more effort a person can be induced to put forth on a chance task, the more he may be deluded into thinking that he can influence the outcome. It might be noted that in Langer's (1975) studies, perceived control was measured before the outcome was known. However, in Wortman's (1975) experiments, subjects found out which prize they were getting and were then asked to indicate how much control they had had. Surprisingly, attempting to select a certain marble resulted in feelings of control even when the subject was unsuccessful and picked the marble that assigned him to the nonpreferred outcome.

Hopefully, follow-up experiments will help to clarify the process by which variables such as choice, foreknowledge, familiarity, and practice influence judgments of control. And since these factors produce an illusion of control in chance settings, they may be even more influential in ambiguous settings—settings where it is difficult to assess the relative contributions of chance or skill. Identifying factors that produce feelings of internal control in performance settings would seem to be worthwhile, since there is good evidence that people with such an orientation work harder and are more likely to persist in the face of failure (see, for example, Dweck, 1975). Surprisingly, little is known about the circumstances under which people believe their task performance is under their control. In addition to the above factors, such parameters as the number of times a person has attempted a task, the percentage of positive and negative outcomes received, and the sequence or pattern of successes and failures (see Jones, Rock, Shaver, Goethals, & Ward, 1968; Langer & Roth, 1976) may strongly affect judgments of personal control (see Wortman & Brehm, 1975, for a more detailed discussion of this issue).

Accidents and disasters. Many studies of "uncontrollable" life events have suggested that people often exaggerate their ability to influence such outcomes. As has been noted previously (see Bucher, 1957; Drabeck & Quarantelli, 1967), individuals seem very uncomfortable with the notion that such outcomes occur by chance. Apparently, people who experience a negative life event seem to prefer to blame themselves for the event than to attribute it to chance factors. This review is not intended to be all-inclusive but is designed to provide some illustrations of reactions to uncontrollable life events.

One striking feature of many accident and disaster studies is the extent to which apparently innocent victims of misfortune appear to experience guilt. The notion that one's fate is caused by past behavior seems to be very prevalent among victims of disease and natural disaster. For example, Abrams and Finesinger (1953) conducted extensive interviews with 60 cancer patients and noted that feelings of guilt were present in 93% of the patients. The tendency of the patients to believe that their cancer was caused by their own misdeeds was striking. Of the 8 cancer patients who had had venereal disease earlier in their lives, all attributed their present disease to their past sexual misdeeds. Similar reactions have been noted in a study of parents with terminally ill children (Chodoff, Friedman, & Hamburg, 1964).

Guilt reactions are also common among victims of natural disasters. According to Lifton (1963), in discussing the psychological effects of the atomic bomb in Hiroshima, "that such feelings of self-condemnation should be experienced by the *victims* of a nuclear disaster is perhaps the most extreme of its many tragic ironies [p. 489]." Guilt also appears to be a very characteristic reaction among women who are raped. Medea and Thompson (1974) have speculated that some women may feel guilty because they had fantasies about rape and may feel that somehow they must have wanted it to happen. Others suffer from guilt because they failed to take all possible steps against their assailants.

Many investigators who have studied the bereavement process have commented on the presence of guilt feelings in the bereaved. In a recent review of studies on grief, Averill (1968) has noted that people who have lost a loved one frequently respond with feelings of guilt for real or imagined transgressions against the deceased. In his classic study of bereaved persons, including people who lost a loved one in the Cocoanut Grove fire mentioned earlier, Lindemann (1944) noted a strong preoccupation with feelings of guilt. He indicated that it is quite common for the bereaved to search the period of time before the death for evidence of failure to do right by the lost loved one.

These investigators have suggested that people exaggerate the extent to which uncontrollable outcomes are caused by their prior misbehavior or mistakes. There is also some evidence to suggest that people attempt to control or avoid such outcomes by engaging in "good" behavior (see Janis, 1951; Kubler-Ross, 1969). According to Janis, "People who are facing the prospect of illness, unemployment, or any extreme form of deprivation, will often attempt to ward off the danger by making sure that they do not deserve to be punished. Evidently, this was one of the dominant types of reaction among the bombed population of Britain [pp. 169–170]." Similarly, Kubler-Ross (1969) has pointed out that terminally ill patients often engage in good and moralistic behavior in an attempt to ward off death. Of course, the tendency to blame or derogate oneself for negative outcomes, or to engage in moralistic behavior to avoid such outcomes, is quite consistent with the just-world hypothesis (Lerner, 1970, 1971, 1975) discussed earlier.

Unfortunately, there are relatively few studies in which victims have been questioned about their assignments of causality for negative events. Therefore, little is known about the conditions under which people blame themselves for these events. It may be informative to investigate whether the factors found to be important in laboratory studies of chance outcomes might also affect guilt reactions to uncontrollable life events. For example, are people likely to feel irrational guilt for an accident if their behavior technically caused it but there was no way they could have foreseen and/or prevented the accident? Let us suppose a person has unexpected car trouble, calls a friend, and asks for a ride back home from the garage. If his friend is hit by a drunken driver and killed on the way to pick him up, will he feel responsible for his friend's death since he "caused" the accident? It may be quite salient to the individual that if he had not initiated the sequence of events by phoning his friend, the incident would not have happened. Many laboratory studies are based on the assumption that this kind of causality produces guilt. In some experiments on guilt and compliance, for example, subjects are tricked into bumping a table and "accidentally" upsetting some cards that another has spent months arranging in order. Again, it may be salient to the subject that if he had not behaved as he did the accident would not have happened. Subjects in these experiments are expected to feel guilty, although most people would regard the outcome as unintentional and probably unforeseeable. There is no direct evidence that such manipulations induce guilt, although they have been shown to produce compliance with an altruistic request (see Cialdini, Darby, & Vincent, 1973, and Regan, 1971, for a more detailed discussion of the guilt-compliance effect).

Some interesting laboratory research by Pallak and his associates has suggested some conditions under which a person may feel responsible for undesirable consequences that occur as a result of his behavior (see, for example, Pallak, Sogin, & VanZante, 1975; Sogin & Pallak, 1976). According to these investigators, if a person freely chooses a course of action and unforeseeable negative consequences occur, he will be disposed to accept responsibility for these consequences. Subjects who make no explicit choice apparently do not take responsibility for unforeseeable consequences that follow from their behavior. The paradigm employed in these studies involved giving subjects choice about performing a dull task (generating random numbers) or leading them to perform the task as a matter of course. Subjects who were told after completing the numbers that their data were useless (an unforeseeable event) felt more responsible for this outcome if they originally had choice.

We should probably be cautious in generalizing from these results to reactions to disasters, but the implications are worth considering. For example, it would be interesting if a person who weighed the alternatives and made a choice as to which route to take while traveling felt more responsible if his car were hit by a drunken driver than a person who took the only available route.

In the study by Jenkins and Ward (1965) mentioned earlier, it was suggested that we may perceive personal causality if we intend a certain outcome and it occurs. Although the author knows of no experimental evidence on this point, clinical case studies abound with examples of people who wished undesirable fates on others and then felt responsible and guilty when these fates occurred. Another factor that may influence a person's assessment of responsibility for an uncontrollable negative outcome is his behavior or intentions at the time the outcome occurs. A person on the way to a self-indulgent shopping spree may feel more responsible if he accidently hits and injures a child than a person on the way to help a sick friend, even though the outcome was unforeseeable and uncontrollable for both. Furthermore, if a person decides to engage in a specific behavior despite the fact that certain negative consequences are foreseeable, he may feel responsible when other negative consequences accrue, even if the consequences that accrue were unforeseeable. A parent who has left his child with a babysitter to go sky diving may feel more responsible if the child is injured in a freak accident than a parent who has left the child to go grocery shopping. Factors that are irrelevant in producing an outcome, but that nonetheless enhance one's responsiblity or feelings of guilt, would seem worth studying.

The Illusion of Freedom

If individuals wish to perceive themselves as possessing control over their environment, it is probably important for them to view their behavior as relatively free from external influences. Indeed, it is difficult for people to interact effectively with the environment unless they can choose how to behave and vary their behavior from one situation to another. If control motivation is important, the view that one's behavior is constrained by personality traits or situational factors is unlikely to be very appealing. In explaining the behavior of others, however, people's desire for predictability and control might well be served by attributing the other's behavior to stable personality traits, or to stable factors in the situation. It is difficult to control others if their behavior is governed by their own personal desires and whims.

In a provocative theoretical paper, Jones and Nisbett (1971) have suggested that the attributions people make for their own behavior may differ from those made by observers. Jones and Nisbett have maintained that there is a pervasive tendency for actors to attribute their actions to situational factors while observers tend to attribute the same actions to stable personal dispositions of the actor. According to Jones and Nisbett, this tendency stems primarily from a differential salience of the information available to the actor and the observer. Presumably, actors' attention is directed outward; they are likely to be especially aware of the situational constraints impinging on them. For observers, the actor's behavior is expected to be the more salient piece of information. Jones and Nisbett have further argued that since observers have limited information about the actor, they are disposed to infer the existence of stable personality traits from the behavior they have seen. Actors, on the other hand, have more information about their behavior in other situations. Therefore, they are likely to be aware of differences in their behavior across situations and thus reluctant to ascribe traits to themselves for this reason.

Throughout their discussion, Jones and Nisbett (1971) have focused primarily on cognitive factors in the attribution process. However, they have not ignored the role of motivational factors completely. They have conceded that the motivation to maintain control may contribute to the tendency for people to avoid making trait attributions to themselves: "... the individual would lose ... freedom to the extent that he acknowledged powerful dispositions in himself, traits that imperiously cause him to behave consistently across situations [p. 14]." But if people are motivated to maintain a sense of personal control, they may wish to minimize the extent to which their behavior is controlled by situational factors as well as personal traits. Jones and Nisbett may be correct in arguing that situational factors are more salient to actors than to observers. But actors may still wish to believe that they can behave as they choose and not as the situation dictates.

During the past few years, many experiments have been designed to examine the causal attributions of actors and observers. A comprehensive review of these studies is obviously not feasible in this chapter. Although the evidence is not entirely consistent (see Gurwitz & Panciera, 1975; Storms, 1973), many investigators have found results which support a perceived-control analysis. In an early interpersonal simulation experiment designed to bear on the dissonance/self-perception controversy (see Bem, 1967), Wolosin (1968) asked subjects who had just consumed a large amount of water to make a tape-recorded speech in which they would either say that they felt very thirsty or very quenched. Both involved subjects and observers were asked to indicate the extent to which the subject felt free to refuse the experimenter's request. Consistent with a perceived-control analysis, subjects rated themselves as having significantly greater freedom to refuse the request than observers rated them as having. Similar results were obtained in an experiment by Bell (1973). Both of these studies imply that relative to observers, actors exaggerate their personal control in a given situation and minimize the extent to which situational factors are shaping their behavior.

Evidence that involved subjects attempt to deny the situational constraints operating on them was also obtained by Miller and Norman (1975). These investigators asked one subject to play a bargaining game with a supposed partner while another subject observed. When asked to indicate how responsible the game situation was for their behavior, actors saw it as less responsible than did observers. Also consistent with a perceived-control analysis is the finding that involved subjects viewed themselves as more responsible for the other player's behavior than did observers. Thus, involved subjects (relative to observers) denied the extent to which their behavior was controlled by the situation and exaggerated the extent to which their behavior influenced others.

Taken as a whole, these studies indicate that people like to believe that their behavior is under their own control and can be modified according to their own needs. Bell (1973) and Miller and Norman (1975) found that people are disposed to view others' behavior as controlled by situational factors, however. It should be noted that these conclusions are not inconsistent with Jones and Nisbett's hypotheses concerning differential information salience and availability. They suggest, however, that motivational factors may be equally important in the attribution process and should not be cast aside by those seeking to understand the process.

Of course, if the desire for control is important to the subject, his willingness to make attributions to personal traits or situational factors will depend on the perceived controllability of these factors. People may be willing to attribute their behavior to situational factors if they believe they can influence the situation; similarly, attributions to traits should not threaten a person's sense of control if the traits are modifiable (for example, "I had the accident because I was careless but can prevent it in the future"). It should also be noted that in attribution experiments, subjects' willingness to endorse a trait or situational attribution may depend upon the way the question is worded (see Bell, 1973; Gurwitz & Panciera, 1975). When people are asked whether the situation forced them to behave a certain way, they may say no, although they might have been willing to acknowledge that they chose to behave a certain way because of the situation.

This analysis is not meant to imply that the motivation to maintain control is the only factor to influence the attribution process or that there are not exceptions to the general tendency to ignore the external determinants of one's behavior. Harvey, Harris, and Barnes (1975) have found that if a person's behavior is extremely negative, he may exaggerate the situational constraints in an apparent effort to ward off personal blame. In many experimental settings, it is difficult to tell whether the particular attributions made by the subjects stem from motivational factors or from errors in information processing (e.g., Nisbett & Borgida, 1975). The results of many experiments are consistent with a perceived-control analysis (e.g., Cialdini & Mirels, 1976), but few studies have been specifically designed to test this view or to discriminate among the various theoretical orientations.

SOME CONSEQUENCES OF ATTRIBUTIONS OF CONTROL

The studies reviewed in this chapter have suggested that people minimize the role of chance in producing various outcomes, exaggerate the relationship between their behavior and "uncontrollable" life events, and tend to be unaware of the extent to which their behavior is controlled by external factors. At this point, it seems appropriate to raise the following question: Why do people make attributions in this manner? Are such attributions adaptive or maladaptive?

In considering this issue, the discussion will center around the psychological concomitants of various control orientations. However, no discussion of this topic is really complete without some mention of the relationship between feelings of personal control and physical well-being. Space limitations prevent a detailed discussion of this literature. However, there is good evidence for a relationship between feelings of helplessness and hopelessness, and various types of disease onset and bodily deterioration (see the APA Task Force on Health Research, 1976, for a review). For example, Schulz (in press) conducted an intriguing experimental study in which members of an old-age home were randomly assigned to receive a college visitor under certain specified conditions. Members who could predict or control their visits from the student were rated as healthier and reportedly took less medication per day after a 6-month period than people who were visited randomly or who received no visits from a student.

What psychological functions might be served by attributions of personal control? Working from the learned helplessness framework, Seligman (1974, 1975) has suggested that enhancing one's feelings of control over environmental factors is adaptive. According to Seligman, it is the attribution that one is helpless that creates problems for the individual. This view is based on a series of impressive laboratory experiments with dogs and rats. Seligman (1974, 1975) and his associates have found that exposure to uncontrollable and inescapable aversive stimulation subsequently interferes with acquisition of escape-avoidance learning. Seligman has used the term learned helplessness to describe this interference with adaptive responding as well as the process which he thinks underlies the behavior: the learning that one's responses and reinforcements are independent. Seligman has suggested that the data from studies of learned helplessness in animals may provide a model of reactive depression in man. He has argued that just as learned helplessness is caused by experience in which reinforcement is independent of one's responses, reactive depression may also have its roots in feelings of loss of control over one's outcomes. (The reader is referred to Wortman & Brehm, 1975; Weiss, Glazer, & Pohorecky, 1974; and Klinger, 1975, for a discussion of the human helplessness literature, some criticisms of the helplessness model, and the presentation of alternative theoretical formulations.)

Learned helplessness theorists have generally assumed that helpless behaviors are maladaptive. In fact, Seligman and his associates (see Seligman, 1974; Seligman, Klein, & Miller, in press) have advocated that our child-rearing processes be geared toward teaching the youngster that he can control his environment. On the basis of studies with animals (see Seligman & Maier, 1967), these investigators have concluded that such "immunization training" will make people more resistant to helplessness effects. In recent years, programs and plans have been developed to change people's control orientation from external to internal (see, for example, de Charms, 1972; Dweck, 1975).

In almost all prior animal and human helplessness experiments, investigators have studied behavior in a situation where the exercise of control was possible. As Wortman and Brehm (1975) have indicated, it is fairly well-established that in situations where it is possible to exercise control, prior experience with lack of control is debilitating, and prior experience with control may be beneficial. However, there are obviously many life situations in which it is impossible to exert control. People may lose their homes and families through natural disaster or may discover that they or their loved ones have a terminal disease. Exaggerating one's causality for such outcomes, or exaggerating the extent to which one will be able to alter them, may actually be maladaptive. Most people would agree that when an outcome is truly uncontrollable, the most adaptive response is to give up and accept the situation. For this reason, Wortman and Brehm (1975) have argued that an accurate assessment of one's potential for control is likely to be more adaptive than an assessment which exaggerates a person's feelings of control. Of course, making such an accurate assessment may often be quite difficult. As was noted earlier, there are many factors that lead individuals to draw erroneous conclusions about their causal powers. Furthermore, there are many outcomes (for example, certain types of cancer) that are not understood well enough to know whether people can influence the course of the outcome by altering their behavior.

However, it is still possible to argue that exaggerating one's causal powers will generally be adaptive if we assume that most of the situations facing people throughout their lives are controllable rather than uncontrollable. Klinger (1975) has maintained that early mastery training may reduce the frequency of depression caused by uncontrollable outcomes. Teaching people early in life to persist longer in the face of failure should prevent premature discouragement and failure by default in all situations where control is possible. However, he has agreed that mastery training is unlikely to reduce the severity of depression when losses or failures occur. In fact, people with exaggerated notions of personal control, or with considerable past experience at controlling the important events in their lives, may find uncontrollable outcomes all the more difficult to accept when they occur. Furthermore, these people may not recognize such outcomes as uncontrollable and may waste considerable effort trying to alter the situation. Langer (1975) has recently suggested yet another possible problem associated with exaggerated notions of personal control. She has pointed out that just as feelings of helplessness with respect to the environment are likely to result in passivity and depression, the illusion of control may be an etiological factor in manic reactions. Langer (1975) quotes Beck (1967) as indicating that the manic patient is "optimistic about the outcome of anything he undertakes. Even when confronted with an insoluble problem he is confident that he will find a solution [p. 93]."

A second question concerns the tendency of people facing such crises to exaggerate their own personal blame. It is certainly more intuitively appealing to believe that people make attributions to protect their self-esteem and sense of personal worth. One would expect that, if anything, people faced with such outcomes would go to great lengths to avoid personal blame and feelings of guilt. A number of questions about this process need to be answered. First, how widespread is this tendency to exaggerate one's own feelings of personal blame for such outcomes? Kubler-Ross (1969) and Klinger (1975) have implied that self-blame is just one of several stages that people go through as they attempt to cope with an undesirable life event. Rubin and Peplau (1973) have suggested that the tendency to blame oneself for an uncontrollable negative outcome may simply represent a "spillover of affect.... People who receive bad lots are unhappy and consequently feel bad about things, themselves included [p. 85]." There may even be a self-presentation aspect to the tendency of people to blame themselves. People may employ this technique because it is a learned response that generally elicits sympathy from others (see Jones &

Wortman, 1973). If so, they may present themselves as being more guilty or blameworthy than they really believe they are.

There is also the possibility that self-blame or devaluation is a response to uncontrollable outcomes adopted only by people with certain personality dispositions. In recent years, a great deal of evidence has emerged to suggest that dispositional factors influence reactions to uncontrollable life events. A detailed discussion of these studies would obviously be impossible, but some brief mention of the most important trends would seem to be in order. Rubin and Peplau (1973) have developed a scale to measure the extent to which people believe in a just world. Several recent studies (see Rubin & Peplau, 1975, for a review) suggest that people who have this belief react differently to victims than people who do not. For example, believers in a just world tend to have more negative attitudes toward underprivileged groups than nonbelievers. Zuckerman (1975) found that people who believe in a just world are more likely to perform an altruistic act prior to final exams, apparently in an effort to increase their own deservingness and corresponding likelihood of doing well on the tests. A second dispositional variable that appears to affect reactions to uncontrollable outcomes is the so-called Type A coronary-prone behavior pattern (see Friedman & Rosenman, 1974). This pattern is characterized by feelings of time urgency, an exaggerated drive to succeed, and general aggressiveness and hostility. Glass and his associates (see, for example, Glass, Snyder, & Hollis, 1974; Krantz & Glass, 1975) have argued that this dimension reflects differential expectations of and needs for control over the environment. Individuals categorized as Type A's seem to respond to preliminary evidence of failure to maintain control by trying even harder, presumably because they find lack of control particularly upsetting and threatening.

An intriguing laboratory experiment by Comer and Laird (1975) also suggests that individuals cope with uncontrollable outcomes in different ways. In this study, subjects were led to believe that they had been randomly assigned to an unpleasant task (eating a worm) or a neutral weight-discrimination task. It was hypothesized that subjects anticipating the unpleasant task would adopt one of three response styles to deal with this outcome: (a) they would change their conception of themselves and decide that they deserved to suffer; (b) they would change their self-view, but in a positive direction, and decide that their willingness to suffer meant that they were brave and altruistic; or (c) they would change their evaluation of the anticipated negative event and decide that eating a worm was not so bad. As predicted, subjects who expected the negative event changed their belief system in one of these three ways, while subjects expecting a neutral outcome did not. About one-third of the subjects "made sense" of their anticipated suffering by altering their evaluation of the worm-eating task; these subjects did not alter their conceptions of themselves. The remaining subjects either inferred that they deserved to suffer or viewed themselves as more brave (and these alternatives were negatively correlated) and did not alter their view of the worm-eating task. Around 10 minutes after subjects had been told what task they would perform, they were told that an error had been made and that they would in fact get to choose whether to eat a worm or do the weight task. The majority of subjects who expected to eat a worm chose to do so or indicated no preference between the two tasks. None of the subjects in the neutral task condition chose to eat a worm, thus subjects changed their beliefs to such a great extent that they then voluntarily chose to suffer.

Comer and Laird were curious as to whether these changes in the subjects' belief system would affect subjects' subsequent behavior. They indicated that subjects who change their view of worms have made a relatively small change in their view of the world, since most people rarely come into contact with worms. However, people who change their conception of themselves have made a change of much greater potential importance. Would people who decided that they deserve to suffer or that they are brave subsequently behave in accordance with this new self-definition? In order to examine this question, the investigators included a group of subjects in the design who expected to eat a worm, but who were then offered a choice between the neutral task and a different negative task (giving themselves painful electric shocks). As predicted, subjects who had changed their views of their bravery or their deservingness to suffer chose the shock, while those who had changed their view of the worm did not.

This study suggests, then, that people adopt different but highly consistent means of dealing with anticipated suffering. However, one conceptual ambiguity should be noted. Most likely for ethical reasons, subjects in the study were told at the onset of the session that they were "free to refuse or to terminate participating at any time . . . if for any reason, you find it difficult to participate." Thus, it is not clear whether the subjects viewed the potential worm-eating as a random, uncontrollable event or whether they felt that by their continued presence, they had implicitly chosen to eat a worm. Obviously, subjects who find that they have cancer or that a loved one has died have no choice about the impending event, and may employ coping strategies different from those used in this study. Nonetheless, the study raises a number of interesting questions. That a victim's reaction to an uncontrollable life event may alter his belief system to a point where he later chooses to suffer is striking. According to Comer and Laird, "As a short-term solution to their immediate problems, such reconstructions are obviously helpful . . . However, if that suffering becomes no longer inevitable, then the results of this study must make us wonder if these adaptations themselves may now become impediments to the amelioration of the suffering [p. 101]."

In later studies, it will be important to determine whether personal attributions for uncontrollable outcomes facilitate or impede the person's ability to cope with such outcomes. Some of the investigators who have noted guilt reactions in their patients have argued that these feelings are destructive and counterproductive and that such feelings interfere with the victim's ability to rehabilitate himself to his fullest capacity (see, for example, Abrams & Finesinger, 1953). Others have stressed that such feelings are quite functional; in discussing the coping responses of parents whose children were dying of leukemia, Chodoff, Friedman, and Hamburg (1964) have stressed the value of such attributions: "It seems that self-blame (and also blame of others) can serve the defensive purpose of denying the intolerable conclusion that no one is responsible, and therefore that neither expiation nor propitiation can undo a malign event which has come about impersonally and meaninglessly [p. 747]."

Becker (1962) has provided a similar analysis in his discussion of bereavement. According to Becker, the most devastating aspect of losing a loved one is that it robs life of meaning. Although shifting blame to the self is painful, Becker maintains that it gives the loss some meaning. Similarly, Averill (1968) has argued that people can make sense of the pain and anguish experienced during bereavement if they view their grief as a form of punishment. To make such a view plausible, they experience feelings of self-condemnation for real or imagined transgressions against the deceased.

It may be painful to blame oneself for such outcomes, but it may be even more painful to view the world as a place where undesirable events happen to innocent people on a random basis (see Lerner, 1975). This is clearly implied in Medea and Thompson's (1974) discussion of guilt among rape victims:

> What appears to be guilt . . . may be the way the woman's mind interprets a positive impulse, a need to be in control of her life. If the woman can believe that somehow she got herself into the situation, if she can feel that in some way she caused it, if she can make herself responsible for it, then she's established a sort of control over the rape. It wasn't someone arbitrarily smashing into her life and wreaking havoc. The unpredictability of the latter situation can be too much for some women to face: If it happened entirely without provocation, then it could happen again. This is too horrifying to believe, so the victim creates an illusion of safety by declaring herself responsible for the incident [pp. 105–106].

If this view is valid, we would expect self-blame to increase as it becomes more likely that people will find themselves in the same situation again. It might also be predicted that people will attribute the outcomes to behaviors that are modifiable (for example, carelessness) rather than behaviors that are fixed and unchangeable (such as low intelligence).

This chapter has probably raised considerably more questions about control motivation than it has answered. Many of the more intriguing phenomena discussed in the chapter, such as the tendency to exaggerate one's blame or guilt for unfortunate life events, have received little systematic investigation. There is a large body of evidence that is consistent with the notion that people have control motivation, but most of this evidence is rather indirect. Furthermore, much of the evidence is equally consistent with other theoretical frameworks, such as a need for justice or a desire for meaning and rationality. Hopefully, future studies will fill some of these gaps. The information gained should not only aid theoretical development, but should be useful in counseling the victims of uncontrollable life events.

ACKNOWLEDGMENTS

Work on this chapter was supported by grants from the National Institute of Mental Health (MH26087-01) and the National Science Foundation (SOC75-14669) to the author. The author is grateful to Fred Bryant, Ronnie Bulman, Sharon Gurwitz, and Jay W. Hillis for their helpful comments on an earlier draft of this paper.

REFERENCES

Abelson, R. P., Aronson, E., McGuire, W. J., Newcomb, T. M., Rosenberg, M. J., & Tannenbaum, P. H. (Eds.), *Theories of cognitive consistency: A sourcebook.* Chicago: Rand McNally, 1968.

Abrams, R. D., & Finesinger, J. E. Guilt reactions in patients with cancer. *Cancer,* 1953, **6**(3), 474–482.

APA Task Force on Health Research. Contributions of psychology to health research. *American Psychologist,* 1976, **31**, 263–274.

Apsler, R., & Friedman, H. Chance outcomes and the just world: A comparison of observers and recipients. *Journal of Personality and Social Psychology,* 1975, **31**, 887–894.

Averill, J. R. Grief: Its nature and significance. *Psychological Bulletin,* 1968, **70**, 721–748.

Beck, A. T. *Depression: Clinical, experimental, and theoretical aspects.* New York: Harper & Row, 1967.

Becker, E. Toward a comprehensive theory of depression: A cross disciplinary appraisal of objects, games, and meaning. *Journal of Nervous and Mental Disease,* 1962, **135**, 26–35.

Bell, L. G. *Influence of need to control on differences in attribution of causality by actors and observers.* Unpublished doctoral dissertation, Duke University, 1973.

Bem, D. J. Self-perception: An alternative interpretation of cognitive dissonance phenomena. *Psychological Review,* 1967, **74**, 183–200.

Brehm, J. W. *A theory of psychological reactance.* New York: Academic Press, 1966.

Bucher, R. Blame and hostility in disaster. *American Journal of Sociology,* 1957, **62**, 467–475.

Chaikin, A. L., & Darley, J. M. Victim or perpetrator? Defensive attribution of responsibility and the need for order and justice. *Journal of Personality and Social Psychology,* 1973, **25**, 268–275.

Chapman, L. J., & Chapman, J. P. Genesis of popular but erroneous psychodiagnostic categories. *Journal of Abnormal Psychology,* 1967, **72**, 193–204.

Chodoff, P., Friedman, S., & Hamburg, D. Stress defenses and coping behavior: Observations in parents of children with malignant disease. *American Journal of Psychiatry,* 1964, **120**, 743–749.

Cialdini, R. B., Darby, B., & Vincent, J. Transgression and altruism: A case for hedonism. *Journal of Experimental Social Psychology,* 1973, **9**, 502–516.

Cialdini, R. B., & Mirels, H. L. Sense of personal control and attributions about yielding and resisting persuasion targets. *Journal of Personality and Social Psychology,* 1976, **33**, 395–402.

Comer, R., & Laird, J. D. Choosing to suffer as a consequence of expecting to suffer: Why do people do it? *Journal of Personality and Social Psychology,* 1975, **32,** 92–101.

de Charms, R. *Personal causation: The internal affective determinants of behavior.* New York: Academic Press, 1968.

de Charms, R. Personal causation training in the schools. *Journal of Applied Social Psychology,* 1972, **2,** 95–113.

Drabeck, T., & Quarantelli, E. L. Scapegoats, villains, and disasters. *Trans-Action,* 1967, **4,** 12–17.

Dweck, C. S. The role of expectations and attributions in the alleviation of learned helplessness. *Journal of Personality and Social Psychology,* 1975, **31,** 674–685.

Friedman, M., & Rosenman, R. H. *Type A behavior and your heart.* New York: Knopf, 1974.

Glass, D. C., Snyder, M. L., & Hollis, J. Time urgency and the Type A coronary-prone behavior pattern. *Journal of Applied Social Psychology,* 1974, **4,** 125–140.

Gurwitz, S. B., & Panciera, L. Attributions of freedom by actors and observers. *Journal of Personality and Social Psychology,* 1975, **32,** 531–539.

Harvey, J. H., Harris, B., & Barnes, R. D. Actor–observer differences in the perceptions of responsibility and freedom. *Journal of Personality and Social Psychology,* 1975, **32,** 22–28.

Hendrick, I. The discussion of the "Instinct to Master." *Psychoanalytic Quarterly,* 1943, **12,** 561–565.

Henslin, J. M. Craps and magic. *American Journal of Sociology,* 1967, **73,** 316–330.

Janis, I. *Air war and emotional stress.* New York: McGraw–Hill, 1951.

Jenkins, H. M., & Ward, W. C. Judgment of contingency between responses and outcomes. *Psychological Monographs,* 1965, 79(Whole No. 594).

Jones, C., & Aronson, E. Attribution of fault to a rape victim as a function of respectability of the victim. *Journal of Personality and Social Psychology,* 1973, **26,** 415–419.

Jones, E. E., & Gerard, H. B. *Foundations of social psychology.* New York: John Wiley & Sons, 1967.

Jones, E. E., Hester, S. L., Farina, A., & Davis, K. E. Reactions to unfavorable personal evaluation as a function of the evaluator's perceived adjustment. *Journal of Abnormal and Social Psychology,* 1959, **59,** 363–370.

Jones, E. E., & Nisbett, R. E. *The actor and the observer: Divergent perceptions of the causes of behavior.* Morristown, New Jersey: General Learning Press, 1971.

Jones, E. E., Rock, L., Shaver, K. G., Goethals, G. R., & Ward, L. M. Pattern of performance and ability attribution: An unexpected primacy effect. *Journal of Personality and Social Psychology,* 1968, **10,** 317–340.

Jones, E. E., & Wortman, C. B. *Ingratiation: An attributional approach.* Morristown, New Jersey: General Learning Press, 1973.

Kelley, H. H. Attribution theory in social psychology. In D. Levine (Ed.), *Nebraska symposium on motivation.* Lincoln: University of Nebraska Press, 1967.

Kelley, H. H. *Attribution in social interaction.* Morristown, New Jersey: General Learning Press, 1971.

Kelley, H. H. *Causal schemata and the attribution process.* Morristown, New Jersey: General Learning Press, 1972.

Kelly, G. A. *The psychology of personal constructs.* New York: Morton, 1955.

Klinger, E. Consequences of commitment to and disengagement from incentives. *Psychological Review,* 1975, **82,** 1–25.

Krantz, D. S., & Glass, D. C. Environmental control and pattern A behavior. Unpublished manuscript, University of Texas at Austin, 1975.

Kubler-Ross, E. *On death and dying.* New York: MacMillan, 1969.

Langer, E. J. The illusion of control. *Journal of Personality and Social Psychology,* 1975, **32,** 311–328.

Langer, E. J., & Roth, J. Heads I win, tails it's chance: The illusion of control as a function of the sequence of outcomes in a purely chance task. *Journal of Personality and Social Psychology,* 1976, **33,** 951–955.

Lerner, M. J. Evaluation of performance as a function of performer's reward and attractiveness. *Journal of Personality and Social Psychology,* 1965, **1,** 355–360.

Lerner, M. J. The desire for justice and reaction to victims. In J. R. Macaulay & L. Berkowitz (Eds.), *Altruism and helping behavior.* New York: Academic Press, 1970.

Lerner, M. J. Deserving versus justice: A contemporary dilemma (Report No. 24). Waterloo: University of Waterloo, 1971.

Lerner, M. J. "Just world" research and the attribution process: Looking back and ahead. Unpublished manuscript, University of Waterloo, 1975.

Lerner, M. J., & Matthews, G. Reactions to suffering of others under conditions of indirect responsibility. *Journal of Personality and Social Psychology,* 1967, **5,** 319–325.

Lerner, M. J., Miller, D. T., & Holmes, J. Deserving and the emergence of forms of justice. In L. Berkowtiz & E. Walster (Eds.), *Advances in experimental social psychology.* New York: Academic Press, 1976.

Lerner, M. J., & Simmons, C. Observer's reaction to the "innocent victim": Compassion or rejection? *Journal of Personality and Social Psychology,* 1966, **4,** 203–210.

Lifton, R. J. Psychological effects of the atomic bomb in Hiroshima: The theme of death. *Daedalus,* 1963, **92,** 462–497.

Lindemann, E. Symptomatology and management of acute grief. *American Journal of Psychiatry,* 1944, **101,** 141–148.

Mallick, S. K., & McCandles, B. R. A study of catharsis of aggression. *Journal of Personality and Social Psychology,* 1966, **4,** 591–596.

Medea, A., & Thompson, K. *Against rape.* New York: Farrar, Straus, & Giroux, 1974.

Michotte, A. E. *The perception of causality.* New York: Basic Books, 1963.

Miller, D. T., & Norman, S. A. Actor–observer differences in perceptions of effective control. *Journal of Personality and Social Psychology,* 1975, **31,** 379–389.

Miller, D. T., & Ross, M. Self-serving biases in the attribution of causality: Fact or fiction? *Psychological Bulletin,* 1975, **82,** 213–225.

Nisbett, R. E., & Borgida, E. Attribution and the psychology of prediction. *Journal of Personality and Social Psychology,* 1975, **32,** 932–943.

Pallak, M. S., Sogin, S. R., & VanZante, A. Bad decisions: The effect of volition, locus of causality, and negative consequences on attitude. Unpublished manuscript, University of Iowa, 1975.

Regan, J. Guilt, perceived injustice, and altruistic behavior. *Journal of Personality and Social Psychology,* 1971, **18,** 124–132.

Rubin, Z., & Peplau, A. Belief in a just world and reactions to another's lot: A study of participants in the national draft lottery. *Journal of Social Issues,* 1973, **29,** 73–93.

Rubin, Z., & Peplau, A. Who believes in a just world? *Journal of Social Issues,* 1975, **31,** 65–89.

Schulz, R. The effects of control and predictability on the physical and psychological well-being of the institutionalized aged. *Journal of Personality and Social Psychology,* in press.

Seligman, M. E. P. Depression and learned helplessness. In R. J. Friedman and M. M. Katz (Eds.), *The psychology of depression: Contemporary theory and research.* Washington, D.C.: Winston-Wiley, 1974.

Seligman, M. E. P. *Helplessness.* San Francisco: Freeman, 1975.

Seligman, M. E. P., Klein, D. C., & Miller, W. R. Depression. In H. Leipenberg (Ed.), *Handbook of behavior therapy*. New York: Appleton-Century-Crofts, in press.

Seligman, M. E. P., & Maier, S. F. Failure to escape traumatic shock. *Journal of Experimental Psychology,* 1967, **74,** 1–9.

Shaver, K. G. Defensive attribution: Effects of severity and relevance on the responsibility assigned for an accident. *Journal of Personality and Social Psychology,* 1970, **14,** 101–113.

Shaw, J. I., & Skolnick, P. Attribution of responsibility for a happy accident. *Journal of Personality and Social Psychology,* 1971, **18,** 380–383.

Smedslund, J. The concept of correlation in adults. *Scandinavian Journal of Psychology,* 1963, **4,** 165–173.

Sogin, S. R., & Pallak, M.S. Responsibility, bad decisions, and attitude change: Volition, foreseeability, and locus of causality for negative consequences, *Journal of Personality and Social Psychology,* 1976, **33,** 300–306.

Starr, B. J., & Katkin, E. S. The clinician as an aberrant actuary: Illusory correlation with the incomplete sentences blank. *Journal of Abnormal Psychology,* 1969, **74,** 670–675.

Storms, M. D. Videotape and the attribution process: Reversing actors' and observers' points of view. *Journal of Personality and Social Psychology,* 1973, **27,** 165–175.

Strickland, L. H., Lewicke, R. J., & Katz, A. M. Temporal orientation and perceived control as determinants of risk-taking. *Journal of Experimental Social Psychology,* 1966, **2,** 143–151.

Veltford, H. R., & Lee, G. E. The Cocoanut Grove fire: A study in scapegoating. *Journal of Abnormal and Social Psychology,* 1943, **38**(Clinical Supplement), 138–154.

Walster, E. Assignment of responsibility for an accident. *Journal of Personality and Social Psychology,* 1966, **3,** 73–79.

Walster, E. "Second guessing" important events. *Human Relations,* 1967, **20,** 239–250.

Ward, W. C., & Jenkins, H. M. The display of information and the judgment of contingency. *Canadian Journal of Psychology,* 1965, **19,** 231–241.

Weiss, J. M., Glazer, H. I., & Pohorecky, L. A. Neurotransmitters and helplessness: A chemical bridge to depression? *Psychology Today,* 1974, **8,** 58–62.

White, R. W. Motivation reconsidered: The concept of competence. *Psychological Review,* 1959, **66,** 297–333.

Wolosin, R. J. *Self- and social perception and the attribution of internal states.* Unpublished doctoral dissertation, University of Michigan, 1968.

Woodworth, R. S. *Dynamics of behavior.* New York: Holt, 1958.

Wortman, C. B. Some determinants of perceived control. *Journal of Personality and Social Psychology,* 1975, **31,** 282–294.

Wortman, C. B., & Brehm, J. W. Responses to uncontrollable outcomes: An integration of reactance theory and the learned helplessness model. In L. Berkowitz (Ed.), *Advances in experimental social psychology* (Vol. 8). New York: Academic Press, 1975.

Wortman, C. B., Costanzo, P. R., & Witt, T. R. Effect of anticipated performance on the attributions of causality to self and others. *Journal of Personality and Social Psychology,* 1973, **27,** 372–381.

Zuckerman, M. Belief in a just world and altruistic behavior. *Journal of Personality and Social Psychology,* 1975, **31,** 972–976.

3

Attribution and Behavior: Social Perception and Social Causation

Mark Snyder

University of Minnesota

Research and theory in social perception and the attribution process are concerned with the processes by which individuals construct causal explanations for behavior and events which they encounter in everyday social interaction. As researchers currently see things, the fundamental questions of one engaged in this information-processing task are two. Is a particular sample of an individual's behavior representative of corresponding inner states, dispositions, or attitudes (dispositional attribution) or a reflection of current social and environmental pressures (situational attribution)? Are the causes of that actor's behavior perceived to be amenable to or beyond self-control? The perceiver's answers to the first question "locate" the perceived causes of an actor's behavior within himself or in the outside social environment. Answers to the second question define the extent to which the perceiver views the actor as responsible for his actions and its consequences.

It is important to realize that perceptions of the situational or dispositional location of cause may be independent of perceptions of self-determination (McGee & Snyder, 1975). Thus, the perceptions "Matthew robbed the store because he had been offered $1,000 to commit the crime" and "Charles robbed the store because he was forced at gunpoint to commit the crime" are both situational attributions, but most perceivers (and our legal-judicial system) would, no doubt, assign less personal responsibility to Charles than to Matthew. We assume that one can and should resist the temptation to earn money illegally but that one cannot and should not resist the desire to stay alive. Similarly, the attributions "Vincent failed his psychology midquarter because he was in a bad mood" and "Michael failed his psychology midquarter because of insufficient effort" are both dispositional perceptions of the causes of behavior, but moods are generally perceived to be less amenable to self-control than are effort and motivation. Although it is all too easy to blur and confuse the distinction between the situational–dispositional and controllability dimensions, it is clear that each dimension focuses on a separate, distinct aspect of our perceptions of the causes of behavior (for a related distinction, see Weiner, Frieze, Kukla, Reed, Rest, & Rosenbaum, 1971).

The outcome of intensive research and creative theorizing has been an impressive array of fairly precise specifications of those social and personal factors which lead perceivers to make the attributions that they do (for example, Collins, 1974; Heider, 1958; Jones & Davis, 1965; Kelley, 1973; Rotter, 1966). Moreover, the notions of attribution theorists have proven equally useful in providing insights about how individuals perceive and understand their own behavior as well as that of other individuals (see, for example, Bem, 1972, and Jones & Nisbett, 1971, for enlightening discussions of the relationship between self-perception and interpersonal perception).

The Perspective of the Perceiver

One hallmark of social perception research is its concern with understanding the construction of causal explanation from the perspective of the perceiver, whether or not his naive theory of causation matches the presumably veridical theory generated by the psychologist as a scientific outside observer. In other words, the attribution process concerns *perceptions* about the causes of behavior and not the *actual causes* of social behavior.

But what of the relationship between attribution and behavior? Do our perceptions of the situational versus dispositional organization of another person's behavior reflect the actual situational or dispositional control of that person's behavior? Within the domain of self-perception, do individuals who tend to organize their self-perceptions in trait or disposition terms also tend to demonstrate more dispositionally controlled social behavior than do others who organize their self-perception in situational terms? The conceptual and empirical links between attributions about the causes of behavior and the actual situational and dispositional determinants of behavior are the central concerns in this chapter.

Several sets of observations seem to suggest that perceptions about the causes of behavior may be independent of the actual causes of behavior. Students of social cognition have time and again noted the importance of both the perceiver and the perceived in the attribution process. Individuals seem to have preferred categories or constructs for organizing their social worlds (for example, Dornbusch, Hastorf, Richardson, Muzzy, & Vreeland, 1965; Kelly, 1955). Thus, what one perceiver might label as altruistic generosity, another might see as ingratiation in the service of manipulative motives. Moreover, the same perceiver may label identical behaviors quite differently when performed by different others. The same item of prosocial behavior could be interpreted as friendliness or kindness in a friend, but ingratiation, manipulativeness, and hypocrisy in an adversary. For example, Regan, Straus, and Fazio (1974) have shown that perceivers attribute identical "good" actions dispositionally for liked others but situationally for disliked others. Similarly, observers view the "bad" actions of those they like as products of situational forces but perceive the same bad actions as reflecting corresponding dispositions in disliked others.

As further evidence of the powerful effect of the perceiver on the outcome of the attribution process, clinical researchers suggest that we may fill in the gaps in our perceptions with information consistent with our preconceived notions about the target of our perceptions. Chapman and Chapman (1967, 1969) have demonstrated that both college students and professional clinicians perceive positive associations between particular Rorschach responses and homosexuality in males, even though these associations are demonstrably *absent* in real life data and in specially constructed experimental stimulus materials. These "signs" are simply those which comprise common cultural stereotypes of gay males. These influences of the perceiver certainly suggest that our perceptions of the causes of behavior might easily stray from accurate portrayal of the true situational and dispositional determinants of behavior.

The Perceiver's "Errors"

In fact, empirical evidence can readily demonstrate that our perceptions about the causes of behavior are often wrong. It is attribution error which time and again characterizes the social perception of participants in experimental investigations of the attribution process. Heider's (1958) observation that we underestimate the impact of social and situational forces and overestimate the role of dispositional causes presumed to underlie the very salient behavior of others in our perceptual field has been repeatedly confirmed (for example, Bem, 1967; Jones & Davis, 1965; Jones & Harris, 1967; Jones & Nisbett, 1971; Kelley, 1967). Social perceivers seem quite willing to perceive dispositions when, in fact, the data (at least in the eyes of the experimenter) do not warrant such inferences.

More intriguing yet is the apparent ease with which our very own self-perceptions, self-attitudes, and self-concepts can be led astray in cleverly contrived misattribution experiments (for example, Davison & Valins, 1969; Nisbett & Schachter, 1966; Nisbett & Valins, 1971; Ross, Rodin & Zimbardo, 1969; Valins, 1966; Valins & Nisbett, 1971). In such misattribution experiments, actors "erroneously" infer dispositional causes for their behavior when in fact the data upon which such explanations are constructed are a product of experimental (situational) manipulations. In truth, the subject's behavior is under situational control at the same time as he makes a dispositional attribution. Such research may, however, reveal very little about the limitations of the perceiver's ability to construct accurate naive analyses of causation. The perceiver may come up with the wrong answers for the right reasons. The information upon which erroneous dispositional attributions are based may look exactly like the patterns of information provided in nature when such attributions are truly warranted.

Social Perception and the "Causes" of Behavior

But when are dispositional or situational attributions actually warranted? As perceivers we tend to fashion our images of ourselves and others largely in dispositional or trait terms (see Schneider, 1973, for a review). Our language, with its rich vocabulary of trait terms combined with awkward and cumbersome means of describing situations, tempts us to talk, if not think, in trait terms. Yet empirical evidence has repeatedly suggested that our intuitions are wrong: rarely do observations within any domain of social behavior reveal the cross-situational consistency and temporal stability which constitute "best evidence" for dispositional or trait inferences (for example, Mischel, 1968). This search for cross-situational consistencies in social behavior has been largely the product of a spirited debate between personality and social behavior theorists. Is behavior controlled by situational factors and hence predictable from characteristics of the surrounding situation, or is it controlled by internal states and dispositions which would facilitate predictions from characteristics of the person, measures of inner states, dispositions, attitudes, and traits? (For excellent reviews of the history of this controversy see Bem & Allen, 1974; Bowers, 1973; Ekehammar, 1974; Mischel, 1968.)

The prevailing situationist view of the causes of human social behavior (see Bowers, 1973, for a review and critique), seems to question the accuracy of the naive person perceiver. For the data of human behavior do not seem to provide a sufficiently solid foundation on which to construct the largely dispositional or trait images which we fashion of ourselves and others. This is not, however, to say that dispositional inferences are never warranted or that social behavior is never under dispositional control. More recent investigations and reconceptualizations of the traits versus situations issue have indicated that at least some of the people some of the time do behave consistently across situations and are predictable from dispositional measures (Bem & Allen, 1974; Snyder & Monson, 1975). What this means is that individuals differ in the extent to which their social behavior is consistent (trait like) or variable (trait free) across social situations. Those whose social behavior is stable across differing situations appear to have "personality traits," and those whose social behavior is predictable from the knowledge of the situational context appear not to have "traits." Thus, an observer may be very wrong if he makes dispositional attributions about some individuals, but he may be quite accurate in ascribing dispositional causes to other individuals. So too might some actors be perfectly accurate in ascribing situational causes to their own behavior. Others would err seriously if they were to do so.

If social behavior differs in the extent to which it is situationally consistent or variable, how do these differences in the presumed situational versus dispositional control of behavior actually relate to social perception? Do those who demonstrate a dispositional organization of their social behavior also tend to perceive the causes of their own behavior in dispositional terms? Similarly, do those whose social behavior is situationally variable or specific tend to ascribe situational causes to their behavior?

The results of several empirical investigations allow us to integrate these perceptual and behavioral concerns. Their outcomes converge on the notion that individuals whose behavior is known to be relatively situationally controlled construe the causes of their behavior in relatively situational terms; by contrast, individuals whose behavior is known to be relatively dispositionally controlled construe the causes of their behavior in relatively dispositional fashion.

Each investigation examined the covariation between separate measures of individual differences in the extent to which persons perceive their behavior in dispositional versus situational terms and diverse methods of identifying situationally and dispositionally controlled behavior. One set of evidence comes from a laboratory study; the other evidence from field–observational settings.

THE LABORATORY INVESTIGATION

The first investigation was the product of research on the interaction between personal and situational determinants of behavior (Snyder & Monson, 1975). It has been suggested that individuals differ in the extent to which their social behavior is consistent (trait like) or variable (trait free) across social situations (Bem & Allen, 1974; Campus, 1974). One interpretation of these findings is that persons differ in the extent to which situational and dispositional factors influence behavior. Moreover, these individual differences can be conceptualized in terms of the social psychological construct of self-monitoring (Snyder, 1972, 1974; Snyder & Monson, 1975). Self-monitoring individuals, out of a concern for the situational and interpersonal appropriateness of their social behavior, are particularly sensitive to the expression and self-presentation of others in social situations and use these cues as guidelines for managing their own social behavior. As a result, these persons show considerable situation to situation discriminativeness or variability in their behavior. By comparison, non-self-monitoring individuals have less concern for the appropriateness of their social behavior. Their behavioral choices in social situations seem to be guided from within by dispositions and other personal characteristics rather than by situational and interpersonal specifications of social appropriateness. As a consequence, these individuals are relatively consistent across situations.

Self-Monitoring and the Control of Behavior

Individual differences in self-monitoring are measured by the Self-Monitoring Scale. The psychometric construction of this instrument and evidence for its convergent and discriminant validity have been presented in detail elsewhere (Snyder, 1972, 1974). Evidence for the relationship between individual differences in self-monitoring as measured by the Self-Monitoring Scale and sensitivity to situationally variable cues to social appropriateness and self-presentation has been documented by Snyder and Monson (1975).

In this study Monson and I demonstrated that the Self-Monitoring Scale may be used to identify persons whose social conformity behavior in a group discussion setting was either very sensitive to situational variation in salience of different peer reference groups or quite independent of such situational cues of social appropriateness of self-presentation. Self-monitoring individuals were highly sensitive to such differences: they were conforming when conformity was the most appropriate interpersonal orientation and nonconforming when reference group norms favored autonomy in the face of social pressures. Non-self-monitoring individuals were virtually unaffected by differences in the discussion group context: presumably their self-presentations were more accurate reflections of their own personal attitudes and dispositions.

Clearly, the self-monitoring construct and the Self-Monitoring Scale can be used to identify individuals who differ in the extent of situational control of their social behavior. Accordingly, Monson and I probed the relationship between self-monitoring and the attribution process.

A Measure of Attribution

We developed a person perception task to assess individual differences in the extent to which persons perceive their behavior in relatively situational versus relatively dispositional terms. Our task involved having the subject judge for himself and for an acquaintance what behavior would be displayed in each of a variety of situations differing in the situational and contextual factors relevant to the display of the diverse traits of generosity, honesty, and hostility. We used measures of the variability (specifically the variance) in the behaviors reported for the self and for an acquaintance to construct an index of the perceived situational versus dispositional organization of each target as perceived by the subject.

These estimates of behaviors across social contingencies are readily interpreted within the person perception framework. To the extent that a person believes that his overt social behavior is a reflection of underlying stable dispositions, he should report little cross-situational variability in that behavior. By contrast, one who considers his social behavior to be caused by or at least molded to fit specific situational contingencies should report considerable variability in his behavior as these factors shift across social settings.

In fact, both theory (Kelley, 1973) and research (McArthur, 1972) indicate that stability over time and setting constitute best evidence for dispositional attributions in interpersonal perception. In the vocabulary of attribution, high self-reported variability in the behavior-rating task designed for this research is an operational definition of situational attribution; low self-rated variability is an operational definition of dispositional attribution.

If, indeed, attributions about the causes of behavior and the actual situational versus dispositional control of behavior do covary, it should be the case that self-monitoring individuals should make more situational self-attributions than should non-self-monitoring individuals. Moreover, an individual's self-perceptions should be relative to those of relevant comparison others, including acquaintances. Self-monitoring individuals ought to make more situational attributions for themselves than for their acquaintances; non-self-monitoring individuals should make more dispositional attributions for themselves than for their acquaintances.

Subjects and Procedure

Our perceivers were 90 undergraduate males enrolled in introductory psychology at the University of Minnesota. We classified subjects as high self-monitoring individuals ($N = 51$) or low self-monitoring individuals ($N = 39$) on the basis of a median (12) split performed on the Self-Monitoring Scale scores collected during a pretest which contained the Self-Monitoring Scale (Snyder, 1974) and the Locus of Control Scale (Rotter, 1966). The Locus of Control Scale, a measure of the extent to which an individual believes that the causes of his behavior are amenable to self-control, was included as a measure of discriminant validity because both the concept and the measure have been related to attribution processes (for example, Collins, 1974). Building upon our previous research which had demonstrated that the Self-Monitoring Scale may be used to identify persons whose social behavior is either very sensitive to situational factors or quite independent of situational cues, we classified those with high self-monitoring scores as situational individuals and those with low self-monitoring scores as dispositional individuals.

Our perceivers all completed the person perception task which concerned their responses in diverse hypothetical situations based upon everyday occurrences. All perceivers read a series of three hypothetical situations, each of which had nine specific contextual variations. The subject was then asked to estimate the probability (from 0 to 10 chances in 10) that he would perform the given behavior in the specific situation described. The subject was also asked to predict what a close friend of his (specified as an acquaintance of the same sex whom he had known for at least a year) would do in the same situations. The three hypothetical situations concerned the behaviors of offering a seat to a middle-age lady on a crowded bus (generosity), returning extra change to a cashier (honesty), and an overt expression of anger to a classmate who had pulled a chair out from under him, causing him to fall to the floor (hostility).

The nine specific contextual variations for each trait involved the presence of members of various reference groups, the presence of various pleasant or unpleasant mood states, and positive or negative previous experiences (for example, "If you were talking to your best friend who was sitting next to you, what is the probability that you ____?" "If you had just finished a quiet afternoon on campus, what is the probability that ____?" "If you had just found out that you had received an A+ on your most difficult midquarter exam, what is the probability that ____?").

Results and Discussion

The variance of the subject's ratings across the variations in situational constraints, either for himself or for his friend, was the major dependent variable. This measure, calculated according to the standard formula for computing the variance, assumes larger values as the subject reports greater situation to situation variability for a target individual. In other words, higher values of this index reflect more situational perceptions, and smaller values of this measure are identified with dispositional attributions. We calculated this measure for each subject for both his self-perceptions and his perceptions of his acquaintance for each of the traits–generosity, honesty, and hostility.

The patterning of the attribution measures provided strong support for our predicted interaction, and an analysis of variance confirmed the statistical reliability of this interaction, $F(1, 88) = 11.27, p < .005$. We performed planned contrasts to specify more precisely the nature of this interaction. High self-monitoring subjects made more situational self-attributions than did low self-monitoring subjects (for high self-monitoring subjects, $\bar{X} = 20.74$; for low self-monitoring subjects, $\bar{X} = 16.61$), $F(1, 88) = 5.81, p < .025$. Furthermore, high self-monitoring subjects made more situational attributions about their own behavior than they did about that of another person (for self, $\bar{X} = 20.74$; for acquaintance, $\bar{X} = 18.08$), $F(1, 88) = 21.09, p < .001$. By contrast, low self-monitoring subjects made more dispositional attributions about themselves than about familiar acquaintances (for self, $\bar{X} = 16.61$, for acquaintance, $\bar{X} = 19.07$), $F(1, 88) = 13.77, p < .001$. We observed identical patterns of results for each trait–generosity, honesty, and hostility–when considered separately. It appears, then, that the hypothesis linking individual differences in self-monitoring to preferences for situational or dispositional attributions has been confirmed.

During the pretest, we had obtained measures of individual differences in internal versus external locus of control (Rotter, 1966). We were, therefore, able to examine the validity of using this measure to predict individual differences in attribution as measured by our experimental task. There was neither a reliable main effect nor interaction between locus of control and the target variable.

Clearly, the first study provided empirical support for the notion that individuals whose behavior is known to be relatively situationally controlled construe the causes of their behavior in relatively situational terms but that individuals whose behavior is known to be relatively dispositionally controlled construe the causes of their behavior in a relatively dispositional fashion.

Self-Monitoring and the Attribution Process

Our research also provided intriguing evidence about the relationship between actor and observer perspectives on the causes of social behavior. Our results indicated that self-monitoring individuals made more situational attributions about their own behavior than they did about that of another person. By contrast, non-self-monitoring individuals made more dispositional attributions about themselves than about familiar acquaintances. The moderating effects of individual differences in self-monitoring on perceptions of selves and acquaintances were such that there was no overall main effect of target in this experiment. Our subjects were no more likely to assign situational causes to themselves than to others. It was only when individual differences in self-monitoring were considered as a moderator variable that divergent perspectives in self-perception and interpersonal perception were evident.

In a theoretical analysis of the differences between self-perception and interpersonal perception, Jones and Nisbett (1971) have suggested that a pervasive tendency exists for individuals to make situational causal attributions about their own behavior and dispositional ascriptions of cause for the identical actions of others. Others do what they do because they are bound by inherent traits, dispositions, and other personal characteristics; we choose our behavior pattern according to particular situational contexts. A series of converging empirical investigations have been offered in support of this hypothesis (Nisbett, Caputo, Legant & Marecek, 1973; Storms, 1973).

The social implications of this hypothesis are considerable. One implication is that clinicians will view their clients' symptoms as expressions of inherent defects, a belief that may seriously hamper the progress of therapy. On the other hand, clients will tend to ascribe their new behavior patterns to the therapist's efforts rather than to themselves. This may undermine any chances for long-term maintenance of behavior change. Similarly, judicial procedures to assign personal responsibility for crimes, efforts to establish equitable social relationships, and attempts at conflict resolution may be jeopardized by such divergent perceptions concerning the causes of behavior.

There is no doubt that the argument for actor–observer differences in perceptions of causality is well reasoned and the empirical evidence suggestive of its validity. Yet we were not able, within our experimental paradigm, to demonstrate that there is a pervasive tendency for individuals to construe their own behavior more situationally than that of others. The fault may lie in the specific nature of our person perception task.

Our behavior ratings made no mention of expectations or attributions concerning the causes of behavior. In our task we simply asked for estimates of behavior in various situations. The task was specifically designed to allow an individual to make situational or dispositional attributions without having, at the same time, to make attributions about freedom of choice or self-determination. Thus, an individual was allowed to report variability in his behavior without having to label it as either adaptive flexibility or capricious whim and without having to label himself as a pawn of circumstances or a shrewd opportunist.

Thus, although self-monitoring individuals perceive more situational variance in their social behavior, it does not necessarily follow that they should locate the causes of their social behavior in the situation. One might perceive this situational variability as reflecting the personal choice of an appropriate self-presentation managed to meet certain situational and interpersonal specifications and contingencies. Or, a situationally variable individual could perceive himself as a chameleon-like or silly-putty creature molded, shaped, or even buffetted by situational forces (see McGee & Snyder, 1975). It may thus be inappropriate to construe perceptions of cross-situational consistency or variability as an operational measure of dispositional versus situational attribution.

However, at least two sets of converging evidence suggest that it *is* appropriate to construe perceptions of cross-situational consistency or variability as an attribution measure. Snyder and Tanke (1975) have found that after forced compliance counterattitudinal essay writing, non-self-monitoring individuals agree with their essays more than do self-monitoring individuals. One interpretation of this occurrence is that non-self-monitoring individuals explain their essay writing behavior dispositionally (they may attribute it to a corresponding attitude) whereas self-monitoring individuals explain their counterattitudinal statement situationally (they may attribute it to requirements of the experimental situation). These results suggest that it is appropriate to interpret self-monitoring differences in perceptions of cross-situational consistency in terms of situational and dispositional self-attributions about the causes of behavior.

But what of the relationship between self-monitoring and the perceptions of the causes of other people's behavior? Our results suggest that self-monitoring individuals make more dispositional attributions about the causes of an acquaintance's generosity, honesty, and hostility than do non-self-monitoring perceivers.

In a study of interpersonal perception, Berscheid, Graziano, Monson, and Dermer (1975) suggested that in an expected interaction context, self-monitoring individuals are more likely than non-self-monitoring individuals to infer dispositions (as measured by extremity and confidence ratings [see Jones & Davis, 1965] on trait dimensions) from observations of behavior of another whom they expect to date socially.

One explanation of these results stresses the functional value of perceiving the behavior of other persons in dispositional trait terms. To do so implies cross-situational consistency and temporal stability of that person's behavior. These characteristics would facilitate prediction and potential influence of the behavior of that individual. It would also allow those who were so motivated to use their expectations (attributions) about that person as a cue to guiding, managing, and monitoring their own social behavior and self-presentation in social interaction with that individual. Such a person is the one identified as the self-monitoring individual by high scores on the Self-Monitoring Scale. For these individuals, perceiving the behavior of another person as dispositionally organized would facilitate use of their perceptions of the behavior of that individual as cues to monitoring their own expressive self-presentation in interaction contexts with that person.

The converging results of Snyder and Tanke (1975) and Berscheid et al. (1975) make more reasonable an interpretation of perceptions of cross-situational consistency and variability in the behavior-rating task of this investigation as indicants of dispositional and situational attributions. It is not clear, however, how to integrate our view of actors and observers with that of Jones and Nisbett (1971), nor is it clear how to fit all of the apparently contradictory sets of data into one conceptual framework.

THE FIELD STUDIES

The results of the laboratory study were certainly encouraging for the hypothesis that links attributions about the causes of behavior with the actual situational and dispositional determinants of social behavior. In fact, they were sufficiently encouraging to motivate us (McGee & Snyder, 1975) to search for further evidence consistent with the hypothesis. We performed two field studies, each designed to demonstrate a correspondence between a measure of individual differences in the extent to which persons perceive their behavior in relatively dispositional versus relatively situational terms and an unobtrusive observation of a criterion behavior which may be either dispositionally or situationally controlled. We conducted these studies in field–observational settings and assessed social perception and social causation with methods different from those used in the laboratory investigations.

The Attribution Measure

In the field studies, we measured individual differences in self-perception of dispositional versus situational organization of social behavior with a measure of self-ascription of traits. Nisbett et al. (1973) have suggested that one result of the tendency to view one's behavior in relatively dispositional terms is that one might view oneself as possessing correspondingly many personality traits. Similarly, a person who sees his behavior as being organized in situational terms should see himself as quite trait free. Nisbett et al. have constructed a questionnaire measure of individual differences in trait ascription. It consists of 20 polar adjectives (for example, realistic–idealistic, cautious–bold, energetic–relaxed), each with the option "depends upon the situation." A subject checks the one of the three alternatives for each trait category which best describes him. The frequency with which individuals endorse the trait-label alternative is a measure of perceived dispositional control of behavior.

The Behavior Measure

In contrast with the laboratory investigation, we assessed individual differences in situational versus dispositional control by means of actual observations of behavior. In choosing a criterion behavior for observation, we sought one which could be reliably categorized as either situationally or dispositionally controlled and which was sufficiently distinct from those traits sampled in the Nisbett et al. (1973) instrument. In addition, we wanted it to be a behavior which could be unobstrusively observed in a nonlaboratory context. One lead in our choice of the criterion behavior was provided by Schachter (1971a, 1971b) in his research on the determinants of eating behavior. In an impressive series of laboratory and field studies, Schachter and his colleagues have demonstrated that the eating behavior of obese and normal weight individuals seems to be influenced by different sets of cues. Obese as compared with normal weight individuals seem to be particularly sensitive to what Schachter called external cues, such as time of day, visibility of the food, and particular taste or textural aspects of food. At the same time, these obese individuals appear to be remarkably insensitive to internal physiological signals of food deprivation and satiation. These physiological cues appear to play a more prominent role in the regulation of the eating behavior of normal weight individuals.

Following Schachter's example that eating behavior may be profitably utilized for the study of the situational versus dispositional control of behavior, we chose as our criterion behavior the use of condiments or seasonings such as salt. We reasoned, quite simply, that an individual who salts his food after tasting it is responding to situational cues provided by the taste of the food ("the food needs salt"). By contrast, the individual who salts his food before tasting it is presumably doing so in response to dispositional cues related to a particular need or desire for salt independent of food-related cues ("I like salt," "I need salt"). Accordingly, the salting behavior of those who salt after tasting the food was considered to be under situational control and the salting behavior of individuals who salt before tasting food was operationally considered to be under dispositional control.

Interestingly, our operational definition converges remarkably well with the wisdom of J. C. Penney. In a radio interview some years ago, the department store tycoon suggested that a good way to spot executive potential is to observe whether an individual salts his food before or after tasting it. Mr. Penney reasoned that any person who salts food without determining that it in fact needed salt might not weigh all the facts before making important executive decisions.

Our hypothesis, then, was that individuals who attribute relatively few traits to themselves would be particularly likely to salt their food after tasting it. By contrast, those who endorse relatively many traits would be particularly likely to salt their food before tasting it. To test our hypothesis, we conducted two studies in field-observational settings.

The First Field Study

Subjects and procedure. In the first study McGee observed 14 females and 16 males ranging in age from 18 to 55 in four different local restaurants. Each person had ordered a meal consisting of a minimum of three courses. McGee selected only subjects who had ordered complete meals, as opposed to those ordering sandwiches or similar items. He did this in order to minimize the effect of kind of food ordered. Furthermore, he chose to observe only individuals seated at tables where salt was readily available. After observing whether the subject salted his or her food before or after tasting it, McGee approached him or her and administered the 20-item trait-ascription questionnaire.

Results. To test our hypothesis that individual differences in perceptions concerning situational versus dispositional causation of behavior would covary with situational versus dispositional control of actual behavior, we dichotomized scores on the trait-ascription measure at the median to form situational ($N = 15$) and dispositional ($N = 15$) self-attribution groups. We then examined the frequencies with which individuals in each of these groups salted their food before or after tasting it. Those who perceived their behavior to be relatively dispositionally controlled were most likely (12 of 15 cases) to salt their food before tasting it, whereas those who perceived their behavior to be situationally determined were most likely (11 of 15 cases) to salt their food after tasting it, $\chi^2(1) = 8.57, p < .005$.

We reached an equivalent conclusion by comparing the mean number of trait ascriptions for those individuals ($N = 16$) who salted before tasting ($\bar{X} = 14.87$

traits ascribed) and those ($N = 14$) who salted after tasting ($\bar{X} = 6.9$ traits ascribed, $t(28) = 5.22$, $p < .005$). Clearly, by either method of statistical assessment, there was a strong and reliable realtionship between self-perception of the situational versus dispositional control of behavior and the situational versus dispositional control of actual behavior. This result is particularly impressive because of the distinctiveness of the criteria chosen to define perception and behavior. Nowhere on the 20 items of the trait-ascription measure is there any mention of any behavior even remotely related to food, its preparation, or its consumption.

Of course, the calculation of significant differences between group means is useful, but does not necessarily yield much information about the power of the questionnaire measure to identify accurately those individuals who will salt before or those who will salt after tasting their food. Tilton (1937) has provided a simple estimate of the percentage of overlap between two distributions. Using Tilton's measure, we calculated the percentage of overlap in number of traits ascribed by subjects who salted before and subjects who salted after tasting to be 32%. Dunnette (1966) has suggested that overlap of less than 45% may confidently be taken as indicating an excellent separation between a measure and dichotomous behavior classification and usually denotes a relationship useful for practical purposes. Another index of the size of the relationship between the perceptual and behavioral measures in this first study is the point biserial correlation of .71 between salting classification and self-ascription scores.

The Second Field Study

As encouraging as the results of the first salting study were for our hypothesis, there were certain shortcomings. The observer who administered the trait-ascription questionnaire was at the same time aware of the salting behavior classification of the respondent. Moreover, no observations were made of the self-perceptions of individuals who did not salt their food. Accordingly, we conducted a modified replication to demonstrate the reliability of our findings and to provide additional information about the nature of our observed relationship. In this study, we employed two observers, one to observe the salting behavior and another to administer the trait-ascription questionnaire. We also included observations of subjects who did not salt their food. In addition, we obtained measures of the subjects' explanations for salting or not salting their food. We expected that subjects who salted their food before tasting it would explain their behavior in terms of characteristics of themselves ("I like salt," "I need salt"). By contrast, we expected that subjects who salted their food after tasting it would explain their behavior in terms of characteristics of the food ("The food needs salt," "The taste needs improvement"). We further assumed that these specific attributions about the causes of salting behavior would covary with more general tendencies toward perception of dispositional versus situational causes of behavior. For example, in the case of nonsalters, we expected that those who ascribed relatively few traits to themselves would explain their nonsalting behavior in relatively situational terms ("The food did not need salt," "The taste was acceptable without salt"), whereas diners who ascribed relatively many traits to themselves would explain their nonsalting behavior in dispositional terms ("I don't like salt").

Subjects and procedure. The procedure for the second study was identical to that of the first investigation, with the exception that the observer who administered the trait questionnaire was unaware of the salting behavior classification of the subject. After each of the 30 subjects (16 males and 14 females ranging in age from 20 to 64 observed in four local restaurants) had completed the trait measure, the observer asked him or her to explain the salting or nonsalting behavior. Salters were asked: "I noticed that you used salt on your food. Why did you salt your food?" Nonsalters were asked: "I noticed that you did not salt your food. Why didn't you salt your food?" The observer transcribed each subject's answers and later coded them into dispositional explanations (those that focused on personal characteristics of the subject) and situational explanations (those that focused on characteristics of the food).

Results. Once again we dichotomized scores on the trait-ascription questionnaire at the median to form situational and dispositional self-attribution groups. Twenty-four of the 30 subjects were categorized as salters. We then examined the frequency with which these individuals salted their food before or after tasting it.

Our results clearly replicated the initial study. Diners who ascribed relatively few traits to themselves were more likely to salt their food after tasting it (9 of 13 cases) than before tasting it; eaters who ascribed relatively many traits to themselves were particularly likely (8 of 12 cases) to salt their food before tasting it, $\chi^2(1) = 4.19$, $p < .025$. In addition, subjects who salted their food before tasting it ($N = 12$) attritubted a mean of 16.92 traits to themselves, whereas subjects who salted their food after tasting ($N = 12$) attributed a mean of only 12.5 traits to themselves. This difference was reliable, $t(22) = 2.65, p < .02$. As in the first study, Tilton's (1937) measure of overlap indicated that there was a considerable degree of predictability between the continuous questionnaire measure and the dichotomous behavioral criterion.

We also examined the relationship, for salters, of salting before versus after tasting food and situational versus dispositional verbal explanations offered for salting behavior. There was a strong association between salting behavior and the explanation one offers for it, $\chi^2(1) = 10.67$, $p < .005$. Our subjects who salted their food after tasting it were particularly likely to explain their salting behavior in terms of characteristics of the food (10 of 12 cases). On the other hand, those who salted their food before tasting it were particularly likely to explain their behavior in terms of aspects of themselves (10 of 12 cases).

Moreover, the verbal explanation offered by salters for salting either before or after tasting their food were remarkably consistent with their more general tendencies to make either trait or situational attributions. Those individuals who ascribed relatively few traits to themselves also tended to explain their salting behavior in terms of characteristics of the food, whereas those who ascribed relatively many traits to themselves seemed to explain their salting behavior in terms of personal characteristics, $\chi^2(1) = 4.19, p < .05$.

What about the relationship between self-perception and explanations of behavior for the 6 subjects who did not salt their food? Three of the nonsalters explained their nonsalting behavior in terms of personal characteristics. All 3 of these individuals scored above the median number of trait ascriptions for the entire sample of 30 subjects. Three of the nonsalters explained their behavior in terms of characteristics of the food. Two of these individuals were below the median number of trait ascriptions for the entire sample. It seems that for these nonsalting individuals, explanations of their behavior in this specific context were congruent with general tendencies to perceive their behavior as being organized either in trait terms or in situational terms.

Discussion

Thus, both field studies provided strong support for the hypothesis that individual differences in self-perceptions of situational versus dispositional control of behavior and individual differences in actual situational versus dispositional control of behavior would covary. We had assumed that individuals who construed their behavior as under dispositional control would ascribe relatively more personality traits to themselves than those who perceived their behavior to be under situational control. We had assumed that individuals who salted their food before tasting it were responding to dispositional characteristics such as needs, tastes, desires, and dispositions; whereas persons who salted their food after tasting it did so in response to taste, texture, and other situational characteristics of the food being eaten. In both field studies, we found that the more traits an individual endorsed the more likely he was to salt his food before tasting it. In the second study we also found that our subjects explained their salting behavior in a fashion consistent with our hypothesis. Diners who salted their food before tasting it explained this in terms of characteristics of themselves; those who salted after tasting explained their behavior in terms of characteristics of the food itself.

CONCLUDING REMARKS

The convergence of the results of the series of studies provide strong confirmation for the hypothesis that self-perception concerning the causation for one's own behavior is actually related to the situational versus dispositional causation of that behavior. Each study assessed individual differences in the situational versus dispositional control of behavior and individual differences in situationally versus dispositionally organized self-perceptions. In the laboratory study, social causation was assessed by scores on the Self-Monitoring Scale; in the field studies, by actual observations of behavior. In the laboratory investigation, social perception was measured by reports of the situational versus dispositional organization of social behavior; in the field studies, by trait ascription. In all cases, there was an impressive correspondence between the perceptual and behavioral measures.

Even though theory and research in person perception have stressed the perceiver's phenomenological views of the world and inferences about cause—whether or not these world views and perceptions are correct—it does turn out that, at least in the contexts studied in this series of investigations, an individual's construction of his or her social reality can be a fairly good match to the actual state of affairs. Of course, perceivers can and do err in their interpretations of behavior—particularly when they deal with limited, biased, or constrained informational inputs.

By necessity, our search for signs of covariation between attributions about the causes of behavior and the actual situational and dispositional determinants of behavior has required a correlational research strategy. Students of psychology learn early in their educational careers of the pitfalls associated with making causal inferences from correlational data. Thus, in the field studies, do those who salt their food before tasting it do so because they have dispositionally organized self-concepts? Or, do individuals who ascribe relatively many traits to themselves do so because their self-concepts are a reflection of this dispositionally controlled behavior? Similarly, in the laboratory study, do self-monitoring individuals make frequent situational self-attributions because they know full well that their behavior is in fact highly sensitive to situational contingencies? Or, do those who for whatever reason believe their behavior to be a product of situational factors proceed to act on this belief and whenever possible guide their social behavior on the basis of relevant situational and interpersonal cues? Which comes first: social perception or social causation?

Answers to these questions are, of course, speculative. Perhaps the pattern of situational and dispositional determinants of our behavior is a self-fulfilling product of our situationally or dispositionally organized self-concepts. Alternately, our self-concepts may be generated by self-perception inferences from observations of the situational versus dispositional organization of our actual behavior. Both the self-fulfilling and the self-perception hypotheses are plausible interpretations of the kinds of covariation between attribution and behavior observed in our investigations. It should be possible to conduct experimental investigations of each hypothesis. To test the self-fulfilling hypothesis, one could experimentally create situational and dispositional self-concepts and observe their effects on the criterion behavior which can be realiably classified as either situationally or dispositionally controlled. To evaluate the self-perception hypothesis, one must produce situational or dispositional behavior and observe the effects of this independent variable on appropriate measures of situational or dispositional self-images.

Although such investigations would tell us what patterns of influence can exist between attribution and behavior, they can never tell us for certain what patterns of causal relationship and mutual feedback and interplay link social perception and social causation in everyday life.

What is certain, however, is that social perception and social causation do go hand in hand. And well they should. Any species for which attribution and behavior were chronically at odds would, no doubt, be a likely candidate for extinction.

ACKNOWLEDGMENTS

Some of the research reviewed in this chapter was supported by grants in aid of research from the Graduate School of the University of Minnesota and by National Institute of Mental Health Grant MH 24998. Preparation of this chapter was supported in part by National Science Foundation Grant SOC75-13872.

The empirical investigations reported in this chapter were conducted in collaboration with Thomas C. Monson (the laboratory study) and Mark G. McGee (the field studies). Paul C. Rosenblatt provided constructive advice and comments on earlier versions of this chapter.

REFERENCES

Bem. D. J. Self-perception: An alternative interpretation of cognitive dissonance phenomena. *Psychological Review,* 1967, **74,** 183–200.

Bem, D. J. Self-perception theory. In L. Berkowitz (Ed.), *Advances in experimental social psychology* (Vol. 6). New York: Academic Press, 1972.

Bem, D. J., & Allen, A. On predicting some of the people some of the time: The search for cross-situational consistencies in behavior. *Psychological Review,* 1974, **81,** 506–520.

Berscheid, E., Graziano, W., Monson, T., & Dermer, M. The initiation of attribution processes: Attention, attraction, and attribution as a function of outcome dependency. Unpublished manuscript, University of Minnesota, 1975.

Bowers, K. S. Situationism in psychology: An analysis and a critique. *Psychological Review,* 1973, **80,** 307–336.

Campus, N. Transituational consistency as a dimension of personality. *Journal of Personality and Social Psychology,* 1974, **29,** 593–600.

Chapman, L. J., & Chapman, J. P. The genesis of popular but erroneous psycho-diagnostic observations. *Journal of Abnormal Psychology,* 1967, **72,** 193–204.

Chapman, L. J., & Chapman, J. P. Illusory correlations as an obstacle to the use of valid psycho-diagnostic signs. *Journal of Abnormal Psychology,* 1969, **74,** 271–280.

Collins, B. E. Four components of the Rotter Internal–External Scale: Belief in a difficult world, a just world, a predictable world, and a politically responsive world. *Journal of Personality and Social Psychology,* 1974, **29,** 381–391.

Davison, G. C., & Valins, S. Maintenance of self-attributed and drug-attributed behavior change. *Journal of Personality and Social Psychology,* 1969, **11,** 25–33.

Dornbusch, S. M., Hastorf, A. H., Richardson, S. A., Muzzy, R. E., & Vreeland, R. S. The perceiver and the perceived: Their relative influence on the categories of interpersonal cognition. *Journal of Personality and Social Psychology,* 1965, **1,** 434–440.

Dunnette, M. D. *Personnel selection and placement.* Belmont, California: Wadsworth, 1966.

Ekehammar, B. Interactionism in personality from a historical perspective. *Psychological Bulletin,* 1974, **81,** 1026–1048.

Heider, F. *The psychology of interpersonal relations.* New York: John Wiley & Sons, 1958.

Jones, E. E., & Davis, K. From acts to dispositions: The attribution process in person perception. In L. Berkowitz (Ed.), *Advances in experimental social psychology* (Vol. 2). New York: Academic Press, 1965.

Jones, E. E., & Harris, V. A. The attribution of attitudes. *Journal of Experimental Social Psychology,* 1967, **3,** 1–24.

Jones, E. E., & Nisbett, R. E. *The actor and the observer: Divergent perceptions of the causes of behavior.* Morristown, New Jersey: General Learning Press, 1971.

Kelly, G. A. *The psychology of personal constructs.* New York: Norton, 1955.

Kelley, H. H. Attribution in social psychology. In D. Levine (Ed.), *Nebraska Symposium on Motivation.* Lincoln: University of Nebraska Press, 1967.

Kelley, H. H. The process of causal attribution. *American Psychologist,* 1973, **28,** 107–128.

McArthur, L. F. The how and what of why: Some determinants and consequences of causal attribution. *Journal of Personality and Social Psychology,* 1972, **22,** 171–193.

McGee, M. G., & Snyder, M. Attribution and behavior: Two field studies. *Journal of Personality and Social Psychology,* 1975, **32,** 185–190.

Mischel, W. *Personality and assessment.* New York: John Wiley & Sons, 1968.

Nisbett, R. E., Caputo, C., Legant, P., & Marecek, J. Behavior as seen by the actor and as seen by the observer. *Journal of Personality and Social Psychology,* 1973, **27,** 154–165.

Nisbett, R. E., & Schachter, S. Cognitive manipulation of pain. *Journal of Experimental Social Psychology,* 1966, **2,** 227–236.

Nisbett, R. E., & Valins, S. *Perceiving the causes of one's own behavior.* Morristown, New Jersey: General Learning Press, 1971.

Regan, D. T., Straus, E., & Fazio, R. Liking and the attribution process. *Journal of Experimental Social Psychology,* 1974, **10,** 385–397.

Ross, L. D., Rodin, J., & Zimbardo, P. G. Toward an attribution therapy: The reduction of fear through induced cognitive–emotional misattribution. *Journal of Personality and Social Psychology,* 1969, **12,** 279–288.

Rotter, J. B. Generalized expectancies for internal versus external control of reinforcement. *Psychological Monographs,* 1966, **80,**(1, Whole No. 609).

Schachter, S. Some extraordinary facts about obese humans and rats. *American Psychologist,* 1971, **26,** 129–144. (a)

Schachter, S. *Emotion, obesity, and crime.* New York: Academic Press, 1971. (b)

Schneider, D. J. Implicit personality theory: A review. *Psychological Bulletin,* 1973, **79,** 294–309.

Snyder, M. Individual differences and the self-control of expressive behavior. (Doctoral dissertation, Stanford University, 1972). *Dissertation Abstracts International,* 1972, **33,** 4533A–4534A. (University Microfilms No. 73-4598)

Snyder, M. The self-monitoring of expressive behavior. *Journal of Personality and Social Psychology,* 1974, **30,** 526–537.

Snyder, M., & Monson, T. C. Persons, situations, and the control of social behavior. *Journal of Personality and Social Psychology,* 1975, **32,** 637–644.

Snyder, M., & Tanke, E. A. Behavior and attitude: Some people are more consistent than others. Unpublished manuscript, University of Minnesota, 1975.

Storms, M. D. Videotape and the attribution process: Reversing actors' and observers' points of view. *Journal of Personality and Social Psychology,* 1973, **27,** 165–175.

Tilton, J. W. The measurement of overlapping. *Journal of Educational Psychology,* 1937, **28,** 656–662.

Valins, S. Cognitive effects of false heart-rate feedback. *Journal of Personality and Social Psychology,* 1966, **4,** 400–408.

Valins, S., & Nisbett, R. E. *Attribution processes in the development and treatment of emotional disorder.* Morristown, New Jersey: General Learning Press, 1971.

Weiner, B., Frieze, I., Kukla, A., Reed, L., Rest, S., & Rosenbaum, R. M. *Perceiving the causes of success and failure.* Morristown, New Jersey: General Learning Press, 1971.

4

Attribution of Freedom

John H. Harvey
Vanderbilt University

> If the concept of freedom appears to the reason as a senseless contradiction, like the possibility of performing two actions at one and the same instant of time, or the possibility of effect without cause, that only proves that consciousness is not subject to reason. It is this unwavering, certain consciousness of freedom—a consciousness indifferent to experience or reason recognized by all thinkers and felt by everybody without exception—it is this consciousness without which there is not imagining man at all . . . [Tolstoy in his postscript to *War and Peace*].

As Tolstoy so elegantly suggests, an "unwavering consciousness of freedom" is a real phenomenon in the experience of people. In the literature of social psychology, evidence has started to accumulate which indicates that this subjective sense of freedom has important consequences, identifiable antecedents, and is associated with other meaningful psychological experiences (see Chapters 2 and 11). People frequently make inquiries and attributions about their state of freedom in various social situations, and their feeling of freedom often represents a matter of considerable import to them (for example, the person who considers how much freedom he/she has in choosing a mate; happiness in the decisional situation is probably associated with some sense of freedom). People also often assess and make attributions about the extent to which others are free.

Philosophers, and to some extent psychologists too, have long debated the issue of the reality of freedom. However, for the purpose of inquiry into people's *perception* of freedom, the answer to whether or not they are really free does not have to be established. Regardless of the reality of freedom, it is an everyday empirical truth that people perceive and can report varying degrees of freedom in their activities.

In another area of considerable controversy, Skinner (1971) in his well-known book *Beyond Freedom and Dignity* has contended that not only is freedom an illusion having no basis in reality, but also that it seriously impedes intelligent rational cultural change. Further, Skinner claims that the literature of freedom (which he says consists of books, pamphlets, speeches) has been designed to "make men 'conscious' of aversive control [for example, punishment] but that in its choice of methods it has failed to rescue the happy slave [p. 40]." Skinner seems to be suggesting that there are forms of control which do not breed escape or revolt but which nevertheless may have harmful consequences; furthermore, it is under these forms of control that an individual is most likely to have an illusory experience of freedom. As an illustration, Skinner notes the government's practice of inviting prisoners to volunteer for possibly dangerous experiments with the promise of shortened sentences. This practice has a controlling effect upon behavior—even as threat of punishment does—and Skinner is asserting that the literature of freedom does not make people aware that such practices, which rely upon positive reinforcement to work, are indeed forms of control.

Skinner's argument about the role of positive reinforcement in fostering an illusion of freedom is provocative. However, some difficult questions may be raised about the more moralistic aspects of his analysis (namely, that people should somehow go "beyond" feeling free since this experience hinders rational cultural planning). We may ask, for example, will people never entertain the feeling of freedom in the new culture envisioned by Skinner? Or will they have some other, perhaps more "rational," feeling as a substitute? Or, if people adequately understand the role of positive reinforcement in their affairs, does it matter whether or not they sometimes feel that they are free? For further critical discussion of Skinner's position, the reader is referred to Wheeler (1973).

In this paper I will briefly review work on the consequences of perceived freedom in attribution theory and provide a more extensive review of theoretical analysis and investigations concerned with the determinants and correlates of perceived freedom.[1]

[1] Although conceptually it seems possible (and perhaps useful) to make distinctions among the concepts of perceived freedom, perceived choice, volition, free will, etc., the small amount of empirical evidence we know about (for example, Harvey, Harris, & Barnes, 1975) suggests that at least the first two concepts listed are perceived as synonymous by average subjects in psychology studies. For purposes of exposition in this paper, these concepts will be considered as referring in a rather general fashion to the same phenomenological state.

A BRIEF REVIEW OF WORK ON CONSEQUENCES OF PERCEIVED FREEDOM IN ATTRIBUTION THEORY

For many years, perceived freedom has been recognized as an important variable in attribution theory. Its centrality as a major variable affecting attributions and related behavior is suggested by practically all of the early, influential attribution theorists (Heider, 1958; Jones & Davis, 1965; Kelley, 1967). Most of these writers seem to have been concerned with the role of perceived freedom in influencing whether attributions are directed "inward," toward the actor, or "outward," toward something in the environment. A general proposition emerging from these statements is that the attribution of freedom may mediate the relationship between the consequence of an overt action and the perceived causal role of the actor in producing the consequence. To the extent that an actor is seen as having been free in taking an action, he will be linked as a causal agent to the consequence of that action. For example, did Patty Hearst *freely* take part in certain illegal activities of the Symbionese Liberation Army (for example, robbery of a bank in California)? If her actions were not freely taken, then her presumed causal role in the production of the negative effects (loss of money by the bank, injuries of bystanders to the robbery) may be deemphasized. However, if she is seen as having acted freely, she will be causally linked to these negative effects, and subsequent reprisal activities by authorities may reflect a consideration of this linkage. Stated succinctly, in real-life affairs, if we are considered free in our acts, we are usually held responsible for the effects of those acts.

Empirically, the foregoing general proposition about the role of perceived freedom in attributional phenomena has been indirectly tested in a number of studies concerned with determinants of the attribution of attitudes. Steiner and Field (1960) conducted a study in which an accomplice expressed prosegregation opinions in a group discussion. In some cases, he was assigned the role by the experimenter, while in other cases, he chose his role with apparent freedom of choice. Steiner and Field found that naive subjects who were members of the discussion groups were more confident in their assignment of prosegregation attitudes to the accomplice in the choice condition than were subjects in the no-choice condition.

Jones and Harris (1967) and Jones, Worchel, Goethals, and Grumet (1971) have examined the effects of perceived freedom on attribution of attitudes within the context of Jones and Davis' (1965) theory of correspondent inferences. The general question asked in this work was: How will an actor's apparent freedom in taking an action affect an observer's inference of the actor's attitude? Similar to Steiner and Field's rationale, correspondent inference theorists suggest that the observer will be interested in making a correspondent (or correct) inference about the actor's true attitudinal disposition and that circumstances surrounding the action–such as the actor's apparent freedom–and the nature of the action will influence the strength of the inference.

The common procedure used in this work concerned with correspondent inference theory involved (a) an accomplice who was serving in the role of a stimulus person either being given freedom of choice or not in taking an attitudinal position which was unpopular, (b) the stimulus person performing an overt action (for example, writing an essay) relevant to the position, and (c) the observer–subject making inferences relevant to the stimulus person's true attitudes. A general finding deriving from these investigations was that attitudes are attributed to be more in line with the overt action under choice than no-choice conditions. This general finding has been replicated in recent research by Miller (1974).

In terms of correspondent inference theory, these data provide support for the assumption that observers gain more information about an actor's predisposition when it is apparent that some action relevant to the predisposition was not taken under coercion or that the actor was not bribed to take the action. A further implication of this research is that the extent to which an actor is considered to have acted freely is especially important when the action in question deviates from normality. For further discussion of perceived choice as a major variable in correspondent inference theory, the reader is referred to Maselli and Altrocchi (1969).

In the sections below, theory and research pertaining to some of the major determinants of perceived freedom will be discussed.

DETERMINANTS OF ATTRIBUTED FREEDOM

Conceptual Analyses

The work of Brehm and Cohen (1962) and Kelley (1967). The theoretical impetus for work on determinants of attributed freedom has come from several sources. Perhaps the first allusion to the study of the determinants may be found in Brehm and Cohen's (1962) extension of dissonance theory. They proposed that volition provides an important source of implication in the arousal of cognitive dissonance and suggested that when inconsistency between cognitions occurs, the greater the volition, the greater the dissonance. In developing this proposition, Brehm and Cohen mentioned some particular conditions in which volition plays a role in the arousal of dissonance.

In his essay on attribution theory, Kelley (1967) assumed Brehm and Cohen's definition of perceived freedom as *the feeling of control over one's own behavior* and extracted from their analysis a set of general conditions conducive to high subjective freedom. These conditions are:

1. when the constraints against a person's leaving a situation (of neutral or negative attractiveness) are low, and the person stays in the situation;
2. when the legitimate forces producing compliance are low, and the person complies;
3. when alternatives are equal in attractiveness, and the person chooses one of them;
4. when the amount of pressure to make a choice between equally attractive alternatives is low, and the person makes a choice;
5. when the strength of illegitimate forces (to comply) is high, and the person complies;
6. when much conscious consideration is given to the choice;
7. when much uncertainty, conflict, and the potentiality of alternative responses exist in making the choice.

In Kelley's analysis, these conditions are explicitly related to self-attribution, although some of them may also be theorized to influence attribution to other. For example, a person's compliance with a request even though illegitimate pressures are known to have been exerted (the fifth condition) may serve as a cue to others that the person must be exercising his own free will in complying. The first five conditions listed above seem to suggest an essentially retrospective, reflective process in which the individual is attempting to understand his own causal role in producing certain actions. A sense of personal causation should be directly related to high-subjective freedom. More of an ongoing feeling-type process is suggested in the last two conditions. Kelley hypothesizes that the individual "feels" more freedom when he deliberates long and/or experiences high uncertainty and conflict in making a choice. Of these conditions outlined by Kelley, only equality of attractiveness of alternatives and uncertainty/conflict have been directly examined in research. This evidence will be reported below after a discussion of other analyses of perceived freedom.

Steiner's (1970) analysis. Theoretical statements by Steiner (1970) and Mills (1970) have stimulated the greatest amount of research on the antecedents of perceived freedom. Our discussion of Steiner's conception will focus only on what he terms decision freedom, which essentially refers to a person's judgment of the comparative gains offered by multiple alternatives. Because it does not necessarily involve a situation in which options are considered and a decision is made, the concept of outcome freedom (defined as an individual's gains that he expects to receive by pursuing desired objectives) in Steiner's analysis will not be of direct concern here.

More specifically, Steiner defines decisional freedom as the volition an individual believes himself to exercise when he (a) decides whether or not to seek a specific outcome or (b) decides whether to seek one outcome rather than another. Steiner hypothesizes that maximum decision freedom should occur when the anticipated investment in an action equals the expected payoff associated with it, or when the expected gain is zero. He defines expected gain as the margin by which the expected payoff (valence × subjective probability of an action) exceeds the investment which the individual feels he must make. Thus, the individual is assumed to develop a perception of freedom after taking into account a number of factors associated with the utility of a particular line of action as opposed to some other line(s) of action.

Steiner's analysis is assumed to hold for both actors and observers (Steiner, Rotermund, & Talaber, 1974). However, as will be discussed later, most of the relevant research has focused on observers' attributions about actors' freedom given certain information about the actors' behavior and the context of its occurrence.

Mills' (1970) analysis. Mills suggests that a person will perceive choice in taking an action to the extent that an alternative action is perceived to exist. Given one possible course of action, another similar kind of action will be perceived as an alternative if (a) the outcome of the action is perceived as different in attractiveness from the outcome of the given action and (b) there is some likelihood that the alternative action will lead to a more desirable outcome than will the given action (that is, there is uncertainty about whether the given action will lead to the most desirable outcome). Both a perceived difference in attractiveness and some uncertainty about outcome must exist for a high degree of perceived choice to obtain.

From these two general propositions, a number of predictions can be derived. For example, Mills' analysis leads to the prediction that a perception of no difference between choice options will inhibit a sense of choice in a decision situation. Further, if the outcomes of potential lines of action are close to one another in attractiveness, uncertainty about which outcome is better should be produced, and a relatively high sense of choice should result. Most of the research testing Mills' analysis has involved actors' attributions about their own choice.

A brief comparison of Steiner's (1970) and Mills' (1970) analyses. Since Steiner's and Mills' analyses have had the most impact on research concerned with antecedents of perceived freedom, it might be informative to point out some apparent theoretical similarities and dissimilarities between the two approaches. The emphasis in Steiner's analysis of perceived decisional freedom is on the individual's rational, sometimes probabilistic assessment of the gains and losses associated with various choice options. A person's sense of freedom is directly linked to his understanding of the payoff or reinforcement matrix associated with a decision. On the other hand, Mills' conception stresses the role of uncertainty and differentiation as mediators of perceived choice. Uncertainty in Mills' conception is conceptually similar to the notion of subjective probability in Steiner's conception. The emphasis on gains and losses is less pronounced in Mills' analysis but does exist implicitly in that uncertainty pertains to a decision maker's quandary about the outcome of various options. Both analyses emphasize the individual's reception and processing of information in decision-making situations, with the burden of these activities assumed to be greater in Steiner's conception. In their analyses, neither Steiner nor Mills is concerned with how perceived freedom may be influenced by strategic and/or self-serving tendencies on the part of the decision-maker (some evidence along this line will be presented later in the paper). As will be mentioned in the next section, rival predictions for the two analyses are possible mainly in the area of consideration of the extent to which outcomes differ in attractiveness for perceived choice to be high.

Research on Determinants of Perceived Freedom

Similarity in attractiveness of options. Harvey and Johnston (1973) and Jellison and Harvey (1973) have examined the derivation from Mills' analysis that people will perceive a greater sense of freedom in their decisions if choice options are close together than if far apart in attractiveness. This prediction follows from the assumption that uncertainty about the attractiveness of outcomes of options will increase if the outcomes of the options become more similar (but not identical). The procedure in these studies involved asking subjects to choose between or among options which were similar in kind but varied in quality. Subsequently, subjects reported their perception of choice in making the decision.

In these studies, the decision involved predicting the football team that would win a game given certain characteristics about the rival teams, and the measure of perceived choice related to the individual's feeling in choosing one team to win as opposed to the other. The characteristics (for example, ratings of a team's defense) were created so as to make teams appear to be either quite comparable (a small difference in attractiveness) or greatly mismatched (a very large difference in attractiveness). It was found in these studies that perceived choice was greater when there was a small difference in attractiveness between choice options than when there was a large difference. It should be noted that no direct evidence about the experiential state of uncertainty was obtained in this work; thus, while the idea that uncertainty mediates the effect of similarity in attractiveness on perceived choice would seem to be a reasonable assumption, it has not been subject to direct empirical test.

Steiner (1970), as well as Brehm and Cohen (1962) and Kelley (1967), seems to suggest a prediction similar to but somewhat different from the above prediction. He suggests that perceived freedom will be high if the choice options are *approximately equal* in presumed benefit. But does approximately equal connote a small difference? If it does, then Steiner's and Mills' analyses make similar predictions for the similarity in attractiveness variable; if it implies equality, the two analyses lead to different predictions. Evidence provided by Harvey and Johnston (1973, Experiment II) and Harvey, Barnes, Sperry, and Harris (1974) indicates that individuals perceive greater freedom when options are close to one another in attractiveness than when they are identical; this evidence is consistent with the implication of Mills' analysis that differentiation between or among options in terms of their attractiveness is a necessary condition for the perception of choice. However, in the first experiment by Harvey and Johnston (1973), it was difficult to create choice options experimentally which would be perceived as exactly identical, and no significant data for this question were obtained. Presumably, people may be quite adroit at constructing differences between or among options despite experimenters' best laid plans to create a sense of equality.

Uncertainty about outcomes of options in making a decision. Uncertainty about which option is superior may directly enhance perceived freedom. Also, it may be operationalized separately from the similarity-in-attractiveness variable. As has already been mentioned, Kelley (1967) predicted that uncertainty might have this effect on subjective freedom, and the prediction was borne out in research by Harvey and Johnston (1973). In this research, actor-subjects were presented with choice options which were described in such a way as to indicate that their associated effects had either a high probability (high certainty) or a relatively low probability (low certainty) of occurring. However, this prediction for the effect of uncertainty on attributed freedom was not supported in a study by Kruglanski and Cohen (1974). These investigators tested the prediction in a situation involving observer-subjects who read about a stimulus person's decisiveness (high certainty) or indecisiveness (low certainty) in making a decision. They found that attributed freedom was greater the more certain the stimulus person was perceived to be in making his decision. According to Kruglanski and Cohen, a decision maker who is perceived to be certain in his choice also will be seen as fulfilling his desires or motives in the situation, and such fulfillment is assumed to connote personal causation (hence, the attribution of freedom to the person will be high).

Thus, the evidence on the effect of uncertainty on perceived freedom is not conclusive. However, there are pronounced differences in the manner in which the uncertainty was operationalized and in the experimental situations designed to investigate this variable. This point is important because it is clear that investigators need to be careful in operationalizing uncertainty. If the concept of uncertainty is equated with the concept of conflict, uncertainty may be defined as a predecisional state involving deliberation about options which cannot be readily eliminated as possibilities (cf. Festinger, 1964). In observer-subject studies, it may be particularly difficult to manipulate uncertainty apart from the cautiousness imputed to the stimulus person in assessing information about choice options or the amount of sheer intellectual difficulty attributed to the stimulus person in evaluating the options. Theoretically, we may wonder whether a person seen as experiencing uncertainty in making a decision is also seen as being cautious or as experiencing conceptual trouble in making distinctions between or among options. Likewise, do these experiences also go together for persons who are actors in decisional situations? And, finally, what effects do the variables of cautiousness and intellectual difficulty, as either ascribed or experienced states, have on perceived freedom? Work seems to be necessary to tease apart these variables and their singular and combinational effects on perceived freedom.

Valence of options. "Imagine that a woman comes to you seeking advice about a major decision in her life. Basically, she is deciding whether to stay married to her inconsiderate, insensitive, and incompetent husband or to get a divorce, an action which she also dislikes because she feels it is a sign of personal failure and will stigmatize her socially" (Jellison & Harvey, 1975, p. 18).

The woman in this example may be conceived as experiencing what Lewin (1935) called an avoidance–avoidance conflict; she is motivated to avoid both outcomes and yet must opt for one or the other alternative. How much freedom would she feel in deciding what to do? The answer is probably very little. In laboratory studies, Harvey et al. (1974) and Harvey and Harris (1975) tested the hypothesis that people who are placed into a decisional situation involving only unpleasant options attribute relatively little freedom to themselves when compared to persons placed into a decisional situation involving only pleasant options (this latter type of situation would correspond roughly to an approach–approach conflict in Lewin's analysis).[2] This prediction was based on the rather common-sense assumption that anticipation of reward is more conducive to feelings of freedom than is anticipation of punishment. Prior to this research on the valence hypothesis, Bramel (1969) proposed a very similar hypothesis with respect to the perceptions of observers. He suggested that a sense of subjective freedom is attributed more to people seeking positive incentives than to those seeking to avoid punishment.

Harvey et al. (1974) and Harvey and Harris (1975) employed a methodology which involved giving individuals descriptions of alternative actions they could take. In order to hold constant the nature of the actions, they made the descriptions very technical and abstruse, but accompanying the descriptions were ratings indicating how pleasant or unpleasant other subjects had found the alternatives to be. The ratings showed that the alternatives either had been perceived as very pleasant or as very unpleasant. Before actually making a decision about which alternative to select, subjects rated their sense of choice in the situation.

[2] Another analogy for the negative–negative and positive–positive decisional situations may be found in Fromm's (1941) concepts of "freedom from" (for example, freedom from economic scarcity) and "freedom to" (for example, freedom to exhibit different life styles), respectively.

The results of these studies provide strong support for the hypothesis that people will perceive more choice when making a selection from positively valenced options than when making a choice from negatively valenced options. Harvey and Harris (1975) also report an unexpected but interesting interaction involving the valence of the options. They found that the previously obtained effect of similarity in attractiveness of options upon perceived choice (with perceived choice greater for similar versus dissimilar sets of options) did not obtain when negative options were available; this previously obtained effect did hold for positive options. No cogent explanation exists as to why similarity in attractiveness may have little effect when options are negative in valence. In fact, on an intuitive level, one might think that in a decisional situation involving negative options, the similarity in attractiveness of the options would be very important to the decision maker (assuming the individual wants to make sure that he selects the "lesser of two evils").

Research on the valence hypothesis seems to offer some support for an implication deriving from Skinner's (1971) argument in *Beyond Freedom and Dignity*. By suggesting that people typically do not recognize the control that positive reinforcement exercises over behavior, Skinner implies that people would feel a greater expectancy of personal control and hence more freedom when confronted by positive options than when confronted by negative options. Of course, Skinner's contention is that this feeling of freedom is illusory; but, as has already been suggested, even if it is an illusion it is a meaningful perception to people and apparently has important effects upon their behavior (see Lefcourt, 1973, for further discussion of this point).

Returning to the earlier mentioning of Lewin's work on decision conflicts, it is suggested that the perception of freedom may play an important role in such conflicts. Presumably, this perception is directly related to the positivity of the options and may also be directly related to the ease of resolution of the conflict.

Number of options. A rather intuitive hypothesis is that the larger the number of choice options, the greater will be the sense of freedom. In terms of one of the mediators already discussed, perceived freedom should be enhanced due to the increased uncertainty about which option is best when many possibilities exist. An alternative line of reasoning is that perceived freedom may reach an asymptote as the number of options increases to a certain point—and then possibly decrease with the addition of still more options. This latter line of reasoning is implied in Toffler's (1970) statement of the complexities of life and decision making in a superindustrialized, bureaucratic society. He suggests that too many options in particular situations may engender a sense of "unfreedom" as opposed to freedom.

Experimental work on the effects of number of options on perceived freedom has shown that the relationship may not be linear under certain conditions. Harvey and Jellison (1974) asked subjects to make selections from a set of options varying in number; the choice options were football teams which subjects were asked to evaluate given certain information. They were then asked to decide which team should be selected to go to a bowl game. Information about the teams was carefully prepared so as to hold constant their relative attractiveness and descriptive nature. Harvey and Jellison found that perceived choice was greater the greater the number of options if subjects were led to believe that they had expeditiously evaluated the options relative to other people who had participated in the study. However, a moderately large number of options (6) was associated with greater perceived choice than was either a small (3) or a very large (12) number of options when the subjects were led to believe that they had spent more time than most other subjects had spent in deliberating over the information.

These data suggest that perceived choice may not be stimulated and may even be diminished when people feel that they possess an overabundance of alternatives. It seems possible that under some circumstances, a very large number of options may inhibit perceived choice because it may give the decision maker a sense of being overburdened by information. Clearly, however, the nature of the alternatives (their complexity, their importance, how distinct they are from one another) and the amount of deliberation necessary to eliminate alternatives must be taken into account when predicting the relationship between perceived choice and number of options. For example, a person searching for an academic position would probably not have the same sense of freedom in making a decision if 10 universities had extended job offers as when confronted by a more mundane choice among 10 different brands of breakfast cereals at the supermarket.

Locus of control of the decision-maker. Locus of control is the one individual difference variable whose relationship to perceived freedom has been directly examined. A prediction for the relationship between locus of control and perceived freedom follows directly from the conceptualization of the locus of control variable. Internals are defined as persons who entertain a pervasive belief that their actions determine their outcomes, while externals are defined as persons who usually believe that they have little control over what happens to them (Rotter, 1966). As has been theorized by Kelley (1967) and reported in empirical work by Harvey and Harris (1975), perceived freedom is directly related to a sense of control over one's behavior. Thus, it follows that internals should perceive more choice than externals in their decision making.

In a set of studies, Harvey et al. (1974) examined this possibility. The general procedure involved asking internals and externals to make selections from sets of options varying in valence and in similarity in attractiveness. Harvey et al. found no general tendency for internals to perceive more freedom than externals. However, as predicted, internals did exhibit more differentiated reactions to the characteristics of the options than did externals, for example, internals perceived significantly more choice when options were close together in attractiveness than when they were identical, but externals did not show a significant difference in response to these two conditions. In formulating a prediction about the differential sensitivity of internals and externals in perceived choice, Harvey et al. argued that such sensitivity may be quite useful to the internal in a decisional situation; it may actually increase the probability that he can act in a manner so as to obtain a desired outcome or avoid an undesired outcome. However, since the external presumably is not concerned with taking actions designed to control his reinforcement, he may be relatively insensitive to whether or not a decisional situation allows a sense of freedom. Locus of control appears to be a dispositional variable which is systematically related to a number of attributional and perceptual tendencies, including attribution of attitudes (Jones et al., 1971)–see Lefcourt (1972) for a review. The work by Harvey et al. (1974) suggests that at least in terms of sensitizing individuals to the amount of choice available to them in their decisions, locus of control appears to be related to perceived freedom.

In the domain of personality variables which may be related to perceived freedom, de Charm's (1968) "origin–pawn" analysis is also relevant. The origin–pawn variable appears to be quite similar to the variable of locus of control. An origin, who is similar to an internal, is postulated to be a person who perceives his behavior as determined by his own choosing, and a pawn, who is similar to an external, is theorized to be a person who perceives his behavior as determined by external forces beyond his control. De Charms (1968) says that the origin–pawn distinction is continuous, not discrete; that is, a person feels more like an origin under some circumstances and more like a pawn under others. In one of the few studies to examine derivations of this conception, de Charms, Carpenter, and Kuperman (1965) reported several findings, including one which showed that persons who had been classified as origins on an origin–pawn measure perceived a hypothetical person to be more free or origin-like in his behavior than did persons who had been classified as pawns. In this study, the hypothetical person had been influenced by a persuasive communication. This work implies that people who generally feel they are free in their behavior may see others as free also. However, this idea needs to be tested in actual behavioral settings and with a greater variety of actions than was the case in the de Charms et al. work.

Variability in reinforcement. Davidson and Steiner (1971) proposed that the manner in which a reinforcing agent administers rewards and punishments is an important source of information concerning the agent's freedom. These investigators assumed that behavior which is routine and highly predictable rapidly acquires an aura of normativeness. It follows from this reasoning that an agent who almost invariably rewards one class of responses and punishes another will be judged to have little freedom to do otherwise. However, by contrast, an agent who administers the same total amount of reward and punishment on a variable schedule should seem comparatively immune to normative constraints. Thus, according to this conception, rewards and punishments can serve not only as reinforcements and incentives, but also as messages concerning the freedom of the agent who administers them.

In the Davidson and Steiner (1971) study, a teacher-accomplice rewarded a learner-subject for correct responses and punished him for incorrect responses on a task either 100 or 40% of the time. As predicted, subjects who received 40% punishment and reward schedules attributed more freedom to the teacher than did subjects who were on 100% schedules. This finding was essentially replicated by Bringle, Lehtinen, and Steiner (1973) with a similar experimental procedure; greater freedom was attributed by subjects on a 50% than on a 100% schedule. Bringle et al. also found that little freedom was attributed to the teacher when reinforcement was costly to the teacher.

Steiner (1972) has made some provocative observations about the implications of this work on variability in real-world situations. For example, he suggests that leaders who reward and punish their followers on a consistent schedule are fairly easy to live with; we can readily predict their behavior. However, the leader who does not reward on a consistent schedule may be very threatening because this person seems so free; accordingly, one must be careful so as not to fall out of favor with such a leader. In this sense, Steiner seems to be suggesting that attributed freedom and the perception of a leader's proneness to capricious acts may go together. But the crucial point with regard to freedom research is that inconsistency over time resulting in the impression of unpredictability in the perception of other's behavior stimulates an attribution of freedom to the other. An interesting question which has not been addressed concerns whether or not actors' perceptions of freedom vary as a function of perceived unpredictability in their own behavior.

Costs in taking an action. The foregoing research by Bringle et al. (1973) indicates that cost may be inversely related to attributed freedom. This finding is generally consistent with Steiner's (1970) analysis. As has already been discussed, Steiner posits that maximum decision freedom should be perceived when a person's anticipated investment approximately equals the expected payoff. And if investment is higher than payoff (that is, net losses accrue), Steiner's analysis implies that little freedom would be attributed. Steiner et al. (1974) proposed that if the cost factor is not controlled or manipulated, a decision-maker may construe the expected values of the outcomes of options differently depending upon how much cost he thinks will be involved in exercising the options.

However, Steiner et al. (1974) found that cost may play a more complicated role in affecting attributed freedom than is suggested above. In this research, observer-subjects read about the business transactions of hypothetical stimulus persons. The transactions varied along a number of dimensions, including how much money the stimulus persons invested to make their transactions. This operation was designed to manipulate the cost factor. In one study, the predicted effect of less attributed choice the greater the cost was found. In another study, it was found that attributed freedom was high when cost (which did not involve any net losses in this situation) interacted with other independent variables apparently to increase the similarity in attractiveness of options. While this research suggests that cost may be an important variable in the attribution of freedom, we may question whether and/or under what conditions actual decision makers in either laboratory or naturalistic settings would be likely to engage in the rather extensive type of cognitive calculus implied in Steiner's (1970) model in assessing their own decisional freedom.

Acting in line with predisposition. Kruglanski and Cohen (1973) proposed that the predisposition of an actor is another variable which affects attributed freedom. In this study, observer-subjects received information about a target person's predispositions and behavior in a specific situation. It was found that greater freedom was attributed to the target person when his behavior was consistent with his presumed predisposition (for example, his attitude was procooperation among human beings, and he acted in a cooperative manner in an interaction setting) than when it was inconsistent. Focusing particularly upon Kelley's (1967) statement as a conceptual base, Kruglanski and Cohen explained this effect in terms of the assumption that freedom will be attributed to a person who is seen as personally responsible for his behavior. Further, they reasoned that behavior which is in line with a person's predisposition (in character) should be more attributable internally to the person than to external factors; and they reasoned that the converse was true for acts which were inconsistent with predisposition (out of character).

More recently, Kruglanski (1975) has reanalyzed the work of Kruglanski and Cohen (1973) within the framework of what he calls the endogenous–exogenous partition. According to Kruglanski, this partition refers to whether an action is judged to constitute an end in itself (endogenous) or as a means to some further end (exogenous). Kruglanski presents arguments and data to suggest that the endogenous–exogenous partition is superior to the frequently invoked internal–external partition in interpreting attribution of causality data. He suggests that the variable in character versus out of character was confounded with the endogenous–exogenous dimension in the Kruglanski and Cohen study. In research reported by Kruglanski (1975) in which these variables presumably were not confounded, it was found that the attribution of freedom is greater for acts which are seen as endogenous than for those which are seen as exogenous, and a nonsignificant result was found for the effect of the in character versus out of character variable on attributed freedom. Although Kruglanski seems to have identified an important new dimension on which attributions of freedom vary, through his research and analysis he does not rule out the possibility that unconfounded variations of ascribed in character versus out of character behavior may affect attributed freedom. It is possible that Kruglanski's (1975) operations to test the effects of this variable, when separated from the endogenous–exogenous variable, may not have been impactful enough to affect attributions of freedom.

Consequences of a decision. This determinant of subjective freedom and the next one to be presented appear to have qualities which set them apart from the other determinants previously discussed in this paper. In contrast to most of the determinants which have been investigated, these two seem to reflect self-serving tendencies on the part of individuals (determinants discussed earlier reflect mainly information-processing and perceptual tendencies). In particular, they appear to be conducive to people's attempts either to employ their attributions so as to assume credit for positive consequences or absolve themselves of responsibility for negative consequences associated with their behavior.

Within the context of examining possible self-serving tendencies, Harris and Harvey (1975) investigated the effect of positive and negative consequences of a decision on individuals' retrospective self-attribution of choice in making the decision. Their procedure involved asking subjects to choose one of two experiments in which another person would presumably participate. They were given information that was very difficult to comprehend about the possible experiments. After making their decision, some subjects discovered via an "unexpected" set of circumstances that the experiment they had selected would be very unpleasant for the other person (the person would receive electric shocks), while other subjects discovered that it would be very pleasant (the person would receive pleasant sensory cues). Subsequent to this discovery, subjects were asked to indicate how much choice they had in making the decision.

In line with predictions based upon the assumption that people sometimes use attributions of freedom to maintain or enhance self-esteem, Harris and Harvey found that subjects in the pleasant consequence condition rated the choice they had in making the decision higher than did subjects in the unpleasant consequence condition. Also, the former subjects rated their choice higher than did subjects in a control condition (where no information about the probable consequence of the decision was obtained), and subjects in the unpleasant consequence condition rated their choice as lower than did these control subjects.

Harris and Harvey concluded that their data provided strong evidence for the operation of a self-esteem or even a dissonance reduction type process in the self-attribution of choice. With respect to the dissonance process, Harris and Harvey suggested that people sometimes may resolve cognitive dilemmas of the dissonance variety—particularly those involving threat to self-esteem (Aronson, 1968)—via retrospective self-attributions of freedom concerning the relevant past acts. This suggestion does not seem to be too unreasonable in light of the events of our time. It is not uncommon to hear alleged wrong doers indicate they are not responsible for the acts in question (which presumably may produce quite a bit of dissonance) because they were simply following orders from "higher ups."

Magnitude of negative consequence of an action: Actor–observer differences. As Jones and Nisbett (1972) have proposed, it appears that actors and observers frequently reach divergent conclusions about the causes of behavior. It also seems likely that these divergencies exist with respect to the perception of freedom. Harvey, Harris, and Barnes (1975) examined actor–observer differences in perceived freedom in a setting involving an actor's taking an action which varied in the magnitude of its negative consequences.

Harvey et al. reasoned that an actor's attributions would not parallel those of an observer in this situation. An actor was expected to employ his attributions so as to serve his need to maintain self-esteem; the more serious the consequence, the greater the need. On the other hand, it was reasoned that in light of the negative consequences associated with the action, observers would feel a need to use their attributions to exercise sanctions against and control over the actor; this need should also be greater the more serious the consequences. Hence, given that perceptions of freedom may be used as a means of ascribing responsibility, it was predicted that an actor would attribute less freedom to himself and that an observer would attribute more freedom to the actor the more serious the consequence of the action.

In the Harvey et al. procedure, actor-subjects (teachers) gave shocks to presumed leaders (accomplices) in the presence of observer-subjects. The shocks were administered for incorrect responses on an associative-learning task. The consequences of the act of giving shocks to the learner were operationalized in terms of the amount of psychological distress shown by the learner on a bogus feedback meter. The learner either displayed a moderately negative or a severely negative distress reaction. Subjects later were asked to indicate how much real choice the actor had in carrying out his job and how much freedom he had to refuse to do what was asked of him.

The data for the measures of perceived freedom and choice for the Harvey et al. study are shown in Table 1. As can be seen from this table, it was found that in both conditions observers attributed more choice and freedom to actors than actors attributed to themselves. These data probably reflect the fact that the actor's behavior, whether in the moderate or severe consequence condition, was negative; thus, the observers may have been demonstrating a general disapproval of the actor's conduct. Further, Table 1 shows that, as predicted, observers attributed more choice and freedom to actors the more severe the consequence of the act, whereas actors attributed less choice and freedom to themselves the more severe the consequence of the act.

These data suggest that there may be situations in which both actors' and observers' attributions are mediated by processes other than information-processing activities such as those articulated by Jones and Nisbett (1972),

TABLE 1
Means for the Measures of Perceived Freedom and Perceived Choice

	Severity of consequence			
	Moderate		Severe	
Dependent measure	Actor	Observer	Actor	Observer
Freedom to refuse	5.2	7.1	3.0	8.3
Choice in fulfilling job	5.3	7.0	2.5	8.8

Note. The higher the score, the greater is the perceived freedom and perceived choice (11–point scales were used). (Adapted from Harvey, Harris, & Barnes, 1975.)

that is, knowledge of actor's past actions and salience of present environmental cues relevant to causality. In these situations, actors may interpret their freedom in terms of their own self-interests, whereas observers may interpret the actor's freedom in a manner suggestive of their acting as an adversary of society's interests.

Harvey et al.'s results show that when an actor's behavior leads to negative effects, actors' and observers' conclusions about the freedom associated with the behavior may diverge in a provocative fashion. Actor–observer differences in perceived freedom appear to represent a viable phenomenon which deserves more theoretical and empirical work. In future work, researchers might extend the range of variables investigated and broaden the focus to include more naturalistic actor–observer situations (for example, parent–child or student–teacher perceptions of freedom in various areas of interaction).

Taken together, the data in the Harris and Harvey (1975) and the Harvey et al. (1975) studies suggest another facet in a multifaceted picture of humans seeking to understand and interpret their state of freedom. People not only make attributions of freedom in a relatively rational, information-processing fashion, but also they sometimes interpret their freedom (and perhaps that of others too) in a rather self-serving manner.

The variables discussed above which have been found to influence perceived freedom in work done to date are summarized in Table 2. This table also indicates the general effects that have been found for these variables.

CORRELATES OF PERCEIVED FREEDOM

In this section, research concerned with the correlates of perceived freedom will be briefly reviewed. Coupled with the previously discussed findings on conse-

TABLE 2
Summary of Some Variables Found to Influence Perceived Freedom

Variable	Finding for variable
Similarity in attractiveness of options	Greater perceived choice for small than for large or zero difference in attractiveness
Uncertainty about outcomes of options	Greater perceived choice the more uncertain the decision maker about the outcomes of the options
Valence of options	Greater perceived choice when the outcomes of the options are positive than when they are negative in attractiveness
Number of options	Greater perceived choice the greater the number of options when decision maker thinks he has expeditiously evaluated options; greater perceived choice for moderate than for small or large number of options when decision maker thinks he has spent more than normal amount of time in evaluating options
Locus of control	Greater differences for internals than externals in response to options varying along dimensions of similarity in attractiveness and valence of outcomes
Variability in reinforcement	Greater freedom attributed to reinforcing agent who rewards and punishes on an intermittent schedule than to one who reinforces on a continuous schedule
Costs in taking an action	Less freedom is attributed to reinforcing agent or to decision maker the more costly the action taken or decision made
Acting in line with predisposition	Greater freedom attributed to person, the more the behavior was consistent with a relevant predisposition
Consequences of a decision	Relatively high self-attribution of choice for decisions having positive consequences and relatively low self-attribution of choice for decisions having negative consequences
Magnitude of negative consequences of an action	Less freedom attributed to self by actors and more freedom attributed to actor by observer the more serious the consequence of an action

quences and antecedents of perceived freedom, this evidence further documents the importance of freedom as a psychological variable.

Perceived Freedom and Perceived Competence

Jellison and Harvey (1973) and Harvey and Jellison (1974) have shown that perceived choice in a decision-making situation is positively related to the perceived gain in knowledge about one's competence in such situations. These investigators explained this relationship by reference to White's (1959) concept of competence motivation. According to White, people have a drive to exercise their mental and physical abilities; thus, they are conceived to seek out situations where they can engage in cognitive analysis and manipulation of the environment. Situations in which perceived choice in decision-making is high, as opposed to those in which it is low, would seem to be most conducive to stimulating cognitive activities. In situations where perceived choice in making decisions is low, little thinking is required (for example, when there is a great disparity in attractiveness between options).

In the studies by Jellison and Harvey and Harvey and Jellison, it was hypothesized (a) that people would report they had gained more information about their competence on a task in conditions in which perceived choice was expected to be high (for example, when the similarity in attractiveness between options was high); and (b) that there would be a positive relationship between perceived choice and feelings of competence. The results provided generally strong support for these hypotheses; the overall correlation for the relationship between perceived choice and competence was .37 in the Jellison and Harvey (1973) study and .46 in the Harvey and Jellison (1974) study. Based upon the results in their research, these investigators have suggested that people may frequently desire to be in situations involving high-perceived choice because they can get maximal information about their competence in these situations.

Perceived Freedom and Perceived Control

In the previously described study by Harvey and Harris (1975), a positive relationship (r = .38) was obtained between perceived choice and expectancy about internal control over own behavior (which in this work refers to a momentary, situationally specific feeling state). Although perceived choice has been defined in terms of perception of control over own behavior (Kelley, 1967), or the potentiality for such control (Ruch & Zimbardo, 1971), the relationship between perceived choice and perceived control has received little empirical attention. Harvey and Harris tested the hypothesis that when an individual is making a decision about which of a set of actions he will take, the degree to which he thinks he will feel control over his behavior in the future situation (the situation which results from his decision) will be directly related to his perception of choice in making the decision.

Harvey and Harris found evidence of a significant positive correlation between perceived choice and expectancy about feelings of internal control over own behavior. Further, it was found that in conditions producing high-perceived choice (when decisions involved positive options), expectancy about feelings of internal control was greater than in conditions producing low-perceived choice (when decisions involved negative options). These data suggest that if it is important for persons to approach tasks with an internal orientation, then care should be taken to ensure that they perceive high choice in their decisions relevant to engagement in the tasks.

Perceived Freedom and Perceived Responsibility

Kruglanski and Cohen (1973), Harvey, Harris, and Barnes (1975), and Wortman (1975) have found positive relationships between perceived freedom and perceived responsibility. Kruglanski and Cohen found evidence for this relationship with observers attributing freedom and responsibility to other persons ($r = .36$). Wortman reported a positive correlation between perceived freedom and perceived responsibility ($r = .61$) in a study in which actors made lottery type drawings, won a prize, and then rated various feelings, including their sense of choice and responsibility in attaining their outcome. Harvey et al. found that this relationship obtained for both observers and actors who were judging their own freedom and responsibility (which in this research was defined as accountability for the consequences of an action). In the Harvey et al. study, the overall correlations for the relationship between perceived freedom and perceived responsibility were over .50 for both actors and observers and did not differ as a function of the attributor.

In attributional conceptions, perceived freedom and responsibility generally are assumed to be closely related. Harris and Harvey (1975) and Harvey et al. (1975) have suggested that feelings of responsibility may underlie attributions of freedom. Other investigators (for example, Collins & Hoyt, 1972) have manipulated perceived freedom as a means of conveying a sense of responsibility. Despite their close relationship, freedom and responsibility may be viewed as separate psychological experiences, the former pertaining more to the decision to act and the latter to the consequences of the act. From an overall perspective, the research on this relationship provides empirical support for the supposed link between perceived freedom and responsibility which frequently has been posited by psychologists, philosophers, and laypeople alike.

Perceived Freedom and Enjoyment

Brock and Becker (1967) asked college students to indicate what they did during a number of time periods for 1 day and also what they could have done during the time periods. After listing these activities, the subjects rated their feelings of personal choice and enjoyment for the various time periods. Brock and Becker found positive correlations between feelings of choice and number of alternatives available for the time periods ($r = .27$) and between feelings of choice and enjoyment of the activity listed for the various periods ($r = .31$).

Harvey and Jellison (1974) also reported evidence about the relationship between perceived choice and enjoyment in a laboratory study where valence of alternatives was held constant. It was found that subjects indicated they enjoyed most decisional situations in which they perceived high choice rather than in situations in which they perceived low choice. Also, the correlation for the relationship between perceived choice and enjoyment was positive but relatively low ($r = .22$).

CONCLUSIONS

In this paper, theory and research which show the extent to which work on the attribution of freedom has progressed have been reviewed. Attributed freedom has been found to be an important determinant of inferences about other people's predispositions. Further, recently investigators have identified a number of variables which influence the attribution of freedom both to self and others. Finally, several variables with which the attribution of freedom has been found to correlate have been discussed. Steiner's (1970) conclusion of only half a decade ago to the effect that researchers have seldom made direct evaluations of individuals' perceived freedom or manipulated many of the factors believed to affect these perceptions no longer seems tenable in light of this fairly substantial amount of evidence.

As has already been discussed in this paper, theoretical accounts of attributed freedom emphasize diverse mechanisms, from the perception of differences to elaborate inferential activities to self-serving tendencies. Conditions under which each of these mechanisms will be operative have been suggested. Taken together, these theoretical accounts do not constitute a comprehensive, refined theory of perceived freedom. However, based upon the amount of empirical work they have already stimulated, it seems likely that fruitful work will continue irrespective of whether a comprehensive theory is developed.

In terms of providing a clearer definition of perceived freedom, it may be useful to reexamine a traditional way of conceiving this experience. Subjective freedom has been defined as a sense of control over one's own behavior (Brehm & Cohen, 1962; Kelley, 1967), and Harvey and Harris (1975) have presented evidence which reveals a positive relationship between perceived freedom and control. However, in this latter research, freedom was defined as the decision maker's feeling at the time of making a choice, whereas control was defined as an expectancy about feelings subsequent to the decision. Given this definition, it is possible to think of instances in which perceived freedom and control do not covary or even may be negatively related. For example, a quarterback may feel quite a bit of choice in deciding upon which play to call, and yet because of the game conditions and the strength of the opponent, he may expect to feel little personal control over the execution of the play. Likewise, the quarterback may perceive hardly any choice when making a decision about a play to call ("it has to be a pass"), but because of his feelings of prowess relevant to the type of play called, he may expect to feel very much in control of his behavior during the execution phase.

Based upon the foregoing reasoning, it would seem useful to treat freedom and control as distinct, though often positively related, experiences. The most clear-cut definitions of perceived freedom would seem to follow ideas articulated by Steiner (1970) and Mills (1970). These approaches emphasize the individual's act of deciding to pursue one alternative as opposed to another and his perception of the extent to which alternative lines of action exist. Important related perceptions, such as those of responsibility and control, may or may not occur contemporaneous to the perception of freedom.

In the course of discussion, several directions for future work have been mentioned. But there are others which seem worthy of consideration. The conditions under which people try to present themselves to others as either free or not free may represent an area where provocative evidence can be obtained. A traditional topic in cognitive psychology concerns the determinants of choice behavior (for example, Luce, 1959). A useful step for future researchers might involve an integration of ideas from the work in cognitive psychology with ideas from more recent research by social psychologists on the perception of choice. In the area of environmental psychology, a question which has been raised but not yet empirically investigated concerns how environmental conditions such as crowding and lack of privacy influence people's sense of freedom (see Proshansky, Ittelson, & Rivlin, 1970, for some theoretical propositions related to this question). Finally, we have little if any systematic information about attributed freedom as a developmental variable or about its role in therapeutic and educational endeavors.

ACKNOWLEDGMENTS

Preparation of this manuscript was supported by grants from the Vanderbilt Research Council and from National Institute of Mental Health (MH 25736). Thanks are extended to William Ickes, Yvette Harvey, Jerald Jellison, and Robert Kidd for critical comments on an earlier draft of this manuscript.

REFERENCES

Aronson, E. Dissonance theory: Progress and problems. In R. P. Abelson, E. Aronson, W. J. McGuire, T. M. Newcomb, M. J. Rosenberg, & P. H. Tannenbaum (Eds.), *Theories of cognitive consistency: A sourcebook.* Chicago: Rand McNally, 1968.

Bramel, D. Interpersonal attraction, hostility, and perception. In J. Mills (Ed.), *Experimental social psychology.* London: Macmillan, 1969.

Brehm, J. W., & Cohen A. *Explorations in cognitive dissonance.* New York: John Wiley & Sons, 1962.

Bringle, R., Lehtinen, S., & Steiner, I. D. The impact of the message content of rewards and punishments on the attribution of freedom. *Journal of Personality,* 1973, **41**, 272–286.

Brock, T. C., & Becker, L. A. Volition and attraction in everyday life. *Journal of Social Psychology,* 1967, **72**, 89–97.

Collins, B. E., & Hoyt, M. F. Personal responsibility for consequences: An integration and extension of the forced compliance literature. *Journal of Experimental Social Psychology,* 1972, **8**, 558–593.

Davidson, A., & Steiner, I. D. Reinforcement schedules and attributed freedom. *Journal of Personality and Social Psychology,* 1971, **19**, 357–366.

de Charms, R. *Personal causation.* New York: Academic Press, 1968.

de Charms, R., Carpenter, V., & Kuperman, A. The "origin–pawn" variable in person perception. *Sociometry,* 1965, **28**, 241–258.

Festinger, L. (Ed.) *Conflict, decision and dissonance.* Stanford, California: Stanford University Press, 1964.

Fromm, E. *Escape from freedom.* New York: Holt, Rinehart, & Winston, 1941.

Harris, B., & Harvey, J. H. Self-attributed choice as a function of the consequence of a decision. *Journal of Personality and Social Psychology,* 1975, **31**, 1013–1019.

Harvey, J. H., Barnes, R. D., Sperry, D. L., & Harris, B. Perceived choice as a function of internal–external locus of control. *Journal of Personality,* 1974, **42**, 437–452.

Harvey, J. H., & Harris, B. Determinants of perceived choice and the relationship between perceived choice and expectancy about feelings of internal control. *Journal of Personality and Social Psychology,* 1975, **31**, 101–106.

Harvey, J. H., Harris, B., & Barnes, R. D. Actor–observer differences in the perceptions of responsibility and freedom. *Journal of Personality and Social Psychology,* 1975, **32**, 22–28.

Harvey, J. H., & Jellison, J. M. Determinants of perceived choice, number of options, and perceived time in making a selection. *Memory & Cognition,* 1974, **2**, 539–544.

Harvey, J. H., & Johnston, S. Determinants of the perception of choice. *Journal of Experimental Social Psychology,* 1973, **9**, 164–179.

Heider, F. *The psychology of interpersonal relations.* New York: John Wiley & Sons, 1958.

Jellison, J. M., & Harvey, J. H. Determinants of perceived choice and the relationship between perceived choice and perceived competence. *Journal of Personality and Social Psychology,* 1973, **28**, 376–382.

Jellison, J. M., & Harvey, J. H. Perceived freedom in everyday decisions. Unpublished manuscript, University of Southern California, 1975.

Jones, E. E., & Davis, K. E. From acts to dispositions. In L. Berkowitz (Ed.), *Advances in experimental social psychology* (Vol. 2). New York: Academic Press, 1965.

Jones, E. E., & Harris, V. A. The attribution of attitudes. *Journal of Experimental Social Psychology,* 1967, **3**, 1–24.

Jones, E. E., & Nisbett, R. E. The actor and the observer: Divergent perceptions of the causes of behavior. In E. E. Jones, D. E. Kanouse, H. H. Kelley, R. E. Nisbett, S. Valins, & B. Weiner (Eds.), *Attribution: Perceiving the causes of behavior.* New York: General Learning Press, 1972.

Jones, E. E., Worchel, S., Goethals, G. R., & Grumet, J. Prior expectancy and behavior extremity as determinants of attitude attribution. *Journal of Experimental Social Psychology,* 1971, **7**, 59–80.

Kelley, H. H. Attribution theory in social psychology. In D. Levine (Ed.), *Nebraska Symposium on Motivation.* Lincoln: University of Nebraska Press, 1967.

Kruglanski, A. W. The endogenous–exogenous partition in attribution theory. *Psychological Review,* 1975, **82,** 387–406.

Kruglanski, A. W., & Cohen, M. Attributed freedom and personal causation. *Journal of Personality and Social Psychology,* 1973, **26,** 245–250.

Kruglanski, A. W., & Cohen, M. Attributing freedom in the decision context: Effects of the choice alternatives, degree of commitment and predecision uncertainty. *Journal of Personality and Social Psychology,* 1974, **30,** 178–187.

Lefcourt, H. M. Recent developments in the study of locus of control. In B. A. Maher (Ed.), *Progress in experimental personality research* (Vol. 6). New York: Academic Press, 1972.

Lefcourt, H. M. The functions of the illusions of control and freedom. *American Psychologist,* 1973, **28,** 417–425.

Lewin, K. *A dynamic theory of personality.* New York: McGraw–Hill, 1935.

Luce, R. D. *Individual choice behavior: A theoretical analysis.* New York: John Wiley & Sons, 1959.

Maselli, M. D., & Altrocchi, J. Attribution of intent. *Psychological Bulletin,* 1969, **71,** 445–454.

Miller, A. G. Perceived freedom and the attribution of attitudes. *Representative Research in Social Psychology,* 1974, **5,** 61–79.

Mills, J. Unpublished analysis of perceived choice. University of Missouri–Columbia, 1970.

Proshansky, H. M., Ittelson, W. H., & Rivlin, L. G. Freedom of choice and behavior in a physical setting. In H. M. Proshansky, W. H. Ittleson, & L. G. Rivlin (Eds.), *Environmental psychology: Man and his physical setting.* New York: Holt, Rinehart, & Winston, 1970.

Rotter, J. B. Generalized expectancies for internal versus external control of reinforcement. *Psychological Monographs,* 1966, **80**(1, Whole No. 609).

Ruch, F. L., & Zimbardo, P. G. *Psychology and life.* Glenview, Illinois: Scott, Foresman, 1971.

Skinner, B. F. *Beyond freedom and dignity.* New York: Knopf, 1971.

Steiner, I. D. Perceived freedom. In L. Berkowitz (Ed.), *Advances in experimental social psychology* (Vol. 5). New York: Academic Press, 1970.

Steiner, I. D. Some antecedents and consequences of attributed freedom. Paper presented at the meeting of the American Psychological Association, Honolulu, 1972.

Steiner, I. D., & Field, W. L. Role assignment and interpersonal influence. *Journal of Abnormal and Social Psychology,* 1960, **61,** 239–246.

Steiner, I. D., Rotermund, M., & Talaber, R. Attribution of choice to a decision-maker. *Journal of Personality and Social Psychology,* 1974, **30,** 553–562.

Toffler, A. *Future shock.* New York: Bantam, 1970.

Tolstoy, L. *War and peace* (C. Bell, trans.). New York: Gottesberger, 1886.

Wheeler, H. (Ed.), *Beyond the punitive society.* San Francisco: Freeman, 1973.

White, R. W. Motivation reconsidered: The concept of competence. *Psychological Review,* 1959, **66,** 297–333.

Wortman, C. B. Some determinants of perceived control. *Journal of Personality and Social Psychology,* 1975, **31,** 282–294.

5

Schedules of Reinforcement: An Attributional Analysis

Stanley Rest

Department of Psychiatry and Behavioral Sciences
University of Oklahoma
Health Sciences Center

Since their initial introduction (for example, Humphreys, 1939; Skinner, 1938), schedules of intermittent reinforcement have been the focus of countless experimental investigations and numerous attempts at theoretical explanation (see Jenkins & Stanley, 1950, and Lewis, 1960, for earlier reviews). Indeed, there are two literatures, one for discrete-trial reinforcement schedules and the other for free-response or operant schedules. With respect to methods, theories, and investigators, only modest overlap exists between the two literatures.

REINFORCEMENT SCHEDULES: DEFINITIONS

Discrete-Trial Schedules

In a discrete-trial reinforcement schedule, a subject is given either a series of presentations of a conditioned stimulus or CS (in the case of classical conditioning) or a series of opportunities to make a response (in the case of instrumental learning). With respect to classical conditioning, the CS (for example, a bell) is followed by an unconditioned stimulus or UCS (for example, a puff of air in the eye) on some percentage of the trials (varying from 0 to 100%). If the response is an instrumental one (for example, running down a runway), then a reinforcement (for example, food) is given to the subject following some of the responses (again, varying from 0 to 100%). A trial may be presented immediately following the previous one, or there may be an interval of time between trials.

In an intermittent or a partial reinforcement schedule, the percentage of reinforcement is greater than 0% and less than 100%. In addition to the percentage of reinforcement, the pattern of reinforcement may be different on two schedules. For example, on a 50% schedule, reinforcement could be administered every other time or on a random basis. A 100% reinforcement schedule is also known as continuous reinforcement, whereas a 0% schedule is known as experimental extinction or continuous nonreinforcement.

Free-Response Schedules

In a free-response reinforcement schedule, there are no trials, as such. Rather, the organism is placed in an environment (for example, a Skinner box) for a given period of time during which reinforcement is available on some basis. A schedule is merely the rule by which certain responses, but not other responses, are followed by reinforcement. The rule for reinforcement may be related to some aspect of responding, to some aspect of time, or to both.

For example, in a fixed-ratio reinforcement schedule, every *n*'th response is followed by reinforcement. In a variable-ratio schedule, reinforcement follows series of responses of varying length, the average of which is *n*. In a fixed-interval reinforcement schedule, the first response after a fixed period of time is reinforced. Following reinforcement, another interval of the same duration begins, during which reinforcement is not available. A variable-interval schedule is similar to a fixed-interval schedule, except that the period during which reinforcement is unavailable varies from reinforcement to reinforcement.

In addition to such simple reinforcement schedules, several more complicated ones have been constructed (see Ferster & Skinner, 1957). For example, reinforcement could be made contingent on the emission of 10 responses after 20 seconds have elapsed since the previous reinforcement.

REINFORCEMENT SCHEDULES: BASIC FINDINGS

Discrete-Trial Schedules

The major finding reported in the discrete-trial reinforcement schedule literature is that resistance to extinction of a learned response is greater when the response is acquired under partial reinforcement than under continuous reinforcement. This finding, verified with different responses, experimental settings, and species, is known as the partial reinforcement effect. Typically, subjects are presented with several trials of a task and given either continuous or partial (usually 50%) reinforcement. After a predetermined number of trials, the subjects are given continuous nonreinforcement. Those subjects who were previously given partial reinforcement almost invariably continue to respond without reinforcement for a longer time. A major thrust in this paper will be the discussion of some of the attempts that have been formulated to explain this finding.

In addition to the partial reinforcement effect, the effect of other variables on resistance to extinction has been investigated. Generally, resistance to extinction has been found to be greater: (a) the smaller the percentage of reinforcement during acquisition (for example, Hulse, 1958), (b) the more random the sequence of reinforcements and nonreinforcements during acquisition (for example, Longnecker, Krauskopf, & Bitterman, 1952), and (c) the larger the number of consecutive nonreinforced trials during acquisition (for example, Capaldi & Stanley, 1965).

As we shall see later, causal attributions have been employed in explanations of the partial reinforcement effect, initially by Rotter and his coworkers (for example, James & Rotter, 1958; Rotter, Liverant, & Crowne, 1961) and more recently by Weiner and his colleagues (for example, Rest, Frieze, Nickel, Parsons, & Ruble, 1972; Weiner, 1972, 1974; Weiner, Frieze, Kukla, Reed, Rest, & Rosenbaum, 1971). In such attributional accounts, the following processes have been examined: (a) how subjects on different reinforcement schedules perceive the causes of their reinforcements and nonreinforcements during the acquisition of a response, (b) how they perceive the complete termination of reinforcement during extinction, and (c) the relationships between their causal perceptions and their subsequent behavior (resistance to extinction).

Free-Response Schedules

Those studying free-response schedules have investigated to some degree such phenomena as the acquisition and extinction of responses under continuous and partial reinforcement, too (for example, Skinner, 1938). The major focus in most operant studies, however, has been on the effect of various intermittent schedules on the performance of an already well-learned response.

One of the most consistent findings of these researchers is that organisms typically respond more rapidly under ratio than under interval reinforcement schedules (for example, Ferster & Skinner, 1957; Reynolds, 1968; Skinner, 1938). Responding under a fixed-ratio schedule is generally rapid and consistent, both within and between reinforcement periods. Responding under fixed-interval reinforcement, however, typically shows a different pattern: Reinforcement is followed by a cessation of responding, known as a postreinforcement pause, followed by a gradually accelerating rate until the next reinforcement, at which point the cycle repeats itself. In addition to such changes in response rate within an interval, there are also rate differences between intervals as well. For example, a subject may respond 200 times in an interval, 15 times during the next interval, and 75 times during the next one.

As will be discussed more fully in a later section, there have been several studies which have examined the relationship between behavior on operant schedules and causal attributions (Baron & Kaufman, 1966; Kaufman, Baron, & Kopp, 1966; Leander, Lippman, & Meyer, 1968; Lippman & Meyer, 1967; Rest, 1974). Three recent experiments which lend support to the thesis that causal attributions mediate the relationship between reinforcement schedules and behavior were conducted by the author. These experiments will be reviewed in somewhat greater detail.

REINFORCEMENT SCHEDULES: THEORETICAL ACCOUNTS

A Generic Model for the Partial Reinforcement Effect

Any explanation of the extinction process must take into account the relationship among variables which affect the rate of extinction. Such variables include the percentage, pattern, and amount of reinforcement, whether reinforcement is given immediately following a response or after a delay, and the amount of effort necessary to make a response. Since percentage of reinforcement is the primary variable studied in the literature, however, it will be the major focus in this section.

A generic model which contrasts the effects of partial and continuous reinforcement on resistance to extinction is depicted in Fig. 1. In this model the process is separated into two stages: acquisition and extinction. Continuous reinforcement, shown at the top, results in some state of the organism, *X*. Partial reinforcement, at the bottom, results in some state, *Y*, which is phenotypically similar yet genotypically different from State *X*. That is, although response strength between the two schedules may be equal following acquisition, when extinction occurs response decrement is slower following partial reinforcement. As we shall see in the next few pages, a wide array of mechanisms underlying the partial reinforcement effect have been proposed.

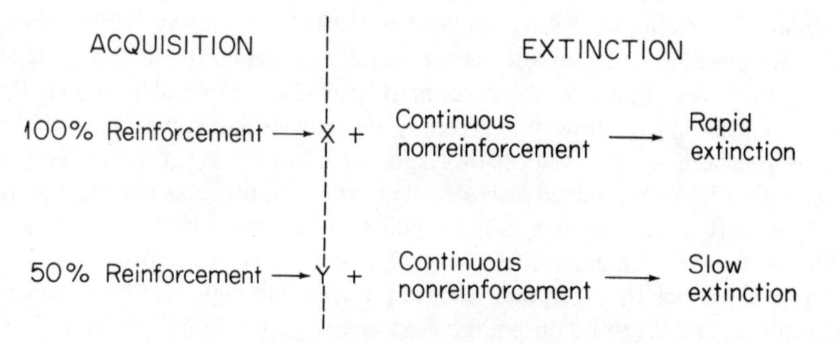

FIG. 1 A generic model of the partial reinforcement effect.

The explanations of the partial reinforcement effect that have emerged in the last four decades may be categorized along two dimensions: whether *mechanistic* or *cognitive* processes are employed to explain the phenomena and whether events which take place during reinforcement or nonreinforcement (*intra*trial events) or events that occur after a sequence of trials (*inter*trial events) are important.

Mechanistic Theories That Emphasize Intratrial Events

The secondary reinforcement hypothesis. Denny (1946) tried to explain partial reinforcement in maze learning in rats by theorizing that, during a reinforced trial, goal-box stimuli acquire secondary reinforcing properties which help maintain the response during withdrawal of the primary reinforcement (experimental extinction). Although secondary reinforcement has been found to play a role in experimental extinction, the partial reinforcement effect has been obtained whether secondary reinforcement cues were present or not. Indeed, as Sheffield (1949) has noted, to explain the partial reinforcement effect by means of secondary reinforcement alone would imply that secondary reinforcement was stronger than primary reinforcement.

Interference theory. In this conceptualization, extinction is viewed as the result of a new response gaining strength and interfering or competing with the original response. Since such a competing response never occurs during continuous reinforcement, when continuous nonreinforcement occurs, the competing response grows strong, and the original response is extinguished. During partial reinforcement, however, the competing response either drops out or comes to evoke the original response. When continuous nonreinforcement occurs, there is less interference with the original response, and extinction is less rapid. In general, the various versions of interference theory (for example, Estes, 1959; Weinstock, 1954) have left ambiguous the nature of the competing response, its source, or both.

Frustration theory. Another derivation of interference theory, known as frustration theory (for example, Amsel, 1967) is by far the most viable of the stimulus–response intratrial explanations of partial reinforcement. Amsel specifies that frustration is the unlearned response to the removal of anticipated reinforcement. Frustration has both drive and cue properties. During acquisition on partial schedules, the subject experiences frustration on nonreinforced trials and such frustration becomes a cue for responding. During extinction, the tendency of partially reinforced subjects to respond in the face of frustration results in slower response decrement. In general, frustration theory gives a plausible account for the partial reinforcement effect and related phenomena. As any theory that is focused on events that happen within a trial, however, frustration theory must ignore variables that are dependent on the particular sequence of reinforced and nonreinforced trials.

Mechanistic Theories That Emphasize Intertrial Events

Stimulus aftereffects theory. Sheffield (1949) suggests that stimuli produced by a response on a given trial persist in some form for a short time, affecting behavior on later trials. In this position resistance to extinction following partial reinforcement is viewed as a product of the conditioning during acquisition of the aftereffects of a nonreinforced trial to the instrumental response. Since the aftereffects are assumed to be short-lived, the increased resistance to extinction should be found only with short intertrial intervals. Although Sheffield (1949) did obtain such results, other investigators have found the effect with spaced, as well as with massed, trials (for example, Weinstock, 1954), thus discounting her hypothesis.

The sequential learning hypothesis. Capaldi and his coworkers (for example, Capaldi, 1966) have suggested that the aftereffects of a reinforced or a nonreinforced trial continue for a long while until replaced by new aftereffects. According to Capaldi (1966), the emphasis of many of the theories of partial reinforcement on nonsequential variables (such as the percentage of reinforcement, the number of reinforced trials, the number of nonreinforced trials) has been misplaced. Instead, Capaldi suggests that sequential variables, such as the number of reinforcement–nonreinforcement transitions, the number of nonreinforcement–reinforcement transitions, the number of successive nonreinforcements (nonreinforcement length), and the number of different nonreinforcement lengths, should be considered. Using such sequential variables, several investigators have made and supported predictions which go beyond the scope of any intratrial theory. As such, the sequential learning approach is one of the most viable explanations of the partial reinforcement effect in the stimulus–response tradition.

Cognitive Theories That Emphasize Intratrial Events

Cognitive dissonance theory. According to Lawrence and Festinger (1962), once an organism comes to expect reinforcement, dissonance is aroused when reinforcement is not obtained. The organism can reduce its dissonance either by changing the behavior that produced it (for example, a rat in a runway situation may stop running) or by changing cognitions about the dissonant relationships (for example, a rat may find other attractions in the goal box).

According to proponents of this approach, a continuously reinforced rat does not experience cognitive dissonance during acquisition and, therefore, does not develop increased attraction towards the goal box. During continuous nonreinforcement, cognitive change is too slow to reduce dissonance. Behavior, therefore, changes. Since stopping running is consistent with receiving no reinforcement, extinction is relatively quick.

On the other hand, a partially reinforced animal experiences nonreinforcement and, therefore, dissonance during acquisition. He continues to respond because quitting is dissonant with the fact that he is rewarded some of the time. The dissonance that arises from the nonreinforced trials is reduced by cognitive change: the rat increases the attractiveness of the situation which leads to greater resistance to extinction.

Lawrence and Festinger (1962) attempted to use cognitive dissonance to tie together the effects of increased effort of the response and delay of reinforcement as well as partial reinforcement on resistance to extinction. Although they presented some data which supported their position, cognitive-dissonance theorists cannot explain many of the findings in the partial reinforcement literature, including results with sequential variables. In addition, the cognitive-dissonance approach assigns theoretical importance to the dissonance aroused by "working for nothing." Longnecker et al. (1952), however, have obtained the partial reinforcement effect for autonomic responses that do not require effort and, therefore, should theoretically not produce any dissonance.

Cognitive Theories That Emphasize Intertrial Events

The expectancy hypothesis. According to proponents of the expectancy hypothesis (for example, Humphreys, 1939), continuous reinforcement leads to an expectation of uniform reinforcement, while partial reinforcement leads to an expectation of irregular reinforcement. Given continuous nonreinforcement, it is easy for continuously reinforced subjects who had expected uniform reinforcement to come to expect uniform nonreinforcement, and if it does not occur, their responding rapidly diminishes. The expectation of irregular reinforcement by subjects given partial reinforcement is more difficult to discard, and they continue to respond.

The expectancy hypothesis has been subjected to much theoretical criticism for being circular and ambiguous. The concept of expectancy, as stated by Humphreys, merely describes the data of partial reinforcement; it does little to explain them.

The discrimination hypothesis. One problem with using the concept of expectancy of reinforcement to explain the partial reinforcement effect is that it appears anthropomorphic when applied to animal behavior. Another problem is that expectancies are mental events that are somewhat difficult to link to observable behavior. The discrimination hypothesis was formulated to solve such problems.

As first introduced by Mowrer and Jones (1945), and later adopted by Bitterman and his colleagues (for example, Bitterman, Fedderson, & Tyler, 1953), this theoretical account states that resistance to extinction is a function of the similarity between acquisition and extinction: the more similar the two schedules, the more difficult is the discrimination between the offset of acquisition and the onset of extinction, and the longer the subject continues to respond. The degree of similarity between acquisition and extinction has been defined in terms of percentage of reinforcement, pattern of reinforcement, and changes in stimuli at the onset of continuous nonreinforcement.

One criticism of the discrimination hypothesis is that there is no specification of the parameters of similarity that are most important. A given pair of acquisition and extinction schedules, for example, could be similar in some ways and different in several other ways. Different predictions would be made about resistance to extinction, therefore, depending on which aspects were chosen for comparison (Lawrence & Festinger, 1962). Another criticism is that, although the discriminability of stimulus conditions by the subject is the crucial variable in the theory, it is always the experimenter who decides, a priori, the degree to which a given set of conditions should be discriminable. Therefore, a measure of acquisition–extinction discriminability independent of resistance to extinction is needed.

Social learning theory. In another cognitive approach to the partial reinforcement effect, social learning theorists (for example, Rotter, 1966) state that the effect of a reinforcement on a person's expectancy depends not only on his reinforcements, but on the way he views the contingencies for his reinforcements as well. For example, James and Rotter (1958) state that, in most studies of partial reinforcement, subjects probably believe that their reinforcements are externally controlled. Those given 100% reinforcement, therefore, perceive their reinforcement as being due to the properties of the task. When continuous nonreinforcement occurs, they perceive that the task has changed, their expectation of reinforcement decreases, and they stop responding. When continuous nonreinforcement occurs for those previously given partial reinforcement, on the other hand, such subjects do not readily discriminate that the task has changed, and they continue to respond. Social learning theorists appear to employ an explanation quite similar to that of proponents of the discrimination hypothesis to account for the partial reinforcement effect under conditions of external control.

Social learning theorists predict that superior resistance to extinction on the part of partially reinforced subjects will not occur, on the other hand, if subjects believe that their reinforcement in a given situation is under internal control. In such a situation, giving subjects continuous reinforcement should lead them to infer that they have a lot of skill at the task, whereas giving subjects partial reinforcement should make them believe they have little skill at the task. During experimental extinction, the former group, still thinking that they are skillful at the task, continue to respond. The latter group's belief that their skill is low should lead to rapid extinction. James and Rotter (1958) and Rotter et al. (1961) found evidence consistent with the above analysis. The partial reinforcement effect was obtained for subjects given chance instructions and reversed for those given skill instructions.

When evaluated critically, the conceptualization of the effect of reinforcement schedules on behavior by social learning theorists appears somewhat ad hoc. As Weiner (1972, 1974) has indicated, proponents of social learning theory have failed to take into account the conflicting information generated by the pattern of reinforcement and the causal instruction. However, despite some flaws in the social learning approach its supporters have made the important point that the way the subject interprets the causes of his reinforcements and nonreinforcements within a reinforcement schedule is a determinant of the rate of extinction.

Causal attribution theory. Like social learning theorists, causal attribution theorists propose that behavior on a reinforcement schedule reflects the organism's beliefs about the causes of its reinforcements and nonreinforcements. As we shall see later, some evidence exists that different reinforcement schedules lead to different causal explanations for outcomes. For example, Jones, Rock, Shaver, Goethals, and Ward (1968) had three groups of subjects observe a stimulus person obtain 50% reinforcement on a task. The pattern of reinforcement, however, was different for each group of subjects. Those subjects who observed a stimulus person obtain more correct responses at the beginning of a series (descending pattern) attributed more ability to the stimulus person than did subjects who observed a random or an ascending pattern.

In addition to between-schedule attributional differences, there is evidence that people respond in different ways to the same reinforcement schedule. For example, two roulette players might each win a small amount during an evening of roulette. One player might attribute his winning to good luck while the other may perceive himself as having the ability to predict correctly the position of the ball.

With respect to experimental extinction, Weiner (1972) has proposed the following three-stage model to account for this process: A given reinforcement schedule (a) provides an organism with information about the causes of his reinforcements and nonreinforcements, which, in turn, results in (b) changes in the expectation of reinforcement which (c) determines how rapidly extinction occurs. According to Weiner, it is the stability or instability of the causal attribution for nonreinforcement which determines, in part, resistance to extinction. If an organism believes that continuous nonreinforcement is due to a factor which is stable over time, such as lack of ability or the extreme difficulty of the task, expectation of reinforcement should decrease rapidly and response decrement will be rapid. If, on the other hand, continuous nonreinforcement is viewed as caused by an unstable factor, such as bad luck or insufficient effort, then expectations of reinforcement should remain high and responding should persist until the attributions have changed.

According to attribution theorists, any antecedent condition that leads to an unstable attribution for nonreinforcement will also lead to resistance to extinction. Similarly, an antecedent factor that leads to a stable attribution for nonreinforcement should result in little resistance to extinction. Weiner et al. (1971) hypothesize that a partial schedule, especially a random one, leads to relatively greater attributions of reinforcement and nonreinforcement to unstable factors, whereas a 100% schedule leads subjects to make stable attributions for reinforcement (for example, "This task is easy" or "I am especially good at this task").

To date, some data which have bearing on an attributional model of the partial reinforcement effect have been collected. Different patterns and schedules of reinforcement have been found to lead to differences in attributions to the factors of ability, effort, luck, and task difficulty (for example, Beckman, 1970; Jones et al., 1968; Weiner et al., 1971). Likewise, changes in expectancy of success have been shown to depend on the attributions made for success and failure. If people attribute their successes or failures to stable causes, then future expectations of success are more greatly influenced than if attributions are made to unstable causes (for example, McMahan, 1973; Weiner, Heckhausen, Meyer, & Cook, 1972).

Besides the experiments that directly test Weiner's model, there are other studies related to an attributional interpretation of partial reinforcement. For example, random partial schedules tend to give greater resistance to extinction than do single alternating schedules (for example, Longnecker et al., 1952). There is evidence that random reinforcement tends to lead to luck attributions (for example, Weiner et al., 1971) and that luck attributions, in turn, result in relatively small changes in expectancy (for example, Weiner et al., 1972). Capaldi's (1958) finding that, given a short acquisition series, resistance to extinction is actually greater for subjects on a single alternating schedule than on a random schedule, is not necessarily troublesome for attribution theorists. It may be that several trials of a random schedule must occur before outcomes are perceived as due to luck.

The finding of Grosslight and his associates (for example, Grosslight & Radlow, 1956) that an acquisition series that ends with a reinforced trial gives greater resistance to extinction than one that ends with a nonreinforced trial is explainable by attribution theory. There is evidence that when rewards occur later in a sequence as opposed to earlier, attributions to effort are greater (for example, Weiner et al., 1972). Such attributions to effort could conceivably mediate the greater resistance to extinction.

As indicated earlier, one criticism of some cognitive accounts is that the operation of cognitions is inferred from the subjects' behavior rather than being measured independently of the behavior the cognitions are supposed to predict. The discrimination hypothesis, for example, lacks an independent measure of similarity between acquisition and extinction from the subjects' point of view. Within attribution theory, however, subjects' own statements of their perceptions of the causes of reinforcement and nonreinforcement have been correlated with their behavior during both acquisition and extinction. Obviously, such an approach is viable only for human subjects, who presumably are able to verbalize their perceptions.

Another criticism of cognitive theorists' approaches to reinforcement is that they use cognitive processes as explanations for behavior, while failing to relate cognitions to prior stimulus events (for example, Skinner, 1969). Such criticism is especially true of the expectancy hypothesis, which does not suggest the mechanism by which sequences of reinforcements and nonreinforcements affect expectancy. Within attribution theory, of course, the links between schedules and expectations are hypothesized to be subjects' beliefs about the causes of reinforcement.

Although Weiner's attributional model of discrete-trial reinforcement schedules is on firmer theoretical ground than some of its cognitive predecessors, a few significant shortcomings remain. It appears that attribution theory can readily explain the resistance to extinction demonstrated by partially reinforced subjects. Such subjects presumably learn during acquisition to attribute their nonreinforcements to lack of effort or to bad luck. When continuous nonreinforcement occurs, they respond more vigorously and/or think that their luck will change.

With respect to continuously reinforced subjects, however, it is not as clear from attribution theory why such subjects extinguish more rapidly. In his model Weiner suggests that both continuous reinforcement and continuous nonreinforcement tend to be attributed to stable factors. That is, when subjects are given 100% reinforcement, they are presumed to attribute their reinforcements to high ability or to task ease. When they are later given continuous nonreinforcement, they are hypothesized as now perceiving themselves as having low ability and/or the task as being extremely hard, with a concomitant decrease in expectation of reinforcement and responding.

The missing step in the chain of events described above, however, is explaining how subjects come to attribute their nonreinforcements to stable factors soon after attributing their reinforcements to stable factors. Perhaps both 100 and 0% reinforcement do indeed cause subjects to make attributions to the task, but the sudden switch from one schedule to the other causes subjects to perceive the task itself as having changed from easy to impossible. Indeed, in one of the studies reported later in the paper, the vast majority of subjects who were given continuous nonreinforcement following 10 minutes of reinforcement at a task attributed their failure to obtain further reinforcement to a change in the task. It would appear that, depending on the situation, Weiner's conceptualization of task difficulty might refer to a stable or to a variable factor.

EXPLANATIONS OF OPERANT SCHEDULES

Skinnerian Explanations

Compared to the discrete-trial reinforcement schedule literature, relatively few theoretical explanations have emerged to account for the effects of operant or free-response schedules on behavior. This state of affairs is due, in part, to the atheoretical orientation of Skinner and his colleagues and, in part, to the relative isolation of operant researchers from the rest of psychology. In general, Skinnerians attempt to explain the effects of reinforcement schedules on behavior with processes that are closely linked to the behavior rather than with hypothetical constructs having no direct behavioral measures. In such accounts, they emphasize conditioning processes such as primary and conditioned reinforcement, response differentiation, and stimulus generalization and discrimination. Differences between responding on ratio and interval schedules are hypothesized by Skinnerians to be due to the differential effects of these processes, operating singly and in combination with one another.

According to Skinner (for example, 1938, 1969) any property of responding that covaries with reinforcement will tend to be repeated over time. On fixed-ratio schedules, in which the number of responses is the only determinant of reinforcement, many properties of responding are reinforced and lead to a subsequent high rate of responding. On fixed-interval schedules, on the other hand, such mechanisms either operate in a different manner or not at all, and a rapid response rate does not occur. Several processes have been suggested as determinants of the higher response rate under fixed-ratio than under fixed-interval reinforcement schedules.

Under a fixed-ratio schedule, a positive feedback mechanism between response rate and reinforcement rate is established. The greater the rate of response, the greater the frequency of reinforcement, which results in a subsequent increase in response rate, etc. On the other hand, the lower the response rate, the less the reinforcement, and responding declines further. This latter mechanism can be used to explain why responses generally cannot be acquired under a high ratio (for example, 50 responses to 1 reinforcement).

On a fixed-interval schedule, on the other hand, there is essentially no correlation between response rate and reinforcement rate and, therefore, no positive feedback mechanism. Neither increments nor decrements in response rate affect reinforcement rate as long as the organism responds at least once per interval.

Reynolds (1968) views the responses emitted on a fixed-ratio schedule as being linked together in a chain. Each response acts as both a conditioned reinforcer for the previous response (thereby increasing the likelihood of the prior response in a chain being emitted in the future) and as a discriminative stimulus for the next response (thereby increasing the probability of its occurrence). Reinforcement of the terminal response of a chain increases the future occurrence of all previous responses. Because the number of responses per reinforcement is generally variable on a fixed-interval schedule, the emission of a response does not act as a conditioned reinforcer for previous responses, and chaining does not occur.

Other properties of responding that tend to covary with reinforcement on fixed-ratio but not on fixed-interval schedules include: (a) the greater likelihood for responses which immediately follow previous responses to be reinforced than those that occur after a delay since the previous response, (b) the number of responses per reinforcement (a constant on a fixed-ratio schedule), and (c) the high response rate itself.

It must be emphasized that the differences in responding under ratio and interval schedules is not merely due to differences in reinforcement rate. Ferster and Skinner (1957) placed two pigeons on a yoked variable-interval, variable-ratio schedule. They found that the bird on the variable-ratio schedule responded at a rate approximately five times as great as the bird on the variable-interval schedule. Thus, even though the number of reinforcements was equal for the two animals, the relationship between reinforcement and response rate was different.

Other Mechanistic Explanations

Other operant researchers have attempted to formulate somewhat more comprehensive theoretical accounts than has Skinner. For example, Schoenfeld and his coworkers (for example, Schoenfeld & Farmer, 1970) have developed a system of classification of reinforcement schedules based on only the dimension of time, instead of one based on both elapsed time and number of responses. Staddon and his colleagues (for example, Staddon, 1972) have formulated the relative proximity principle, in which reinforcement is viewed as selecting both stimulus and response properties of behavior. Anger (1956) and Morse (1966) view performance on reinforcement schedules as due, in large part, to the differential reinforcement of interresponse times, which are subject to the same principles as are other operants.

Cognitive Explanations

Proponents of the mechanistic explanations summarized above have tended to explain behavior on reinforcement schedules as a function of reinforcement contingencies set up by the interaction of the subject's behavior and the rule for reinforcement defined by the schedule. There have been several studies of operant behavior in humans, however, in which researchers have focused on the relationship between behavior and the subject's *cognitive representation* of such reinforcement rules (that is, reinforcement hypotheses).

Such reinforcement hypotheses, when manipulated by use of instruction, have frequently been shown to exert powerful effects on behavior. Kaufman et al. (1966), for example, varied the reinforcement hypotheses for subjects on a fixed-interval schedule. They found that response rate was different for subjects given different hypotheses. Subjects given variable-ratio instructions responded the most, with subjects given fixed-interval instructions responding the least. The variable-interval instruction group and a no-instruction control group were intermediate.

In another study (Baron & Kaufman, 1966), subjects were reinforced on a five-component fixed-interval schedule. A group of subjects who were given accurate instructions about the rule for each component displayed more optimal behavior (they made fewer nonreinforced responses) than did a group merely told the nature of the response. Similarly, Weiner (1970) reported a study in which subjects who were instructed when to expect reinforcement to stop, responded less during continuous nonreinforcement than did a group not given such instructions. This finding might appear quite trivial to the more cognitively oriented reader. It must be noted, however, that the effect on behavior of instructing a subject that reinforcement will cease is somewhat difficult for Skinnerians to explain.

In other studies the hypotheses about the causes of reinforcement formulated by the subjects themselves were measured. For example, Leander et al. (1968) reinforced subjects on one of several fixed-interval schedules without instruction. A correspondence between response rate and the perceived cause of reinforcement was found. Subjects who believed that their reinforcements were based on time responded considerably less than did those who believed that reinforcement was based on responding. In an earlier study, Lippman and Meyer (1967) found a strong instruction effect for subjects given fixed-interval reinforcement as well as a high correspondence between postresponding verbalizations and response rate.

The thesis that reinforcement hypotheses are a determinant of response rate is given some indirect support from studies in which subjects were reinforced on a schedule which followed a certain rule and then were switched to another schedule in which reinforcement followed a different rule. DeCasper and Zeiler (1972) reported a study with children who were initially reinforced on a ratio schedule and then tested on a fixed-interval schedule. Response rates, which were high under the ratio schedule, remained high under the interval schedule. Other children who were initially reinforced on a differential low-rate schedule, in which interresponse times greater than a certain interval were reinforced, were able to respond more "appropriately" when switched to a fixed-interval schedule. Similar results with adults were reported by Weiner (1964).

The researchers cited above all utilized a cognitive variable (the subjects' awareness of their reinforcement contingencies), either by manipulation, measurement, or inference, to explain the effect of reinforcement schedules on response rate. None of these researchers, however, have attempted to integrate their findings into a larger cognitive framework. Causal attribution theorists offer a potentially viable framework.

An important principle in attribution theory is that of covariation. That is, one of the sources of data that people use to make inferences about the causal relationship between two events is the covariation of the events over time. The underlying point of an attributional approach to reinforcement schedules is that response rate is largely a function of the perceived causal relationship between responding and reinforcement.

An attributional analysis of the effect of reinforcement schedules on response rate would suggest the following process: Depending on the particular rule for reinforcement followed by a reinforcement schedule, responding at a given rate and pattern will lead to a particular rate and pattern of reinforcement. For example, rapid responding on a fixed-ratio reinforcement schedule will lead to a high rate of reinforcement, whereas slow, irregular responding will lead to a low, irregular rate of reinforcement. The pattern of responses and reinforcements are analyzed by the organism to determine their degree of covariation. The degree to which responses and reinforcements are perceived as covarying is a determinant of subsequent responding: the greater the perceived covariation, the greater the responding; the less the perceived covariation, the less the responding.

The difference between ratio and interval responding is readily explainable given attribution theory. On a fixed-ratio reinforcement schedule, response rate and reinforcement rate covary. The perception of such a covariation between responding and reinforcement is essentially what is meant by an attribution to effort (for example, Weiner et al., 1971). Attribution theorists suggest that the high response rate in a fixed-ratio schedule is mediated by the subject's perceptions that his reinforcements are due to his efforts.

In contrast, for a fixed-interval reinforcement schedule, there is virtually no correlation between response rate and reinforcement rate. Reinforcement rate is largely a function of the length of the interval, a factor outside the organism's control. Responding on such a schedule would lead an organism to make an external attribution. Such an external attribution is presumed to mediate the effect of the fixed-interval schedule on the subsequent low rate.

In the next section. a series of three studies is summarized in which I tested the hypothesis that reinforcement schedules operate, in part, by giving subjects information about the causes of their reinforcements. Three kinds of questions will be considered: (a) How do people behave on various reinforcement schedules? (b) What do people on various reinforcement schedules say about the causes of their reinforcements? (c) What is the relationship between what people say and how they behave on various reinforcement schedules?

OVERVIEW OF THE PROCEDURES FOR THE EXPERIMENTS

The subjects in the experiments were all college-age males, either at the University of California, Los Angeles (Experiments 1 and 2) or the University of Oklahoma (Experiment 3). Basically, similar procedures were followed in all the experiments. Subjects were tested individually in a "human Skinner box," which was a room containing only a chair and a spring scale mounted on a shelf. The subjects' task was to pull the cord with a minimum of 25 lbs. pressure for $.02 per reinforcement. Reinforcement was made available according to some schedule of partial reinforcement. In Experiments 1 and 2, subjects were given instructions (either accurate or inaccurate) about the causes of their reinforcements before they were exposed to the particular reinforcement schedule. In Experiment 3, no such instructions were given. In addition, on one or more occasions, subjects in each study rated the importance of various factors in their performance: their ability or lack of ability at the task, their effort or lack of effort, the ease or difficulty of the task, and good or bad luck.

Experiment 1

In Experiment 1, subjects were randomly given either effort instructions ("The machine is set up so that you can exert a great influence on the amount of money you make") or noneffort instructions ("The machine is set up so that you can exert only a small influence on the amount of money you make"). One-half of the subjects in each of the instruction groups were randomly given reinforcement on a fixed-ratio schedule (for example, 45:1). The other half of the subjects in each of the groups were randomly given fixed-interval reinforcement (for example, one reinforcement for the first response after 30 seconds had elapsed). After 20 minutes at the task, subjects rated the importance of the four attributional factors in their performance.

The results from Experiment 1 indicated some answers for all three questions raised on page 112.

1. Response rate was related to type of reinforcement schedule. Subjects who were given ratio reinforcement pulled significantly more times than those given interval reinforcement (the average responses per minute was 106.7 versus 73.7). Furthermore, this difference became significantly greater over time.

2. Subjects on different reinforcement schedules made different causal attributions for their reinforcement. Those on fixed-ratio schedules tended to perceive effort as a significantly more important cause and luck as a significantly less important cause than did those given fixed-interval reinforcement.

3. The relationship between causal attributions and response rate was examined in two ways: experimentally and correlationally. Subjects instructed that their efforts would be a major determinant of their reinforcements pulled insignificantly faster than those who were told that their efforts would play only a minor role. The failure to obtain a difference in responding due to causal instruction may be a result of subjects learning after only a few minutes exposure to the schedule whether or not effort was indeed an important determinant of reinforcement. As new information was assimilated over time, the effect of the schedule became stronger and that of the instruction weaker.

In addition, correlations between causal attributions and response rate were computed. For subjects given ratio reinforcement, response rate was positively correlated with the perceived importance of effort, $r = .35$, the perceived importance of ability, $r = .27$, and the perceived importance of task difficulty, $r = .34$, while negatively correlated with the perceived importance of luck, $r = -.24$. For subjects given interval reinforcement, however, there were no significant relationships between response rate and causal attribution.

Alternative Explanations of the Data

In this study, reinforcement schedules were found to lead to differences in causal attributions as well as to differences in response rate. In addition, relationships between causal attributions and response rate were obtained for subjects given fixed-ratio reinforcement. The data, while consistent with the attributional explanation of the effect of reinforcement schedules on behavior (see pp. 111–112), are not necessarily conclusive. Since subjects made causal attributions after their performance on the schedule, one cannot conclude unequivocally that causal attributions must have mediated the effect of reinforcement schedules on response rate. Attributions may have indeed exerted a causal influence, but the design of Experiment 1 was not an adequate test of this hypothesis.

Another possible explanation of the results is that subjects' attributions were a function, rather than a cause, of their behavior. Rather than viewing subjects on reinforcement schedules as analyzing their patterns of reinforcements and responses, which, in turn, influences their subsequent behavior, response rate is viewed as being the result of some other process. In such a view, causal attributions are conceived of as being made by the subjects after they have responded, partly to explain or justify their behavior. Such attributions do not affect future responding. Such a model is consistent with Bem's (1972) approach to the relationship between cognitions and behavior. Bem essentially views people as observing their own actions and then inferring what they must have been thinking.

For example, subjects on fixed-ratio schedules, after responding at a relatively high rate, decide that their reinforcements must have been strongly influenced by their efforts. Subjects on fixed-interval schedules, after responding at a low, irregular rate, decide that their reinforcements were unrelated to their efforts, but due instead to factors such as good or bad luck. Although good or bad luck, in the sense of a random occurrence, could not actually have affected reinforcement rate under either schedule, it is understandable that subjects on fixed-interval schedules may either have concluded accurately that their reinforcements were based primarily on the passage of time and used the term luck to express this conclusion, or they may have perceived no covariation whatsoever between their reinforcement rate and any other factor and used luck to express this lack of covariation.

Still another possible explanation of the results is that, although reinforcement schedules influence both behavior and attributions, there is no direct relationship between the two variables. In this approach, response rate is viewed as an operant influenced by such factors as stimulus discrimination, secondary reinforcement, and differential reinforcement of interresponse times. Causal attributions are viewed as verbal operants, also shaped by reinforcement contingencies. More specifically, a subject on a ratio schedule will come to respond relatively rapidly as well as to verbalize that his efforts are a determinant of his reinforcements. Likewise, a subject on an interval schedule will come to respond relatively slowly as well as to say that his reinforcements are due to good or bad luck. There may be a correspondence between responding and attributions, but no causal connection is presumed to exist. Such an explanation reflects a philosophical view known as mind–body parallelism and is consistent with the theoretical position of Skinner (for example, 1969).

Experiment 2

In Experiment 2, attributions were obtained from subjects twice during their performance instead of after their performance. It is possible, therefore, to examine subjects' response rates both before and after they were asked to explain their reinforcements. Response rates calculated for different segments of performance can then be compared to attributions made at different times, yielding information about the causal relationship between cognitions and behavior.

Proximity- and Direction-Rule Correlations

According to Hume (1965), we infer that an event, A, causes another event, B, if the two events occur in temporal proximity and if Event B follows, but does not precede, Event A. These two statements will be referred to as the proximity rule and the direction rule, respectively. With respect to Experiment 2, the proximity rule would suggest that an individual's statement of the causes of his reinforcements at a given point in time should correlate more highly with his behavior near that point in time than with his behavior at a point temporally distant. The direction rule, on the other hand, would suggest that a subject's attributions at a given point in time should correlate more highly with his subsequent than with his prior rate of response.

To test these hypotheses, I computed correlations between attributions made at 2 minutes and at 8 minutes with the beginning (first 2 minutes), middle (next 6 minutes), and the ending (last 2 minutes) response rates. For example, according to the direction rule, attributions at 8 minutes should correlate more highly with the ending than with the middle response rate. The results of Experiment 2 are highly consistent with the proximity and the direction rules. Of 32 possible comparisons between different pairs of correlations, 29 were in the direction predicted from the two rules.

The pattern of the direction-rule correlations obtained in this experiment support the Weiner-type model. Weiner (1972) would argue that since attributions have a causal influence on response rate, they should correlate more highly with subsequent than with prior behavior. Conversely, a derivation from Bem's approach would probably state that, since attributions are inferred from behavior, they would correlate more highly with prior than with subsequent rates of responding. The data were opposite to such a prediction. It is not clear whether in a derivation from Skinner's analysis one would predict that verbal operants would correlate more, less, or equally highly with prior, as compared to subsequent, responding.

Changes in the Perceived Importance of Causal Attributions as Related to Changes in Behavior

Another source of data concerning the causal relationship between atrributions and behavior involves correlations between changes in attributions and changes in response rate. Table 1 presents correlations between changes in response rate among the three periods (beginning–middle, beginning–ending, and middle–ending) and changes in the perceived importance of ability, effort, task difficulty, and luck between 2 and 8 minutes of responding. Particularly noteworthy are

TABLE 1
Correlations between Attribution Change Scores and Change Scores among Various Segments of Responding: Experiment 2[a]

Response rate change score[b]		Attribution change score[c]			
First segment	Second segment	Ability	Effort	Task difficulty	Luck
Beginning	Middle	+.12[d]	+.12	+.04	–.25**
Beginning	Ending	+.20*	+.21*	+.20*	–.33**
Middle	Ending	+.20*	+.22**	+.29**	–.36**

[a]Taken from Rest, 1974.
[b]Response rate per minute during first segment minus response rate per minute during second segment.
[c]Attribution at 2 minutes minus attribution at 8 minutes.
[d]$N = 80$ for each correlation.
*$p < .05$.
**$p < .01$.

the correlations between attributional change scores and changes in average response rate between the middle and ending periods. The correlations were positive for ability, $r = .20$; effort, $r = .22$; and task difficulty, $r = .29$; and negative for luck, $r = -.36$. That is, if a subject's ratings of the importance of ability, effort, or task difficulty tended to increase over time, his subsequent response rate also tended to increase. If, however, a subject's rating of the importance of luck tended to increase, his subsequent response rate tended to decrease.

From a Bemian approach one would predict that attributional changes would correlate with prior changes in response rate. With one exception, correlations between prior rate changes (beginning–middle periods) and attributional changes were not significant. Thus, support for Weiner's attributional model is given by the result that attributional changes tended to be related to subsequent but not prior changes in response rate.

Experiment 3

In both schedules examined in the previous studies, reinforcement depended on subjects' responding to some extent. Although only on fixed-ratio schedules is there a perfect correlation between response rate and reinforcement rate, at least minimal responding is necessary on fixed-interval schedules. Consequently, many subjects given fixed-interval reinforcement may have "superstitiously" exaggerated the importance of their efforts.

It was decided, therefore, to examine subjects on fixed-time reinforcement schedules. Subjects on such schedules receive reinforcement at constant intervals, whether they respond or not. It was hypothesized that subjects on a fixed-time schedule would readily learn that responding and reinforcement were completely independent and cease responding.

Results showed a significantly lower response rate for subjects given fixed-time reinforcement than for those given fixed-ratio or fixed-interval reinforcement. Unlike the first two experiments, however, significant differences in responding between subjects given ratio and those given interval reinforcement were not found.

In addition, attributional differences were found among groups reinforced on various schedules. Compared to subjects given fixed-interval or fixed-time reinforcement, subjects given fixed-ratio reinforcement tended to perceive reinforcement as due more to their efforts and the difficulty of the task. On the other hand, subjects given fixed-interval or fixed-time reinforcement tended to see ability and luck as more important than did subjects given fixed-ratio reinforcement. (The fact that ability was given a low rating by subjects who rated effort and task difficult highly, and vice versa, was most likely an artifact of the method of measuring attributions used in the third study. The factor of ability was compared to the factors of effort and task difficulty but *not* to luck because luck and ability differ on both the locus of control and stability dimensions. Since both task difficulty and effort tended to be rated highly, ability was given a spuriously low rating.) There were no significant differences in causal attributions between fixed-interval and fixed-time subjects.

Taken as a whole, the results of the three studies summarized above lend some support to the notion that reinforcement schedules operate by giving subjects information about the causes of their reinforcement. In their subsequent behavior, subjects reflect the relationship they perceive between their efforts and their reinforcement.

As indicated earlier, proponents of the mechanistic view of learning have tended to stress the automatic action of reinforcers in shaping behavior (for example, Skinner, 1938; Thorndike, 1932). In such views, learning takes place independent of the learner's "understanding" of the situation. Proponents of cognitive views (for example, Saltz, 1971; Tolman, 1932), on the other hand, stress the information value of a reinforcer in changing an organism's expectancies of reinforcement and, therefore, his behavior. What is learned essentially is that a relationship exists between a stimulus and a response in the case of classical conditioning or between a response and a reinforcer in the case of operant conditioning.

In his early approach to the understanding of reinforcement schedules (for example, 1938,1963), Skinner appeared to rule out cognitive variables completely. His major criticism of the cognitive theories of the time was that such approaches attempted to explain behavior as cognitively mediated but failed to explain the sources of the cognitions. Attribution theory, on the other hand, explicitly investigates the stimulus antecedents as well as the response consequences of cognitions. Thus, it appears that attribution theorists are attempting to develop more clearly defined empirical rules than found in many of the earlier cognitive approaches.

In addition to such changes within cognitive psychology, some reinforcement theorists seem to be modifying their extremely anticognitive positions. Skinner (1969) himself has altered his view of the role of internal events in a direction closer to that expressed by cognitive theorists. He states "that the behavior of a person *who has calculated his chances or considered his consequences* is usually more effective than that of one who has *merely been exposed to the unanalyzed contingencies* of the schedule [pp. 121–22, my italics]."

In some situations, a verbal statement of the reinforcement contingencies may have as powerful an effect on behavior as exposure to the contingencies themselves. For example, many people use seat belts, not because they have had direct experience with the negative consequences of not wearing them (which occur on an intermittent basis) but because they have been exposed to a statement about a rule. This is not to say that rules are equally effective as direct experience in all situations. Perhaps all children need to burn themselves at least once in order not to touch a hot stove.

REFERENCES

Amsel, A. Partial reinforcement effects on vigor and persistence. In K. W. Spence & J. T. Spence (Eds.), *The psychology of learning and motivation* (Vol. 1). New York: Academic Press, 1967.

Anger, D. The dependence of interresponse times upon the relative reinforcement of different interresponse times. *Journal of Experimental Psychology,* 1956, **52,** 145–161.

Baron, A., & Kaufman, A. Human free-operant avoidance of "time-out" from monetary reinforcement. *Journal of the Experimental Analysis of Behavior,* 1966, **9,** 557–565.

Beckman, L. J. Effects of students' performance on teachers' and observers' attributions of causality. *Journal of Educational Psychology,* 1970, **61,** 76–82.

Bem, D. J. Self-perception theory. In L. Berkowitz (Ed.), *Advances in experimental social psychology* (Vol. 6). New York: Academic Press, 1972.

Bitterman, M. E., Fedderson, W. E., & Tyler, D. W. Secondary reinforcement and the discrimination hypothesis. *American Journal of Psychology,* 1953, **66,** 456–464.

Capaldi, E. J. The effect of different amounts of training on the resistance to extinction of different patterns of partially reinforced responses. *Journal of Comparative and Physiological Psychology,* 1958, **51,** 367–371.

Capaldi, E. J. Partial reinforcement: A hypothesis of sequential effects. *Psychological Review,* 1966, **73,** 459–477.

DeCasper, A. J., & Zeiler, M. D. Steady-state behavior in children: A method and some data. *Journal of Experimental Child Psychology,* 1972, **13,** 231–239.

Denny, M. R. The role of secondary reinforcement in a partial reinforcement situation. *Journal of Experimental Psychology,* 1946, **36,** 373–389.

Estes, W. K. The statistical approach to learning theory. In S. Koch (Ed.), *Psychology: A study of a science* (Vol. 2). New York: McGraw–Hill, 1959.

Ferster, C. B., & Skinner, B. F. *Schedules of reinforcement.* New York: Appleton–Century–Crofts, 1957.

Grosslight, J. H., & Radlow, R. Patterning effect of the nonreinforcement–reinforcement sequence in a discrimination situation. *Journal of Comparative and Physiological Psychology,* 1956, **49,** 542–546.

Hulse, S. H., Jr. Amount and percentage of reinforcement and duration of goal confinement in conditioning and extinction. *Journal of Experimental Psychology,* 1958, **56,** 48–57.

Hume, D. *Essential works.* New York: Bantam Books, 1965.

Humphreys, L. G. The effect of random alternation of reinforcement on the acquisition and extinction of conditioned eyelid reactions. *Journal of Experimental Psychology,* 1939, **25,** 294–301.

James, W. H., & Rotter, J. B. Partial and 100% reinforcement under chance and skill conditions. *Journal of Experimental Psychology,* 1958, **55,** 397–403.

Jenkins, W. O., & Stanley, J. C., Jr. Partial reinforcement: A review and critique. *Psychological Bulletin,* 1950, **47,** 193–224.

Jones, E. E., Rock, L., Shaver, K. G., Goethals, G. R., & Ward, L. M. Pattern of performance and ability attribution: An unexpected primacy effect. *Journal of Personality and Social Psychology,* 1968, **10,** 317–340.

Kaufman, A., Baron, A., & Kopp, R. M. Some effects of instructions on human operant behavior. *Psychonomic Monograph Supplements,* 1966, **1**(11), 243–250.

Lawrence, D. H., & Festinger, L. *Deterrents and reinforcements.* Stanford, California: Stanford University Press, 1962.

Leander, J. D., Lippman, L. G., & Meyer, M. E. Fixed-interval performance as related to subjects' verbalizations of the reinforcement contingency. *Psychological Record,* 1968, **18,** 469–474.

Lewis, D. J. Partial reinforcement: A selective review of the literature since 1960. *Psychological Bulletin,* 1960, **57,** 1–28.

Lippman, L. G., & Meyer, M. E. Fixed-interval performance as related to instructions and to subjects' verbalizations of the contingency. *Psychonomic Science,* 1967, **8,** 135–136.

Longnecker, E. D., Krauskopf, J., & Bitterman, M. E. Extinction following alternating and random reinforcement. *American Journal of Psychology,* 1952, **65,** 580–587.

McMahan, I. D. Relationships between causal attributions and expectancy of success. *Journal of Personality and Social Psychology,* 1973, **28,** 108–115.

Morse, W. H. Intermittent reinforcement. In W. K. Honig (Ed.), *Operant behavior: Areas of research and application.* New York: Appleton–Century–Crofts, 1966.

Mowrer, O. H., & Jones, H. Habit strength as a function of the pattern of reinforcement. *Journal of Experimental Psychology,* 1945, **35,** 293–311.

Rest, S. *An attributional interpretation of reinforcement schedules.* Unpublished doctoral dissertation, University of California, Los Angeles, 1974.

Rest, S., Frieze, I., Nickel, T., Parsons, J., & Ruble, D. Effects of chance versus skill instructions, schedule of reinforcement, and locus of control on resistance to extinction. *Proceedings of the 80th Annual Convention of the American Psychological Association,* 1972, *7,* 257–258. (Summary)

Reynolds, G. S. *A primer of operant conditioning.* Glenview, Illinois: Scott, Foresman, 1968.

Rotter, J. B. Generalized expectancies for internal versus external control of reinforcement. *Psychological Monographs,* 1966, **80**(1, Whole No. 609).

Rotter, J. B., Liverant, S., & Crowne, D. P. The growth and extinction of expectancies in chance controlled and skilled tasks. *Journal of Psychology,* 1961, **52,** 161–177.

Saltz, E. *The cognitive bases of human learning.* Homewood, Illinois: Dorsey Press, 1971.

Schoenfeld, W. N., & Farmer, J. Reinforcement schedules and the "behavior stream." In W. N. Schoenfeld (Ed.), *The theory of reinforcement schedules.* New York: Appleton–Century–Crofts, 1970.

Sheffield, V. F. Extinction as a function of partial reinforcement and distribution of practice. *Journal of Experimental Psychology,* 1949, **39,** 511–526.

Skinner, B. F. *The behavior of organisms.* New York: Appleton–Century–Crofts, 1938.

Skinner, B. F. Behaviorism at fifty. *Science,* 1963, **140,** 951–958.

Skinner, B. F. *Contingencies of reinforcement: A theoretical analysis.* New York: Appleton–Century–Crofts, 1969.

Staddon, J. E. R. Temporal control and the theory of reinforcement schedules. In R. M. Gilbert & J. R. Millenson (Eds.), *Reinforcement: Behavioral analyses.* New York: Academic Press, 1972.

Thorndike, E. L. *The fundamentals of learning.* New York: Bureau of Publications, Teacher's College, Columbia University, 1932.

Tolman, E. C. *Purposive behavior in animals and men.* New York: Century, 1932.

Weiner, B. *Theories of motivation.* Chicago: Markham, 1972.

Weiner, B. Achievement motivation as conceptualized by an attribution theorist. In B. Weiner (Ed.), *Achievement motivation and attribution theory.* Morristown, New Jersey: General Learning Press, 1974.

Weiner, B., Frieze, I., Kukla, A., Reed, L., Rest, S., & Rosenbaum, R. M. *Perceiving the causes of success and failure.* Morristown, New Jersey: General Learning Press, 1971.

Weiner, B., Heckhausen, H., Meyer, W. U., & Cook, R. E. Causal ascriptions and achievement behavior: A conceptual analysis of effort and reanalysis of locus of control. *Journal of Personality and Social Psychology,* 1972, **21,** 239–248.

Weiner, H. Conditioning history and human fixed-interval performance. *Journal of the Experimental Analysis of Behavior,* 1964, **7,** 383–385.

Weiner, H. Instructional control of human operant responding during extinction following fixed–ratio conditioning. *Journal of the Experimental Analysis of Behavior,* **1970, 13,** 391–394.

Weinstock, S. Resistance to extinction of a running response following partial reinforcement under widely spaced trials. *Journal of Comparative and Physiological Psychology,* 1954, **47,** 318–322.

6

The Self-Perception of Intrinsic Motivation

Michael Ross

University of Waterloo

Psychologists and the educators consider the distinction between intrinsic and extrinsic motivation to be important to an understanding of human behavior. A person is described as intrinsically motivated if he performs an activity for its own sake and extrinsically motivated if he performs an activity solely as a means to an end, for example, to obtain a reward or to avoid a punishment. My major purpose in this chapter is to investigate the determinants and consequences of self-perceptions of intrinsic and extrinsic motivation.

To begin, let us consider the relevance of attribution theory to this problem, Attribution theory is not really a formal "theory," but rather a set of empirical generalizations about causal inferences made by people in their everyday lives. The theory has had a major impact on social psychology because of the unique orientation provided and the kinds of questions raised. With respect to self-perceptions, attribution theory is concerned with the person's understanding of the causes of his own behavior, rather than the actual causes. From this perspective, then, we are led to consider whether the person views his own behavior as intrinsically or extrinsically motivated. These self-perceptions are assumed to have important implications for behavior. A person who attributes his behavior to external consequences is likely to act as if he were extrinsically motivated. Similarly, if he perceives himself as performing an activity for its sake, he is likely to behave as if he were intrinsically motivated, whatever the true cause of his behavior. In succeeding portions of this chapter when I describe a person as intrinsically or extrinsically motivated, I am referring to the person's own view of the source of his motivation. For the present purposes, at least, the true source of the person's motivation is unimportant.

An attribution theory approach to intrinsic motivation raises two important questions: (a) What causes a person to perceive himself as intrinsically or extrinsically motivated?[1] and (b) What are the consequences of these differing attributions? To answer the second question first, psychologists and educators tend to agree that intrinsic motivation is somehow more desirable, though the reasons for this assertion are not often elucidated. The most obvious advantage is that the intrinsically motivated person finds his work inherently satisfying. If an individual is working solely for the sake of some extrinsic goal, however, he may find the work, itself, "neutral or even disagreeable" (Heider, 1958, p. 126). It is important to point out, however, that enjoyment should not be equated with intrinsic attribution. For example, a person may like an activity because he gets paid for it. In this sense, one may like an activity and yet not be interested in it for its own sake. On the other hand, the extrinsically motivated person should, in general, be less likely than the intrinsically motivated person to persist at the task in the absence of external reward, coercion, or surveillance. In a similar manner, Brazelton (1974) noted the importance of the intrinsic–extrinsic distinction with respect to children's motivation: Intrinsic motivation

> . . . has ever increasing rewards, and gets constant refueling from the sense of mastery in the child that accompanies each step. Motivation injected by those around him is more likely to need ever increasing amounts of fuel (from outside) to fire the system. At a certain point, the need outgrows the supply [pp. 190–191].

In this chapter, I describe a series of experiments that were designed to provide an answer to the first question, what are the determinants of the self-perception of intrinsic and extrinsic motivation? Since 1970, there has been an increasing amount of research conducted on this topic. Because of space limitations, I will concentrate on the research carried out in my own laboratory. Where possible, though, I will point out how our results replicate, extend, or contradict the findings of others.

THE EFFECTS OF TASK-CONTINGENT REWARDS ON INTRINSIC MOTIVATION

Most of the researchers in this area have studied the effects of rewards on intrinsic motivation. In 1950, Harlow reported that monkeys' enthusiasm for a puzzle-solving activity waned after the activity had become associated with a food reward. While it may seem overly anthropomorphic to discuss monkeys'

[1] For ease of communication, I refer to intrinsic and extrinsic motivation as if they were completely dichotomous states. In reality, though, most behaviors are probably perceived as both intrinsically and extrinsically motivated. Consequently, it would be more accurate to consider whether the person views himself as *predominately* motivated by intrinsic or extrinsic factors.

"self-perceptions," these results are at least consistent with a derivation from attribution theory. An attributional analysis suggests that a person will be more likely to perceive himself as extrinsically rather than intrinsically motivated if he is provided with a salient reward for engaging in an activity (Bem, 1967, 1972; Kelley, 1967, 1973).

This hypothesis has also received support from a fine series of studies conducted by Lepper and his colleagues with nursery school children. For example, Lepper, Greene, and Nisbett (1973) had children draw with magic markers (an enjoyable activity for these children) under one of three experimental conditions: (a) they anticipated and received a reward in return for doing the drawing; (b) they received the reward unexpectedly after finishing the drawing; or (c) they were neither promised nor given a reward. Intrinsic interest in working with the magic marker was assessed 1 or 2 weeks later in a free-play situation in which a number of other activities were available, and no rewards were given. During this period subjects who had been in the expected-reward condition were less likely to play with the magic marker than were subjects in the remaining two conditions which did not differ. Presumably, the children who had anticipated a reward attributed their initial performance of the activity to that reward; as a result, they did not persist in the activity during the subsequent free-play period when no reward was available. They had lost interest in performing the activity for its own sake. Presumably, unexpected rewards did not affect intrinsic interest because the initial performance could not reasonably be attributed to the unanticipated reward.

The detrimental effect of expected rewards on intrinsic interest has been replicated using very different activities and rewards, with subject populations ranging from preschool children to college students. In general, individuals promised and given a reward for engaging in an enjoyable activity are less interested subsequently in that activity than individuals not promised or given a reward (for example, Calder & Staw, 1975; Deci, 1971; Kruglanski, Friedman, & Zeevi, 1971).

Like most researchers in this field, my own interest in intrinsic motivation stemmed from practical as well as theoretical considerations. Extrinsic rewards are powerful aids to learning. However, the research findings suggest that the employment of rewards may have an unintended, deleterious effect on the child's intrinsic interest in the activities with which rewards are associated. A greater understanding of the conditions which do or do not produce a decrement in intrinsic interest may ultimately enable the judicious use of rewards, so that their benefits but not their dangers are realized.

Most of my own research has been conducted with children. The principal advantage to employing young children as subjects in this research area is that some of the most important practical implications of the research relate to the use of contingent reinforcements in education. Nevertheless, there are hazards facing the social psychologist who has only conducted experiments with adults and suddenly decides to perform research on children. Children are simply not college sophomores and cannot be treated as such. For over a year, Ken Bowers and I collaborated on research on intrinsic motivation in children under 10 years of age and obtained data that looked as if it had been generated from a random numbers table. Ken's previous research experience had concerned hypnosis and creativity in adults; my own focused on social perception and attitude change in college students. In retrospect, I can see that we were often insensitive to the age of our subjects in terms of the kinds of experimental situations we devised. We probably scared or, at the very least, confused the hell out of them. Our abject failures led Ken, quite understandably, back to his previous research interests. I continued and almost all of the research that I describe below was conducted following this initial period of training.

A recurrent theme in my research program concerns the attempt to specify the conditions under which rewards do or do not cause a decrement in intrinsic motivation. In initial efforts I focused on the salience of the reward. An attributional analysis does not imply that extrinsic motivation will be inferred whenever external rewards are anticipated. Since all of the possible causes of a behavior need not be salient at a given time (Kelley, 1972, 1973; Kiesler, Nisbett, & Zanna, 1969), it should be possible to modify the inferences that a person will draw with respect to his behavior. The more salient or figural external consequences become, the more likely the person should be to regard them as the reason for his behavior and to perceive his actions as extrinsically motivated.

When I examined previous research in which a decrement in intrinsic motivation had been obtained as a function of reward, I discovered that the reward had been made highly salient to the subjects. For example, Deci (1971) reported that his subjects felt "inequitably overpaid" while Lepper et al. (1973) described the reward in glowing terms to the children at the very point at which they were required to commit themselves to performing the activity.

I conducted two experiments to test the effect of varying the salience of an anticipated reward on subjects' intrinsic interest in an activity (Ross, 1975). The subjects were preschool children between 3 and 5 years of age. In both the salient and nonsalient reward conditions, the child was told that he would receive a prize if he played a drum for a few minutes; however, he would not discover the nature of the prize until he had finished playing. In the salient reward condition, the child was told that the reward was located under a box in front of the drum. I assumed that the box would act as a constant, visible reminder of the forthcoming reward. In the nonsalient reward condition, the child was not told that the prize was under the box, and, hence, the box did not act as a cue for the reward. In a control condition, no prize was promised or awarded. The children played with the drum for several minutes. Subjects in the two reward conditions were then given their prizes, assorted candies. The children's intrinsic interest in the drum was assessed during a subsequent 5-minute free-play period in which no rewards were forthcoming, and they could choose to play with the drum or engage in a number of other activities.

The results suggested that the salience of a reward is an important determinant of intrinsic motivation. During the free-play period, children in the nonsalient reward condition played the drum for a significantly longer time period, $\bar{X} = 183.3$ seconds, than did the children in the salient reward condition, $\bar{X} = 81.7$ seconds. Comparisons with the control mean, 138.5 seconds, indicated that these results principally reflected a decrease in intrinsic interest due to the high salience of the reward. Children in the salient reward condition showed significantly less interest in the drum than did control subjects. Results of the control and nonsalient reward conditions did not differ reliably.

In a second experiment, I incorporated a different reward and a different manipulation of salience in order to test the generalizability of the initial results. Preschool children were promised two marshmallows if they would play the drum for a few minutes. For some subjects the reward was made highly salient: they were asked to think about the forthcoming marshmallows while playing the drum. (This rather strange requirement was introduced under the rubric of having subjects play the "think about" game while they thumped the drum.) In a second condition, subjects were asked to think about an unrelated topic (snow) while playing the drum. I assumed that thinking about snow would distract subjects from thinking about the reward, and, hence, its salience would be reduced. In a third condition, the subjects were promised the marshmallows but not given any instructions to ideate while playing the drum (nonideation condition). Predictions about the results in this condition could not be made with confidence. Although subjects were not asked explicitly to think about the reward, they were not prevented from doing so. Moreover, the reward was possibly more salient than in the previous experiment, since subjects were aware of the type of reward that they would obtain (and pretesting had indicated that the reward was very popular), whereas in the first study subjects were given only a promise of some kind of prize. In a final control condition, the subjects were neither promised nor given a reward.

In this second experiment, I was attempting to manipulate directly subjects' attention to the reward while they engaged in the target activity. I assumed that the amount of intrinsic interest in the drum that they subsequently displayed would vary inversely with the amount of attention that they had focused on the reward. As in the initial experiment, intrinsic interest was assessed in a subsequent 5-minute free-play period. As expected, the distraction condition yielded significantly more interest in the drum during the free-play period than did the think-reward condition (mean duration of play, 195.6 and 84.5 seconds, respectively). Although, the nonideation condition tended to produce more play with the drum, $\bar{X} = 96.9$ seconds, than did the think-reward condition, the difference was small and not significant. This suggests that many subjects in the nonideation condition were thinking about the reward on their own initiative. In the no-reward control condition, there was significantly more play with the drum, mean = 201.1 seconds, than in the think-reward and nonideation conditions. On the other hand, amount of play with the drum in the control and distraction conditions did not differ significantly. This supports the evidence from the first experiment that an anticipated reward need not reduce intrinsic interest.

Assuming that at least some children find schoolwork intrinsically interesting, the practical implication of these results is clear: teachers who employ rewards to enhance learning should not constantly draw the children's attention toward the rewards. The less the children think about the external consequences, the less likely their intrinsic interest in the activity associated with the reward is to suffer. Nevertheless, the reward cannot (at least initially) be totally deemphasized if it is to have any positive impact on learning and performance. As Bowers (1974) has noted, the very conditions which may be most conducive to learning (highly salient rewards) may be least conducive to intrinsic interest. Thus, the teacher may be forced to make certain trade-offs. He may wish to tolerate a reduction in the efficiency of reward procedures in order that at least some of the children in the classroom continue to find learning inherently satisfying.

ALTERNATIVES TO THE ATTRIBUTION EXPLANATION

Delay of Gratification

The results of the above two experiments appeated to offer substantial support for the attributional analysis of self-perceptions of intrinsic motivation. Nevertheless, it is possible to view the data from these studies and preceding research on intrinsic motivation in children from an entirely different perspective (Ross, 1975; Ross & Karniol, 1974). The reward conditions in most intrinsic motivation studies incorporate a period of time between the promise of the reward and its attainment. That is, children are offered a reward, but it is not given until they have performed the target activity. Children's ability to wait for a reward has been studied in research on delay of gratification. Mischel (1974) has found that such a waiting period causes young children to experience an aversive state of frustration. Conceivably, in the intrinsic motivation studies, the frustration generated by having to wait for the reward becomes associated with the interpolated activity and thereby makes the task somewhat aversive. It may be argued that as a result of this temporal association, children promised and given a reward should be less likely to engage subsequently in the reinforced activity than children who had not been promised or given a reward.

Some additional research evidence is consistent with this alternative hypothesis, for example, the Lepper et al. (1973) finding that an unanticipated reward did not produce a decrease in intrinsic interest. As this procedure eliminates delay of gratification, no decrease would be predicted. Moreover, the data that I reported on the effects of reward salience are consistent with evidence indicating that increasing the salience of a reward makes the delay period more frustrating for the child (Mischel & Ebbesen, 1970; Mischel, Ebbesen, & Zeiss, 1972).

The delay of gratification explanation seems in at least one respect more and in another respect less plausible than the attributional account. One positive aspect of the delay of gratification hypothesis is that it requires less cognitive work on the part of the child. Can a child as young as 3 or 4 engage in the causal reasoning posited by attribution theory? Some of Piaget's (1970) analyses can be interpreted to suggest that such young children are simply incapable of thinking in these terms. Piaget argues that the young child's thought processes are egocentric and that, as a result, he is unable to shift attention from one aspect of a situation to another or to deal with several aspects of a situation simultaneously. According to this conceptualization, the preschool child should be hard pressed to consider multiple plausible causes for a behavior and to infer that one was more likely to have caused the behavior than another. Yet an attributional analysis of the intrinsic motivation studies would suggest that the young child considers multiple plausible causes in precisely such a fashion. On the other hand, the delay of gratification hypothesis fails the test of parsimony. Rewards have been shown to decrease intrinsic motivation in adults (Calder & Staw, 1975; Deci, 1971), and it is unlikely that they would find these short delay periods frustrating. It is possible, though, that the effects of rewards on intrinsic motivation warrant different explanations in young children and adults.

The delay of gratification hypothesis is sufficiently compelling to merit an experimental test that would contrast it with the attribution hypothesis in a situation that yielded differential predictions. The delay of gratification analysis would imply that the temporal association of the task with the frustrating delay period is the cause of the subsequent decreased interest in the activity. In contrast, the attribution hypothesis would suggest that the contingency between the task and the reward is critical. Accordingly, an experiment was devised to identify the important element: association with the delay period or reward–task contingency (Ross, Karniol, & Rothstein, in press).

The subjects were children in the first, second, and third grades. After playing with the children for a few minutes, the experimenter informed them that he had some work to finish in another room. In the task-contingent reward condition, the children were told that they would receive a reward if they drew pictures with magic markers while the experimenter was absent. In the wait-contingent reward condition, the children were promised the same reward for waiting for the experimenter's return, rather than for performance of the interpolated drawing activity. In the no reward control condition, the children were neither promised nor given a reward for waiting or for drawing. As in the previous experiments, the children's intrinsic interest in the drawing activity was assessed during a subsequent free-play period which in this study lasted 15 minutes.

The delay of gratification hypothesis predicts that subjects in the wait-contingent and task-contingent reward conditions should show less interest in the drawing activity than subjects in the control condition during the free-play period. According to the delay of gratification analysis, a decrement in interest would occur in the two reward conditions because subjects performed the activity initially while awaiting a reward. The attribution hypothesis predicts that this decrease should occur solely in the task-contingent condition and that interest in the wait-contingent and control conditions should not differ. It is only in the task-contingent condition that the subject may perceive the reward as a cause of his willingness to engage in the drawing activity.

The attribution interpretation was supported by the results of this experiment. Children in the wait-contingent reward and control conditions did not differ significantly in the amount of interest that they displayed toward the drawing activity during the free-play period (wait-contingent $\bar{X}$, 379.4 seconds; control $\bar{X}$, 286.7 seconds), and they showed significantly more interest than children in the task-contingent reward condition, $\bar{X}$ = 153.2 seconds. The data strongly suggest that rewards must be made contingent on performance of the target activity for the decrement in intrinsic interest to occur, a finding that is consistent with an attributional analysis.

Task Performance and Intrinsic Motivation

Several investigators have offered a second alternative to the attribution explanation that would account for the detrimental effects of rewards on intrinsic interest in both adults and children (Calder & Staw, 1975b; Reiss & Sushinsky, 1975). They reasoned that the promise of a reward affects task performance during the period in which rewards are anticipated but not yet available. Such alterations in performance are, then, assumed to have an impact on the subjects' intrinsic interest in the activity. Thus, Calder and Staw suggested that the introduction of task-contingent rewards may cause the subject to increase the amount of effort that he puts into an activity. As a result, rewarded subjects may become more fatigued or satiated with the task than nonrewarded subjects. Reiss and Sushinsky argued that anticipated rewards may distract subjects from the experimental activity or cause them to engage in hasty, low-quality play. As a consequence, subjects in the task-contingent reward conditions may find the activity less enjoyable.

The critiques by Calder and Staw and Reiss and Sushinsky generate two major questions: (a) Do anticipated rewards affect task performance? and (b) If so, are these differences in performance causally related to changes in intrinsic interest?

The answer to the first question appears to be yes. For example, Lepper and his colleagues (1973) have found in some of their studies that children in the expected reward conditions produce lower quality drawings prior to receiving the reward than do children in the no reward conditions. Similarly, Kruglanski et al. (1971) reported that subjects who anticipated a reward were less "creative" prior to obtaining the reward than were subjects in a no-reward condition. While a number of other experiments have not obtained differences in prereward behavior (for example, Ross, 1975), it is difficult to interpret negative results. Conceivably, these researchers employed insensitive or inappropriate assessments. It is always possible to argue that additional measures of task performance may well have produced positive results. In any case, there is ample evidence that anticipated rewards can affect task performance.

The second question concerns the relation between these differences in performance and intrinsic interest. There are two sources of evidence that suggest that the performance differences do not readily account for the effects of rewards on intrinsic interest. First, reward-induced decrements in interest have occurred in studies in which no differences due to anticipated rewards were obtained on task performance (Calder & Staw, 1975b; Green & Lepper, 1975; Ross, 1975). This evidence is not compelling, though, for again the negative results on task performance may simply indicate inadequate measurement techniques. Perhaps more convincing evidence is provided from the results of an individual difference measure employed in the study by Ross et al. (in press). Prior to the experiment, they administered a test of ability to delay gratification (adapted from Mischel & Gilligan, 1964) to their subjects. Ability to delay gratification was found to influence the quality of the drawings subjects made while awaiting their reward. Not surprisingly, subjects who were less able to delay gratification are poor artists since in the no-reward control condition the their drawings were of poorer quality than were those of high delay of gratification subjects. These results do not imply merely that subjects who are unable to delay gratification are poor artists since in the no reward control condition the drawings of low delay of gratification subjects were, if anything, better than those of high delay of gratification subjects. These results indicate that the experiment incorporated an adequate measure of task performance and that only subjects who were less able to delay gratification showed a decrement in performance while awaiting a reward. Consequently, the task-contingent reward should reduce the intrinsic interest of low delay of gratification subjects only if the relation between extrinsic rewards and intrinsic interest is mediated by task performance. In fact, ability to delay gratification was not related to subsequent interest in the activity during the free-play period, $F < 1$. Moreover, the correlation between the quality of the drawings that subjects made prior to obtaining the reward and the amount of time that they engaged in the drawing activity during the subsequent free-play period was negligible, $r = .19$. In summary, anticipated rewards may affect task performance. However, objective differences in task performance do not appear to be causally related to changes in intrinsic interest.

I do not wish to overstate the case, however. I am not arguing that distraction, hurried play, or for that matter frustration will never cause a reduction in intrinsic motivation. It is conceivable that with larger rewards (which might act as more potent distractors or frustrators), more complex activities, or tasks that are more clearly skill-oriented, a relationship between objective task performance and intrinsic motivation would be obtained. Moreover, the research that I will turn to next indicated that *self-perceptions* of task mastery may well affect intrinsic motivation.

THE EFFECTS OF PERFORMANCE-CONTINGENT REWARDS

From the above research it appears that salient rewards which are contingent on task engagement reduce intrinsic interest in an activity. Nevertheless, this decrement in intrinsic interest may not be inevitable. A number of theorists have argued that people are motivated to exercise effective control over their environment and derive intrinsic satisfaction from those activities at which they feel competent (de Charms, 1968; Hunt, 1965; Nuttin, 1973; Smith, 1968; White, 1959). Yet a person often uses extrinsic or socially defined standards to assess his degree of mastery at a task (Festinger, 1954; Smith, 1968). From this perspective, extrinsic factors which define one's performance as effective should facilitate intrinsic interest. Rewards may be expected to function in this way when their attainment is a function of the individual's success at the activity. Such rewards have cue value: their reception signifies the individual's degree of competence at the task.

Note that in the experiments that I have described previously rewards were not employed in this manner. In these studies, rewards were dispensed contingent on undertaking the task rather than on any specific performance requirements. Rewards which are allocated without regard to quality of performance have little or no cue value. Such rewards may be particularly injurious to intrinsic motivation since they are appreciated in their own right but do not arouse the intrinsic satisfaction associated with effective control. As a result, there is simply a means–end relation between the activity and the reward. On the other hand, when rewards are contingent upon degree of success, their attainment makes figural the person's competence. The reward is still likely to be appreciated for its own sake and to be perceived as a deserved outcome for performing the activity. The person may be less likely, however, to view the reward as the *sole* reason for performing the activity, and, thus, to discount his intrinsic interest in the activity. The converse is more probable: the activity may be perceived as the raison d'etre for the reward. The cue value of a reward may thus be maximized when rewards are made contingent on the person's degree of success, rather than on task engagement per se.

Several investigators have suggested that the impact of rewards on intrinsic motivation may depend on their meaning with respect to feelings of effective control (for example, Deci, 1974; Lepper & Greene, 1974; Salancik, 1974). Deci (1974) argued that perceptions of effective control are reduced through the use of tangible rewards and enhanced by the use of social reinforcement. This analysis is likely to be true when the individual learns less about his competence on a task from the tangible reward than from the social reinforcement. Yet the same information can be communicated from tangible rewards as from social reinforcement when the rewards are contingent on the quality of performance. As a result, such rewards should not be inimical to, and perhaps should even enhance, intrinsic interest.

Karniol and Ross (1975a) conducted two experiments in which the same tangible rewards were given either contingent or noncontingent on the subjects' degree of success at an enjoyable activity. In the first experiment, college students worked at a concept-formation task for which they either received no money (control condition) or money contingent solely upon undertaking the task or contingent upon successful solutions. By means of bogus feedback, we led all subjects to believe that they had solved the same number of problems. Also, the amount of money received by subjects in the two reward conditions was identical ($1.50). Intrinsic interest was assessed from subjects' evaluations of the experimental task and from their reported willingness to perform the task on subsequent occasions when no rewards were available.

The results supported the hypothesis that intrinsic interest is affected by the cue value of the reward. Subjects who were rewarded on the basis of their alleged successes indicated greater liking for the task and volunteered to participate in more future sessions for no reward than did subjects whose reward was not linked to quality of performance. Comparisons with the control condition revealed that rewards reduced liking for the task only when they were not contingent on performance level.

In the second experiment we tested the generalizability of the initial results by employing a younger subject population (ages 4 to 9), a nonmonetary extrinsic reward (marshmallows), and a behavioral measure of intrinsic motivation (persistence at a concept-formation task during a 6-minute free-play period in which no rewards were available and other toys were present). In order to test further the hypothesis that the impact of performance-contingent rewards is dependent on perceptions of task mastery, we also included a manipulation of the degree to which subjects succeeded at the activity. One-half of the subjects in each experimental condition learned (via bogus feedback) that that they had performed either better or worse then average. It was predicted that in the success condition, performance-contingent rewards would enhance intrinsic interest when compared to rewards allocated irrespective of quality of performance. In the failure condition, however, this result was not expected, for subjects cannot perceive that they possess a significant degree of task mastery.

The data were generally consistent with these hypotheses. When subjects allegedly succeeded, those in the performance-contingent and no-reward control conditions played with the concept-formation task for a significantly longer time period during the free-play session than did subjects in the noncontingent reward condition, $\bar{X}$s = 247, 260.5, and 126.4 seconds, respectively. These results paralleled the results obtained in the initial experiment. When subjects failed, however, those in the performance-contigent condition displayed somewhat *less* interest in the concept-formation task than did subjects in the noncontingent reward condition, $\bar{X}$s = 155.5 and 255.3 seconds, respectively. In summary, then, the data from the two experiments indicate that rewards which are contingent on the quality of performance yield more intrinsic interest than noncontingent rewards when subjects perceive that they possess some degree of mastery over the activity.

One other point should be made with respect to these studies. If performance-contingent rewards arouse effectance motivation, it may be expected that more intrinsic interest would be shown in that condition than in the no-reward control condition. In the first study, subjects in the performance-contingent condition did, in fact, volunteer for significantly more future sessions than did subjects in the control condition. The lack of a significant difference in the second study may be attributable to the explicit performance information that was provided. Subjects knew independently of any information yielded by the rewards exactly how well they had performed. As a result, the cue value of the reward may have been reduced. The performance-contingent reward may still make salient the person's degree of success and thus arouse effectance motivation. Nevertheless, a performance contingency in which the reward provides the sole outcome information would possibly further enhance intrinsic motivation.

PRACTICAL IMPLICATIONS: TOKEN ECONOMY PROGRAMS

In recent years, token economy programs have become an increasingly popular means of modifying children's classroom behavior (Kazdin & Bootzin, 1973; O'Leary & Drabman, 1971). Children in these programs are selectively reinforced for emitting desired behaviors, for example, increased attention in class and improved grades. These programs have proven to be extremely effective in changing children's academic and social behavior. Recently, though, a number of investigators have questioned the use of such programs because of their possible negative impact on children's intrinsic motivation (Lepper & Greene, 1974; Lepper et al., 1973; Levine & Fasnacht, 1974). Levine and Fasnacht observed,

> . . . as applied to token economies, the results of these studies (on intrinsic motivation) indicate the trepidations that practitioners should have before instituting a token

program. There should not be an attempt to use external rewards for behaviors that are of some intrinsic interest. If children are engaged in classroom activities, adding rewards for learning should have the effect of increasing the learning while the rewards are used. However, the longer range consequences of using external rewards are clear, the token rewards will lead to a decrease in interest [pp. 817–818].

It is my contention, however, that at least on the basis of present evidence, these comments are overly pessimistic. In the following section, I will examine research that directly tests the effects of token economy programs on intrinsic motivation.

O'Leary and Drabman (1971) noted that a token economy program usually includes:

(a) a set of instructions to the class about the behaviors that will be reinforced, (b) a means of making a potentially reinforcing stimulus–usually called a token–contingent upon behavior (the token is merely a symbol, for example, a gold star or plastic chip, which can be exchanged for the actual reward), (c) a set of rules governing the exchange of tokens for backup reinforcers such as prizes [pp. 379–380].

Reiss and Sushinsky (1975) obtained evidence that a short-term token economy program does not decrease intrinsic interest. These investigators trained kindergarten children to discriminate among three songs by reinforcing them for listening to one of them. The training procedure consisted of 10 reinforcement trials, such that listening to the target song for 10 consecutive seconds earned the first token (a poker chip), but the child was required to listen for an additional 15, 20, 35, 20, 60, 30, 100, 80 and 90 seconds to earn tokens 2 to 10. The children had been informed that if they earned 10 tokens, they would be able to buy a toy. The children's interest in the songs was assessed in a 10-minute posttest, 48 hours later, in which they could listen to any of the three songs, and no rewards were forthcoming. Reiss and Sushinsky found that the children evidenced a marked preference for the song that had been associated with reward. Reiss and Sushinsky concluded that the token reinforcement schedule increased rather than a decreased intrinsic interest in the target song.

Unfortunately, the results of this experiment are difficult to interpret for the following two reasons. First, Reiss and Sushinsky failed to include a control condition in which subjects were exposed to the target song and not rewarded. Comparisons with such a baseline condition would have indicated whether the token reinforcement schedule per se produced an increment in intrinsic interest. As it is, the preference for the reinforced song may merely indicate that subjects liked the target song, not because they had been reinforced, but because they had listened to it more. (Zajonc, 1968, found that mere exposure to a stimulus can increase liking for it.) Second, it is possible that the children expected that they might receive additional rewards for listening to the target song during the posttest. The tokens had originally been dispersed at variable intervals and, conceivably, during the posttest the children believed that they were merely experiencing a long delay before the first token. As a result, they may have switched to the other songs only after they had given up any hope of receiving tokens. Thus, the results of this study are inconclusive.

Ross and Bowers (1974) conducted two experiments to test the impact of short-term token economy programs on intrinsic motivation. The first experiment was designed to determine whether a decrement in intrinsic motivation would be obtained if the tokens and rewards were dispensed irrespective of the subject's quality of performance. In short, would the reduction in intrinsic interest produced from salient rewards that have no cue value (Karniol & Ross, 1975) occur in a token reinforcement paradigm?

The subjects were third-grade children, and the tokens were made contingent solely on their willingness to perform the experimental activity. The children worked at a booklet of follow the dots and received a token for each page that they completed. Children in the reward condition were told that they would receive a prize–a bag of colored marbles–when they had obtained nine tokens. Children in the control condition received tokens but neither anticipated nor received a reward. Acquisition of the tokens was dependent merely on the children's performing the activity. There was no relation between competence at the activity and reward attainment. In fact, the children were told: "Don't worry if you make a mistake, just finish the page." Intrinsic interest in the dot task was assessed in a subsequent 5-minute free-play period in which no rewards were forthcoming and alternative activities were available.

The results were consistent with those obtained in previous research. Subjects in the reward condition tended to show less interest in the dot activity ($p < .08$) than did subjects in the no reward control condition (the means were 207.8 and 270 seconds, respectively).

The second experiment differed in only one respect form the first: token and reward attainment indicated high-quality performance at the dot task. Subjects received tokens in the form of cards on which the words "good job" were written. They were told that the good job cards indicated that they "didn't make any mistakes." Subjects in the reward condition were told that if they received four good job cards they would win a bag of marbles. Within the space of 15 minutes, they were awarded the four cards. Control subjects were also given the four cards, but no tangible rewards were promised or given in exchange for them.

In this experiment, the reward was linked explicitly to subjects' degree of success at the activity. In the Karniol and Ross (1975a) studies, such performance-contingent rewards did not cause a decrease in intrinsic motivation. In the token economy paradigm used by Ross and Bowers (1974), the performance-contingent reward actually significantly enhanced interest in comparison to that shown by subjects in the no reward condition (the means were 159.6 and 65.6 seconds, respectively).

The major implication of these results is that token economy programs may increase rather than undermine intrinsic interest when the tokens have high cue value with respect to the children's competence at the task. Nevertheless, two caveats are in order. First, this study employed only a short-term token economy program. While there is no reason to predict that the results would fail to hold for a longer term program, the issue is of such practical importance that it certainly merits testing. Second, the present results are probably generalizable only to programs that yield high quality performance (see Karniol and Ross, 1975a). If the child is aware that he is doing poorly, performance-contingent rewards are unlikely to enhance his intrinsic interest. However, with token economy programs, at least, this may not be an important consideration, for herein lies the strength of the programs. As noted earlier, they have proven to be very effective in modifying behavior.

CHILDREN'S USE OF THE DISCOUNTING PRINCIPLE

In the above discussion I have attempted to specify some of the factors which influence the realtionship between rewards and intrinsic motivation. The research that I have summarized has been based on attribution theory. Yet, there is a curious gap between the theory and the research. The theory focuses solely on cognitive processes, while the research almost always assesses only overt behavior. I have finessed this problem by making the tacit assumption that the behavioral indicant of intrinsic motivation is controlled by, or reflects, the hypothesized cognitive events. Is this assumption valid? Is it plausible to assume that young children engage in the logical, causal analyses posited by attribution theory?

One approach to answering these questions is to examine directly young children's inferences with respect to a behavior which may be prompted by either intrinsic or extrinsic causes. Attribution theory implies that the child should disregard or discount the intrinsic cause when a salient extrinsic cause could reasonably serve as a sufficient explanation for the behavior.

In a recent study, Smith (in press) suggests that young children do *not* systematically engage in these cognitive processes. Kindergarten children were presented with pairs of stories about one child who was either rewarded for playing with or ordered to play with a toy and another child who clearly played with the same toy of his own volition. Smith found that the children responded randomly to the question: "Which child wanted to play with ______?" Thus, these subjects did not appear to discount the intrinsic cause (wanting to play) when an extrinsic cause in the form of a reward or order was present. Smith reported that discounting occurred with greater frequency as the age of the children increased (kindergarten, second grade, fourth grade), and that fourth grade children did not differ from university students on this measure.

The finding that kindergarten children do not employ the discounting principle is perplexing, since in a number of the studies described in the present chapter children within the same age range behave *as if* they are discounting the intrinsic cause for their own behavior in the presence of a reward (for example, Lepper et al., 1973; Ross, 1975). In analyzing Smith's experiment, Karniol and I speculated that the kindergarten children responded randomly because they did not understand the stories as well as the older children did. Since Smith did not test for story comprehension, this hypothesis cannot be ruled out on the basis of his data. Consequently, we conducted a study in which we attempted to be as certain as possible that children of the various ages understood and remembered the stories (Karniol & Ross, 1975b).

The subjects were children in kindergarten, first and second grades. Each child was asked to listen carefully to the stories which were played on a tape recorder. After the subject heard one story he was asked to repeat it. If he were unable to do so, the story was replayed (up to three times) until the child could correctly repeat it (the data from four kindergarten and two first-grade children were omitted from the analyses because they were unable to repeat one or more stories even after multiple replays). Each subject was exposed to four story pairs adapted from those used by Smith (in press). In two story pairs Child A was promised a reward for playing with a certain toy, while Child B was not, as in the following example:

> Judy was home and there were two of her toys there, a ball and a puzzle. Judy's mother said that Judy could have some cake if she played with the ball. And Judy played with the ball. Remember, Judy's mother said that Judy could have some cake if she played with the ball.
>
> Suzie was home and there were two of her toys there, a ball and a puzzle. And Suzie played with the ball.

In the two remaining story pairs Child A was ordered to play with the toy by his mother, while Child B was not. One-half of the subjects were exposed first to the order–story pairs and then the reward–story pairs; whereas, the remaining subjects received the reward–order sequence. After each story pair the subject was asked: Which child really liked to play with ______?[2] The subject was then asked to explain his response.

The data indicated that use of the discounting principle increased with age and that most kindergarten children did not systematically engage in discounting (see Table 1). These findings are consistent with the results of Smith's research. In contrast to Smith's data, however, Karniol and Ross (1975b), in whose study recall of the experimental stories was ensured, observed that the majority of kindergarten children were not responding randomly. Our data suggested that

[2] Karniol and Ross actually conducted two studies in which the wording of this question varied slightly. In the first experiment subjects were asked, "Which child really *wanted* to play with ______?" In the second experiment subjects were asked, "Which child really *liked* to play with ____ ?" The switch from "want" to "like" appeared to increase somewhat young children's use of the discounting principle. The two experiments lead to the same general conclusions, however, and only the second experiment is described here.

TABLE 1
Children's Use of the Discounting Principle[a]

Grade	Reward–story pairs			Order–story pairs		
	Discount	Inconsistent	Additive	Discount	Inconsistent	Additive
Kindergarten	32.1	28.6	39.3	50.0	39.3	10.7
Grade 1	63.0	11.1	25.9	55.5	22.2	22.3
Grade 2	70.0	10.0	20.0	80.0	5.0	15.0

[a]The figures in the table represent the percentage of subjects choosing the unconstrained target person on both story pairs (discount), one story pair (inconsistent), or on neither of the story pairs (additive) in response to the question: "Which child really liked ______________?"

many kindergarten children do make reliable causal inferences for a behavior which may be motivated by either intrinsic or extrinsic causes. However, two quite distinct heuristics appeared to be used to infer causality in this instance. Some of the kindergarten children seemed to employ systematically an additive model in which more liking is inferred when both intrinsic and extrinsic factors can plausibly explain the occurrence of a behavior. When the extrinsic cause was a reward, slightly more than one-third of the kindergarten children consistently inferred liking in this manner. An additional third of the kindergarten children appeared to employ the multiplicative model which is implied from the discounting principle when the extrinsic cause was a reward; that is, more liking was consistently attributed in the absence of an extrinsic cause. The remaining kindergarten children responded inconsistently and are perhaps best categorized as responding randomly.

The two different heuristics may also be inferred from the children's explanations for their selections. For example, children who consistently chose the rewarded target person as really liking the toy tended to explain that he liked the toy *because* he was rewarded for playing with it. On the other hand, children who consistently selected the nonrewarded story character explained that the person who was rewarded played with the toy in order to get the reward, whereas the nonrewarded person played with the toy because he liked it.

The proportion of children employing either heuristic cannot be specified exactly at any age level, for it appears to depend, in part, on the particular extrinsic factor involved. In both the Smith (in press) and Karniol and Ross (1975b) studies the younger children were more likely to employ the additive heuristic when the extrinsic cause was a reward rather than an order. However, the Karniol and Ross data do reveal a developmental transition, with use of the additive model declining and the use of the multiplicative model increasing with age of the child.

One purpose of the Karniol and Ross (1975b) study was to determine whether children who participated in the research on intrinsic motivation were capable of the logical, causal analyses posited by attribution theory. The results indicated that approximately one-third of the kindergarten children and over 60% of the first-grade children consistently employed the discounting principle when the extrinsic cause was a reward. These results, coupled with Piaget's (1948) speculation that children can employ more sophisticated processes when judging their own behavior than in judging the behavior of others, suggest that a reasonable number of children participating in the intrinsic motivation research were capable of employing the attributional analysis.

On the other hand, the Karniol and Ross (1975b) results do not demonstrate that children in the typical study of intrinsic motivation actually engage in these cognitive processes. Such a demonstration would seem to be a mandatory step in the effort to validate the attributional analysis of these studies. One approach suggested from the Karniol and Ross (1975b) data would consist of assigning the story pairs to children of kindergarten age and selecting groups of children who systematically employ either the additive or multiplicative rules. These children could then be employed as subjects in a study of intrinsic motivation modeled after Lepper et al. (1973), with the expectation that the individual difference variable would affect the results. Those children who demonstrate that they are capable of the cognitive processes assumed by attribution theory (that is, the multiplicative rule) should be more likely to behave as if they were discounting the intrinsic cause for their own behavior in the presence of a salient extrinsic reward. Research along these lines is currently in progress.

SUMMARY

In this chapter, I have described a series of experiments relating rewards to intrinsic motivation. The theoretical rationale for this research was derived from the discounting principle posited by attribution theorists and from a consideration of the influence of rewards on self-perceptions of task mastery. The data suggest that a reward reduces intrinsic motivation in the activity with which it is associated when the reward is (a) salient and (b) allocated irrespective of the person's degree of success at the activity. The data also indicate that rewards which are contingent upon quality of performance are not injurious to and may even enhance intrinsic motivation when the person is able to assume that his performance is reasonably satisfactory. While these data are congruent with attribution analyses, there is no substantive evidence that young children are engaging in the cognitive processes posited by attribution theorists. However, the results of a study in which children's use of the discounting principle was directly assessed were consistent with the assumption that a substantial number of the children are employing these attribution processes. Further, two alternative explanations—delay of gratification and differential task performance—that have been offered to account for the experimental results are not supported by the data presently available.

The practical implications of the results center on the negative consequences of engendering self-perceptions of extrinsic rather than intrinsic motivation. The research reported in this chapter, specifies some of the conditions under which rewards may be employed without any apparent loss in intrinsic motivation. While much more research needs to be conducted to delineate the full impact of rewards on self-perceptions, it appears entirely possible at this juncture that the detrimental effects of rewards on intrinsic motivation may be limited to a rather narrow set of circumstances.

ACKNOWLEDGMENTS

The writing of this chapter was facilitated by Canada Council Research Grant s74-0696. I would like to thank the principals, teachers, and students whose cooperation enabled me to conduct the research described in this chapter.

REFERENCES

Bem, D. J. Self-perception: An alternative interpretation of cognitive dissonance phenomena. *Psychological Review,* 1967, **74** 183–200.

Bem, D. J. Self-perception theory. In L. Berkowitz (Ed.), *Advances in experimental social psychology* (Vol. 6). New York: Academic Press, 1972.

Bowers, K. S. The psychology of subtle control: An attributional analysis of behavioural persistence. *Canadian Journal of Behavioural Science,* 1975, 7(1), 78–95.

Brazelton, T. B. *Toddlers and parents: A declaration of independence.* New York: Delacorte Press, 1974.

Calder, B. J., & Staw, B. M. Self-perception of intrinsic and extrinsic motivation. *Journal of Personality and Social Psychology,* 1975a, **31,** 599–605.

Calder, B. J., & Staw, B. M. The interaction of intrinsic and extrinsic motivation: Some methodological notes. *Journal of Personality and Social Psychology,* 1975, **31,** 76–80. (b)

de Charms, R. *Personal causation: The internal affective determinants of behavior.* New York: Academic Press, 1968.

Deci, E. L. Effects of externally mediated rewards on intrinsic motivation. *Journal of Personality and Social Psychology,* 1971, **18,** 105–115.

Deci, E. L. Cognitive evaluation of rewards and feedback. Paper presented at the meeting of the Eastern Psychological Association, Philadelphia, 1974.

Deci, E. L. Intrinsic motivation, extrinsic reinforcement, and inequity. *Journal of Personality and Social Psychology,* 1975, **22,** 113–120.

Festinger, L. A theory of social comparison processes. *Human Relations,* 1954, **7,** 117–140.

Greene, D., & Lepper, M. R. An information-processing approach to intrinsic and extrinsic motivation. Paper presented at the meeting of the American Psychological Association, Chicago, September 1975.

Harlow, H. F. Learning and satiation of response in intrinsically motivated complex puzzle performance by monkeys. *Journal of Comparative and Physiological Psychology,* 1950, **43,** 289–294.
Heider, F. *The psychology of interpersonal relations.* New York: John Wiley & Sons, 1958.
Hunt, J. McV. Intrinsic motivation and its role in psychological development. In D. Levine (Ed.), *Nebraska Symposium on Motivation.* Lincoln: University of Nebraska Press, 1965.
Karniol, R., & Ross, M. The effects of performance contingent rewards on intrinsic motivation. Unpublished manuscript, University of Waterloo, 1975. (a)
Karniol, R., & Ross, M. A developmental study of causal inferences in the presence of multiple sufficient causes. Unpublished manuscript, University of Waterloo, 1975. (b)
Kazdin, A. E., & Bootzin, R. R. The token economy: An evaluative review. *Journal of Applied Behavior Analysis,* 1973, **5,** 343–372.
Kelley, H. H. Attribution theory in social psychology. In D. Levine (Ed.), *Nebraska Symposium on Motivation.* Lincoln: University of Nebraska Press, 1967.
Kelley, H. H. *Causal schemata and the attribution process.* Morristown, New Jersey: General Learning Press, 1971.
Kelley, H. H. The process of causal attribution *American Psychologist,* 1973, **28,** 107–128.
Kiesler, C. A., Nisbett, R. E., & Zanna, M. P. On inferring one's beliefs from one's behavior. *Journal of Personality and Social Psychology,* 1969, **4,** 321–327.
Kruglanski, A. W., Friedman, I., & Zeevi, G. The effect of extrinsic incentive on some qualitative aspects of task performance. *Journal of Personality,* 1971, **39,** 606–617.
Lepper, M. R., Greene, D., & Nisbett, R. E. Undermining children's intrinsic interest with extrinsic rewards: A test of the "overjustification hypothesis." *Journal of Personality and Social Psychology,* 1973, **28,** 129–137.
Lepper, M. R., & Greene, D. Overjustification and intrinsic motivation: Some implications for applied behavioral analysis. Paper presented at the meeting of the Eastern Psychological Association, Philadelphia, 1974.
Levine, F. M., & Fasnacht, G. Token rewards may lead to token learning. *American Psychologist,* 1974, **29**(11), 816–820.
Mischel, W. Processes in delay of gratification. In L. Berkowitz (Ed.), *Advances in experimental social psychology* (Vol. 7). New York: Academic Press, 1974.
Mischel, W., & Ebbesen, E. B. Attention in delay of gratification. *Journal of Personality and Social Psychology,* 1970, **16,** 329–337.
Mischel, W., Ebbesen, E. B., & Zeiss, A. R. Cognitive and attentional mechanisms in delay of gratification. *Journal of Personality and Social Psychology,* 1972, **21,** 204–218.
Mischel, W., & Gilligan, C. Delay of gratification motivation for the prohibited gratification and responses to temptation. *Journal of Abnormal and Social Psychology,* 1964, **69,** 411–417.
Nuttin, J. R. Pleasure and reward in human motivation and learning. In D. E. Berlyne & K. B. Madsen (Eds.), *Pleasure, reward, preference: Their nature, determinants, and role in behavior.* New York: Academic Press, 1973.
O'Leary, D. K. & Drabman, R. Token reinforcement programs in the classroom: A review. *Psychological Bulletin,* 1971, **75,** 379–398.
Piaget, J. *The moral judgment of the child.* London: Routledge & Kegan Paul, 1948.
Piaget, J. *Genetic epistemology.* New York: Columbia University Free Press, 1970.
Reiss, S., & Sushinsky, L. Overjustification, competing responses, and the acquisition of intrinsic interest. *Journal of Personality and Social Psychology,* 1975, **31,** 1116–1125.
Ross, M. Salience of reward and intrinsic motivation. *Journal of Personality and Social Psychology,* 1975, **32,** 245–254.
Ross, M., & Bowers, K. S. The effects of token reinforcement programs on intrinsic motivation. Unpublished manuscript, University of Waterloo, 1974.

Ross, M., & Karniol, R. The effects of rewards on intrinsic motivation: A frustration theory analysis. Paper presented at the meeting of the Eastern Psychological Association, Philadelphia, 1974.

Ross, M., Karniol, R., & Rothstein, M. Reward contingency and intrinsic motivation in children: A test of the delay of gratification hypothesis. *Journal of Personality and Social Psychology,* in press.

Salancik, G. R. Interactive effects of performance experiences and extrinsic reward on intrinsic evaluations: Personal causation or personal accomplishment. Paper presented at the meeting of the Eastern Psychological Association, Philadelphia, 1974.

Smith, M. C. Children's use of the multiple sufficient cause schema in social perception. *Journal of Personality and Social Psychology,* in press.

Smith, M. B. Competence and socialization. In J. A. Clausen (Ed.), *Socialization and society.* Boston: Little, Brown, & Co., 1968.

White, R. W. Motivation reconsidered: The concept of competence. *Psychological Review,* 1959, **66,** 297–333.

Zajonc, R. B. Attitudinal effects of mere exposure. *Journal of Personality and Social Psychology,* 1968, **8,** (Monograph), 1–29.

7

Attribution Processes and Emotional Exacerbation of Dysfunctional Behavior

Michael D. Storms
Kevin D. McCaul

University of Kansas

In the *Annals of Popular Neuroses,* a chapter is needed for all those cases in which worrying about a problem only makes it worse. The first casestudy reported in this category would feature our overweight colleague who talks incessantly about his problem of overeating. The more he describes his inability to control food intake, which he does invariably at every department cocktail party, the more distraught he gets and the more hors d'oeuvres he pops into his mouth. This fellow's self-stoking syndrome is not unlike that of another colleague who bemoans her tennis game. Her serve, for no predictable reason, has a tendency to fall apart in the midst of a match. She gets angry and frustrated, curses her athletic inadequacy, and double-faults her way through the rest of the set. Then again, perhaps first honors for self-defeating anxiety should go to our friend with a writer's block. Just earlier today he was sitting at his typewriter trying to get started on a book chapter for a fast approaching deadline. The basic idea for the chapter was clear in his mind, but the opening paragraph would not come to him. Lighting cigarette after cigarette, his anxiety soared uncontrollably. Words and sentence fragments flew through his mind—none of them made any sense, none of them were worth writing down. He realized, once again, how poor a writer he is, how poor a thinker he is, and how the basic idea for his chapter will seem simple-minded and inane to his critical readers.

The above complaints fit a common pattern that may characterize a number of different emotional syndromes. In this pattern an individual becomes aware of some undesirable aspect of his own behavior, overeating or underwriting, which he views as faulty or unattractive. From that individual's point of view, the behavior further implies a deeper lying problem, a basic lack of ability or self-control, or a pathological and embarrassing inadequacy. Anxiety, guilt, and self-deprecation follow from these inferred negative dispositions. Then, ironically, this anxiety promotes an increase in or exacerbation of the unwanted behavior.

In the present chapter we will develop a model of the emotional exacerbation of dysfunctional behavior. Initially, we will review recent work in attribution theory and emotional behavior. From this background, we will offer the simple, two-part proposition that: (a) attribution of unwanted, dysfunctional behavior to the self leads to an increased emotional state (which can loosely be called "anxiety"), and (b) this anxiety may serve to increase the frequency or intensity of the dysfunctional behavior.

In the second section of the chapter, we will examine both parts of the exacerbation model in greater detail. First, we will discuss some of the potentially harmful aspects of self-attributions which may contribute to increased anxiety, including perceptions of inadequacy and loss of self-control. In addition, we will point to the importance of situational stress which may enhance the anxiety provoking properties of these attributions. Second, we will discuss the types of dysfunctional behavior which are most likely to be increased by anxiety. These include physiological responses that are affected by arousal properties of anxiety, habitual behaviors that are affected by drive properties of anxiety, and performance behaviors that are affected by distraction properties of anxiety.

In the final section of this chapter, we will examine a particular dysfunctional behavior, stuttering, and report on a demonstration of its exacerbation with self-attribution and situational stress.

ATTRIBUTION AND DYSFUNCTIONAL BEHAVIOR

Initial research on causal perceptions and emotions suggested that emotional states and associated behaviors could be modified by changing an individual's attributions for symptoms associated with emotionality. In the now classic example, Schachter and Singer (1962) demonstrated that symptoms of arousal, such as increased heartrate and butterflies in the stomach, will lead to heightened emotional states and behaviors if those symptoms are attributed to emotional stimuli in the person's environment. However, those same symptoms, if attributed to a relatively nonemotional stimulus, such as a drug injection, will not increase the individual's emotionality. In fact, subsequent research demonstrated that an individual's emotionality can be *decreased* by misattributing naturally occurring emotional symptoms to a nonemotional source (Nisbett & Schachter, 1966; Ross, Rodin, & Zimbardo, 1969). This research suggested an important implication of attribution theory, namely, that pathological emotional conditions might be affected by attributions a patient makes about the causes of his symptomology. Attribution processes may be operative in the etiology of dysfunctional emotional behaviors, and reattribution might be useful in treating emotional disorders. To test these possibilities, Storms and Nisbett (1970) focused on the emotional behavior of insomniacs. In their study they gave insomniacs placebo pills to take at bedtime. In one condition, the pills were described as producing behaviors symptomatic of insomnia, including alertness, increased heartrate, and increased body temperature. It was reasoned that subjects taking these pills would be able to reattribute their naturally occuring bedtime arousal to the "pep pills," would experience less emotionality at bedtime, and would fall asleep more quickly. In another condition, insomniacs were given placebo pills described as "sedatives." The authors reasoned that this manipulation would lead subjects to attribute greater arousal to emotional cues as a result of taking a drug that was supposed to reduce their symptoms but which, in fact, had no physiological effects. Such attributions may result in increased insomnia. A comparison of the time subjects reported going to bed and the time they reported falling asleep on two nights without the pills and on two nights with the pills showed strong support for the hypotheses. Subjects who took "pep" pills reported a decrease in their insomnia averaging 12 minutes per night, while subjects on "sedative" pills reported an increase in their insomnia of 15 minutes per night.

These findings are consistent with the hypothesis that attributions about the symptoms of emotional states will influence the subsequent intensity of those emotional states. Furthermore, during debriefing interviews with the insomniac subjects, some important observations emerged which suggested that attributional processes contributed to their problems. First, it was striking that each subject had adopted the common, self-descriptive label of "insomniac." It was noticeable that each subject described himself as an individual who had some problem with sleeping, and each subject used the insomnia label as though it implied the possession of a particular psychological condition. Secondly, there were strong similarities among subjects' subjective experiences with insomnia. Most subjects reported something like the following sequence as occurring on any given sleepless night. Initially upon retiring they would begin to wonder whether it would be another "problem" night. They would begin looking for telltale symptoms of insomnia, such as feeling too hot or being too sensitive to noise. Then, with the advent of any of these symptoms, their initial worries would be confirmed. Their still active minds would begin imagining the consequences of another sleepless night: fatigue, lowered alertness in class or at work, poor performance on a test, and so on. As more time passed without sleep, their worries expanded. Next, they would speculate about their "condition" of insomnia, whether it might have some psychological meaning or might be a deeper symptom of some deeper neurotic disorder. With repetition of this experience night after night, these individuals would come to view themselves as chronic insomniacs. As they described themselves to the experimenter, it became apparent that these individuals viewed their irregular sleeping patterns as a serious personal problem, perhaps as a sign of pathology.

Although these interview data from the insomnia study are anecdotal, they strongly suggest a cyclical pattern in the development of insomnia. Individuals may observe an unwanted aspect of their own sleeping behavior and attribute its occurrence to an internal, dispositional condition. This attribution, and its implications for other undesirable dispositions, may increase the individual's anxiety at any manifestation of insomnia. This anxiety would then exacerbate the symptomatic behavior. Given this view of the nature of their insomnia, the experimentally induced changes in subjects' reported sleep seem reasonable. On experimental nights, subjects who took "arousal" pills were told, in effect, that their insomnia was being caused by a drug. On those nights, subjects did not have to view their symptoms as evidence of inadequacy or pathology. The drug attribution would have "short-circuited" any exacerbating worry about their condition and would have resulted in less delay in sleep onset. On the other hand, subjects who took pills that were supposed to relax them were led to expect that their insomnia symptoms should be reduced. When they failed to experience any reduction after taking the placebo drug, they may have concluded that they were experiencing a particularly bad bout of insomnia—so bad that the drug's relaxation effects were being overridden by their own inner turmoil. As a result, they may have worried even more than usual about their condition and consequently had even more difficulty falling asleep. In short, the attributional effects of the pill manipulations may have been to increase or reduce the amount subjects worried about their insomnia on that particular night.

Storms and Nisbett (1970) further suggested that the two key elements in their analysis of insomnia, self-attributions of pathology and increased anxiety, may also characterize many other emotional disorders, including male impotence, stuttering, extreme shyness, and excessive awkwardness in athletic situations. In a more recent article, Valins and Nisbett (1971) add supportive evidence to this proposition. Drawing on a number of clinical and case studies, they point out the important role of attributions and anxiety in a variety of emotion related behaviors, including homosexual panic, schizophrenia, depression, and phobias. In many instances, Valins and Nisbett observe, patients will be aware of their own symptomatic behavior before seeking out a therapist and will arrive at dispositional self-diagnoses of abnormality, inadequacy, or delusional thinking. These attributions frequently lead to increased anxiety that can fuel more of the unwanted emotional behavior and that can impede successful therapeutic progress. Successful therapies often must begin by convincing the patient to drop his self-diagnosis and to reattribute his symptoms to specific external, situational stressors (cf. Davison, 1966). Although additional therapy may be needed to help the individual cope with those external conditions, the initial dissolution of self-attributions stops the self-feeding emotional cycle and relieves a large part of the patient's anxiety.

In addition to the evidence from these casestudies, support for the exacerbating effects of self-attributions is found in sociological analyses of deviant behavior (see Becker, 1964; Lemert, 1951, 1967; Merton, 1959; Schur, 1971; Tannenbaum, 1938). These authors propose a social labeling process which leads an individual from his first, isolated deviant act to a full-scale deviant identity and lifestyle. According to these theorists, individuals and persons in social institutions respond to an initial deviant act by identifying and labeling the deviant publicly. These reactions may include interpersonal behaviors such as name-calling, stereotyping, and approbation; or organizational responses such as institutional confinement, psychological treatment, and criminal rehabilitation (Schur, 1971). These social reactions serve to reinforce a deviant self-identity in the differentiated individual. Social reactions to deviant behavior not only inform society that the individual is different, they make his differences clear to the individual himself. As social labeling increases, Schur (1971) asserts, "there is a tendency for the actor to define himself as others define him [p. 70]." In other words, these reactions lead to dispositional self-attributions. They impute pathological and disordered traits to the individual and convince the individual to adopt a deviant self-label. The inducement of deviant self-attribution is portrayed in Lemert's (1967) description of a treatment program for stutterers.

> We may safely say that going to a speech clinic in all cases confronts the individual with a clear-cut societal definition of the stuttering self. The association with other stutterers and with speech cases in the clinic situation has a clear implication for self and role, as well as the knowledge that other students or members of the community know the function of the clinic. One well-known clinic, at a Middle Western college, makes it more or less of a prerequisite for treatment of adult stutterers that they make frank avowals in speech and behavior that they are stutterers. This is done by having the stutterers practice blocks in front of mirrors, exaggerate them, copy one another's blocks, and have or fake blocks in public situations. While there are several objectives behind this procedure, one of its chief consequences is to instill an unequivocal self-definition in the stutterer as one who is different from others [p. 159].

Finally, social labeling theory contends that these changes in the individual's self-concept produce increased deviant behavior. Thus, deviance is exacerbated by self-attributions.

Storms and Nisbett's research with insomnia, Nisbett and Valins' review of emotional disorders, and social labeling theorists' analysis of deviance suggest a common, descriptive model which may characterize a number of dysfunctional behaviors. In terms of this model, an individual observes some unwanted or uncomplimentary aspect of his own behavior for which he makes a dispositional, self-attribution. This self-attribution often takes the form of inferences about real or imagined inadequacies, psychological disorders, character flaws, lack of self-control, personality deficits, deviant tendencies, and so on. These negative views of the self give rise in turn to a variety of unpleasant emotions–anxiety, guilt, frustration, perhaps even self-hatred. These emotions, then, promote increases in emotional behaviors, some of which may be the very same dysfunctional behaviors which began the whole process. For convenience, we can refer to this syndrome as exacerbation and propose two essential steps in the model. First, attribution of dysfunctional behavior to negative dispositions in the self produces an increased emotional state. Second, increased emotionality exacerbates the occurrence of some dysfunctional behaviors. This model should be applicable to any emotional disorder in which the primary symptomatic behavior is increased by anxiety. In such cases, attributionally triggered anxiety about the symptoms should produce more of the same symptoms.

SELF-ATTRIBUTIONS, ANXIETY, AND EXACERBATION

The proposed exacerbation model focuses on two hypothesized relationships: the effects of self-attributions on anxiety levels and the effects of anxiety on certain dysfunctional behaviors. In this section we would like to examine each step in the model more closely with an emphasis on the similarities between the exacerbation model and other social psychological theories. The link between self-attributions and anxiety will be discussed in comparison to objective self-awareness theory (Duval & Wickland, 1972) and to various theories of perceived loss of control (Brehm, 1966; Glass & Singer, 1972a, 1972b; Seligman, Maier, & Solomon, 1971). In addition, it will be suggested that real or imagined situational stress may interact with negative self-attributions to produce even more anxiety.

The link between anxiety and dysfunctional behavior will be discussed in comparison to habit–drive theory (Spence & Spence, 1966) and attentional theories of test anxiety (Wine, 1971). Our intent in these comparisons is eclectic; each theory will be examined for possible contributions to a better understanding of exacerbation phenomena and for possible mechanisms by which exacerbation occurs. The exacerbation model is not posed as a competitor to these other theories. It is instead a descriptive model of certain emotional syndromes to which these and other theories may apply.

Self-Attributions and Anxiety

It is proposed that an important characteristic common to exacerbation phenomena is a self-attribution which leads to negative affect. The individual becomes preoccupied with the implications of his behavior for psychological disorders or inadequacies, and this produces an emotional state of anxiety. This condition has many similarities to the state of objective self-awareness proposed by Duval and Wicklund (1972; see Chapter 8). According to their theory, an individual is in one of two conscious states: either he is thinking about himself (objective self-awareness) or his attention is directed outward toward his environment (subjective self-awareness). Many factors determine which state is being experienced at any particular time, one of the most important being the distinctiveness of one's own behavior (Wicklund, 1975). To the extent that one's own behavior stands out, is noticeably different, or is unique, the individual is pushed toward a state of objective self-awareness. That is, simply, his own distinctive behavior leads him to think about himself. This theorem would suggest that behaviors which characterize exacerbation cycles may produce states of objective self-awareness. Insomnia, stuttering, or sexual impotence, for example, are unusual behaviors both in terms of their general infrequency in the population, and perhaps in terms of their infrequency in the past life of the individual himself. This perceived distinctiveness, whether real or not, should produce increased objective self-awareness.

Objective self-awareness theory goes on to propose certain definite consequences of this self-focused state. In particular, Duval and Wicklund propose that objective self-awareness usually leads to increased negative affect about the self. Self-focusing, they claim, results in the individual comparing his own behavior to internalized standards of ideal behavior. Since actual achievements seldom match idealized aspirations, it is assumed that the objectively self-aware person will find shortcomings within himself and will experience negative affect. Several researchers have supported this hypothesis. Ickes, Wicklund, and Ferris (1973), for example, report three studies in which subjects are put into a self-focused state (by the use of mirrors or tape recordings to make them more aware of themselves) and asked to fill out questionnaires about discrepancies between ideal self-concepts and real self-concepts. In each study, these and other measures of self-esteem were lowered by the objective self-awareness state. Additional investigators have demonstrated that individuals in this state perceive themselves to be less in control of their performance at perceptual motor tasks (Duval & Ritz, cited in Duval & Wicklund, 1972) and take more responsibility for negative outcomes (Duval & Wicklund, 1973; Wicklund & Duval, cited in Duval & Wickland, 1972).

It is evident that objective self-awareness theory is directly applicable to exacerbation phenomena. In the present model, we propose that exacerbation occurs when an individual makes self-attributions about his behavior and experiences negative emotional states as a result. Objective self-awareness theory provides a possible mechanism by which this occurs. The individual who observes his own distinctive behavior is quite likely placed in a state of self-focused attention. This state will lead to unfortunate comparisons between the observed behavior and ideal standards. The objective self-awareness state enhances a tendency to make dispositional self-attributions of inadequacies and lack of control. Emotionally, the state is aversive, unpleasant, and affectively negative.

If the objectively self-aware individual is psychologically upset by dispositional attributions of inadequacy, he may be even more traumatized by implications that he has lost self-control. In terms of exacerbation cycles, an examination of examples and casestudies cited earlier in this chapter suggests that one of the most damaging effects of self-attributions is a perceived loss of control. One of the self-deprecating cognitions that occurs to the insomniac is that he has lost control of his sleeping functions. The stutterer may attribute to himself a loss of control over his own tongue. The impotent male may think that he has lost control over his sexual functioning. Loss of control may be a frequently occuring cognition in exacerbation phenomena, and it may be one of the primary contributors to increasing anxiety.

In a recent article on loss of control, Wortman and Brehm (1975) point to a considerable body of evidence that perceived loss of control can cause profound psychological upset.

> There is good reason to believe that exposure to uncontrollable outcomes can result in profound psychological upset. Many kinds of maladaptive behavior have been attributed to feelings of helplessness with respect to one's environment. For example, investigators have argued that the helplessness stemming from feelings of lack of control is an important factor in the development of such psychiatric disorders as depression and schizophrenia (Bateson, Jackson, Haley, & Weakland, 1956; Seligman, 1974, in press). Cofer and Appley (1964), Janis, (1958), and Janis and Leventhal (1968) have maintained that feelings of helplessness interfere with the ability to respond adaptively in stressful situations. Feelings of helplessness have also been proposed as a precusor of physical disease (Engle, 1968; Schmale, 1971). Some investigators (Greene, Goldstein, & Moss, 1972; Richter, 1957) have suggested that the perception of inability to exert control over one's environment can even result in sudden death from coronary disease or other factors. Feelings of lack of control have also been viewed as a cause of many types of antisocial behaviors [p. 278].

While Wortman and Brehm point out important theoretical distinctions between different loss of control models (for example, reactance theory [Brehm, 1966] versus learned helplessness theory [Seligman, Maier, & Solomon, 1971]), their review underlines a common finding in all of this research. Perceived loss of control creates an aversive motivational state. Perhaps the clearest empirical demonstration of this aversive state is provided in the urban stress studies of Glass and Singer (1972a, 1972b). In this research, subjects are typically exposed to noxious noise while performing a puzzle solving or proofreading task. This uncontrollable stress is shown to debilitate performance on the task during the noise exposure and even after the noise is terminated. In one variation on this basic condition, subjects are told that, if they wish to, they can push a button to terminate the noise. The experiment is set up in such a way, however, that subjects usually choose not to terminate the noise, and all subjects actually receive the same amount and schedule of the noxious stimulation. Just the perception of possible control over the noise causes a significant decrement in the stress-producing effects of the noise. Subjects who think they can terminate the sound show less deleterious aftereffects on the proofreading tasks. As Glass and Singer (1972a) suggest, "Presumably, the helpless group experienced not only the aversiveness of unpredictable noise but also the 'anxiety' connected with their felt inability to do anything about it [p. 462]." These researchers clearly suggest that perceived lack of control over an unwanted behavior increases anxiety about that behavior, consistent with the exacerbation model.

The research on objective self-awareness and loss of control supports the notion that self-attributions of inadequacy or of loss of control will produce anxiety. However, it seems likely that the relationship between attributions and anxiety is somewhat more complex. In particular, the potential consequences of inadequacy and loss of control should influence the amount of anxiety produced by these attributions. If the individual is in a situation where highly stressful outcomes will result from inadequate or uncontrolled behavior, he may react more anxiously than if the potential outcomes are less stressful. In other words, anxiety may not stem directly from negative self-attributions, but may result from anticipations of stressful consequences that will befall the individual because of his inadequacy or lack of control.

In a study recently completed in our own laboratory, we suggest this might be the case (Storms & Leviton, 1975). In the context of a study on driving ability, female subjects were tested on a pursuit rotor task over several trials while listening to noises delivered over headphones. They were told that the task was related to driving ability and were also told one of two things about the noise. External-attribution subjects were informed that the noise was generated randomly and was meant to simulate the distracting sounds of city traffic. Self-attribution subjects were told that the noise would signal whenever their performance fell below preset standards and was meant to simulate the sounds of other cars honking at them for making a driving error. However, in an effort to avoid the alternative explanation that self-attribution subjects would perform worse simply because they received erroneous feedback, these subjects were told that actual time of the noise onset was determined by a number of factors, including their performance, time elapsed in the trial, and computerized criteria which would change across trials. In fact, the noise was delivered randomly, and no subject had control over its frequency. In addition, the volume of the noise was varied. Half of the subjects heard quite loud, stressful noise, and half heard very soft noise. We used the Palmar Sweat Index and actual performance on the pursuit rotor task as measures of anxiety. Unfortunately, large variability in subjects' ability at the task made it difficult to assess the meaning of changes on the performance measure. However, while all subjects improved at the task over trials, the rate of improvement was significantly influenced by the attribution manipulation. Subjects who attributed the noise (loud or soft) to their own performance improved at the slowest rate. Noise level alone did not affect performance nor was there an interaction between noise level and attributions.

Results of the Palmar Sweat Index and other, self-report measures were more conclusive. Both loud noise and self-attributions increased subjects' anxiety. Furthermore, a significant interaction was obtained showing that loud noise and self-attributions increased anxiety more than each of those factors considered alone. Self-reports of satisfaction with performance showed the same effects. Attributions and noise level interacted significantly; loud noise/self-attribution subjects were least satisfied with their performance.

These results suggest that anxiety levels are directly related to control attributions and to situational stress factors. But more importantly, the two variables can interact to produce magnified increases in anxiety. Self-attributions which imply lack of control in important stressful situations may produce the worst negative affect states. In terms of exacerbation cycles, this would suggest that insomniacs, for example, might experience a greater exacerbation of their problems the night before an important test or on any night when they need sleep the most. Impotent males may find it most difficult to perform for females whom they most want to impress or please sexually. Stutterers, likewise, may stutter worse when they are saying something important to an audience that they wish to persuade or that will be critical of their speech. In all of these cases, negative self-attributions are damaging, but the degree of the anxiety is multiplied by situational factors relevant to the self-attribution.

Anxiety and Exacerbation

In our discussion to this point, we have sought to document the effects of self-attributions on emotional states such as anxiety. The importance of anxiety in exacerbation cycles, it is proposed, is the role it plays in producing more of the unwanted dysfunctional behaviors that give rise to self-attributions. Only behaviors which are subject to the exacerbating effects of anxiety can be described in terms of the exacerbation model. It is important, therefore, to determine what types of behavior are increased by anxiety. Unfortunately, anxiety is a very complex concept and has generated an equally complex literature. Nevertheless, we will examine how the following three aspects of anxiety may exacerbate dysfunctional behavior: (a) physiological arousal, (b) drive properties, and (c) distraction.

Physiological arousal which may accompany anxiety can directly exacerbate some behaviors. For example, insomnia is quite likely directly related to excessive physiological arousal at bedtime (Monroe, 1967). If the self-attributions of insomniacs (perhaps accompanied by situational factors) produce more physiological arousal, there will almost certainly be a decrease in ability to fall asleep. Similarly, male impotence is known to be affected by arousal from the sympathetic nervous system. Too much sympathetic activity produces both inadequate erections and premature ejaculation, the most common symptoms reported by impotent males who seek treatment (Masters & Johnson, 1970).

It is unlikely, though, that all emotional syndromes which are exacerbated by anxiety are cases of physiological arousal acting directly on the behavior of the organism. In most instances, the effects of anxiety are probably psychologically mediated. The best theory of such a relationship between anxiety and behavior is the drive theory formulated by Spence and Spence (1966). According to this theory, anxiety acts as a drive state which will energize behavior according to its habit strength. The higher the level of anxiety, the more likely dominant, habitual responses will be emitted over any other responses. In other words, high anxiety in a particular situation will increase the likelihood that the individual will continue to act as he always has. Old habits are reinforced and new, possibly more adaptive, behaviors are more difficult to learn. The adoption of new response patterns under high anxiety is unlikely. It is quite unlikely that you will successfully change the style of your serve in the midst of an important tennis match. By similar reasoning, any habitual emotional response is more likely to occur when anxiety is high. The excessively shy individual is apt to invoke his habitual pattern of social interaction, withdrawing from others, when he is most anxious. If the source of his anxiety is a concern over that very behavior, the fact that he always withdraws at parties, the anxiety should produce a continuance of his typical emotional response and an exacerbation of his shyness syndrome.

If the drive properties of anxiety can be generalized to emotional behaviors, as we have suggested, exacerbation cycles should occur whenever an individual has a dominant emotional response to a particular situation. Paradoxically, when the individual becomes concerned about changing that emotional response, if his concern takes the form of increased anxiety, he is even less likely to succeed. Storms (1973), commenting on the treatment of alcoholics, suggests that this is exactly what happens in some therapy programs. When by various therapeutic techniques the alcoholic is made more aware of the unattractiveness of his drunken behavior, he becomes more anxious. Unfortunately, his habitual response to anxiety is to reach for another drink.

Another possible way in which anxiety may promote increased emotional behavior is suggested by a special case of the drive model. Spence and Spence (1966) have suggested that one particular competing response may be prevalent in such performance behaviors as test taking, namely, "covert verbalizations reflecting self-depreciation [p. 308]." In situations where performance relies on attention to task-relevant cues, such as test taking, these self-depreciating verbalizations will distract the individual from attending to important task-relevant cognitions. Mandler and Watson (1966) demonstrated this process. Their subjects were led to believe that they would not be able to finish, and thus would fail, one of two intelligence tests. When subjects were not told on which test they would not have enough time, their performance was debilitated. Mandler and Watson contend that this occurred when subjects' attention to the tests was disrupted by distracting ruminations about the time running out.

Anxiety may promote the effects of distraction in another way. Wine (1971), drawing on the earlier work of Easterbrook (1959), proposes that anxiety narrows the range of cues which can be attended to at any one time. Even if an individual can attend both to distracting cognitions and to enough necessary task-relevant cues under low anxiety, under high anxiety task-relevant cues are likely to fall outside the narrowed scope of attention. Wine supports this proposition in a review of the test-anxiety literature. She finds that test "clutchers" are more likely to report being distracted by self-depreciating cognitions during a test (Ganzer, 1968) and that their total range of attention narrows such that they become fixated on these irrelevant cognitions (Wachtel, 1968). Wine suggests a further complication to the distraction effects of anxiety. Not only does increased anxiety produce an increment in irrelevant, self-deprecatory cognitions, it also limits attention to those cognitions to the exclusion of other cues which may be important for changing the unwanted behavior.

There are important similarities between Wine's formulation of test-anxious behavior and our model of exacerbation. If an individual makes a self-attribution for some unwanted behavior, that inference may accomplish two things. It may increase anxiety, and it may provide a repertoire of self-deprecating cognitions. In future situations, the increased anxiety would trigger increased rumination over the self-conscious cognitions to the exclusive of any cognition necessary for changing that behavior. Thus, the individual is doubly damned. His attribution not only enhances counterproductive anxiety, it also provides self-defeating cognitions for that anxiety to feed upon.

Given the effects of the arousal, drive, and distraction properties of anxiety, it is possible to describe three categories of dysfunctional behavior which may be subject to exacerbation. First, behaviors which are comprised of specific physiological functions may be disrupted by anxiety. This category would include sleeping, sexual functioning, and perhaps certain motor tasks that require precise muscular movements. Second, behaviors which are habitual, well-learned responses may be increased by anxiety. This category may include various addictions, alcoholism, overeating, and perhaps other typical emotional responses to stressful situations. Third, behaviors which require attention to appropriate cues may be affected by anxiety. This category would include driving and other perceptual-motor tasks, and test taking and other highly cognitive activities. In short, any behavior may be subject to an exacerbation cycle if it (a) can lead to negative self-attributions and (b) is further increased by the anxiety those self-attributions may cause.

STUTTERING AND EXACERBATION

The preceding review of clinical and social psychological research suggests that a variety of dysfunctional behaviors develop and are sustained by an exacerbation cycle. We have proposed two factors which play a determinative role in exacerbation. Namely, we have suggested that the attribution of undesirable behaviors to dispositional qualities of the self creates anxiety and that this anxiety subsequently increases the occurrence of the negative behavior. In this section of the paper, we wish to apply the exacerbation model to a behavior—stuttering—not previously discussed and to demonstrate experimentally the development of this dysfunctional behavior via exacerbation.

Stuttering would appear to be a likely candidate for application of exacerbation concepts. Perceptions of inadequacy and loss of control and a high degree of anxiety may be common experiences among stutterers. Van Riper (1971) has noted that:

> Perhaps the most common of the advanced stutterer's feelings is fear. Fear is the expectation of unpleasantness. This expectation ranges from vague doubt to complete certainty. The stutterer may be afraid of social penalty and stigma; he may dread listener loss and rejection . . . but most of the stutterers with whom we have worked have told us there is another more basic fear—the expectation of communicative inability and verbal impotence. What they fear most is the momentary loss of self control. [This lack of control] is traumatic to the basic integrity of the self [p. 158].

In addition, stuttering may be an ideal behavior with which to demonstrate the origins of an exacerbation cycle. Stuttering can be an emotional behavior that grows out of normal speech disfluency (Bloodstein, Alper, & Zisk, 1965; Lemert, 1967), and may become a syndrome as a result of exacerbation processes. According to Lemert (1967), stuttering "appears to be exclusively a process-product in which . . . normal speech variations, or at most, minor abnormalities of speech . . . can be fed into an interactional or evaluative process and come out as . . . stuttering [p. 56]."

Our model suggests the following speculation about the role of exacerbation in stuttering. We would propose that stuttering may arise out of the normal speech disfluencies which all individuals display. At some point, however, a speaker may become aware of his disfluency. Either by direct communication of the label "stutterer" or by other signs of his failure to meet standards for proper speech, the individual adopts a negative self-attribution for his behavior. He infers inadequacy and lack of control over his speech. The negative emotion associated with the label "stutterer" and the accompanying self-attribution creates anxiety. This anxiety is likely to be highest in situations which have relevance to the attribution, such as speaking before an evaluative audience. Paradoxically, then, this anxiety produces more of the unwanted behavior, exacerbating the problem to the point where the speaker develops severe and persistent speech disfluency.

This scenario provides the basis for a study recently completed in our laboratory (Storms & McCaul, 1975). The experiment dealt with normal speakers and their naturally occurring disfluencies. Subjects were labeled as disfluent speakers, formed a self- or situational-attribution for their disfluency, and recorded a speech under conditions designed to produce high or low situational stress.

Forty-four male subjects were asked to make two tape recordings of their speech. The first of these was a practice recording, during which subjects were taped for 3 minutes as they responded to a written list of topical questions (for example, "What was the last movie that you saw, and did you like it?"). This first recording was used as a pretest estimate of individual disfluency rates as well as a basis for the manipulations. After each subject completed his recording, the experimenter made a number of comments about his speech. Specifically, he commented briefly and favorably on the subject's volume, tone of voice, and sentence length. He noted, however, that the subject had displayed a very high number of disfluencies such as "uhs," pauses, repetitions, and stammers. At this point, some subjects were told that their disfluency rate was a "normal" result of situational factors such as being in an experiment, while other subjects were led to make a dispositional attribution for their disfluencies. These latter, self-attribution subjects were told that their disfluency was a symptom of their own personal speech pattern and ability. The experimenter then played back the practice recording for the subject and pointed out the various types of disfluencies as they occurred.

Subjects subsequently recorded a second tape under conditions designed to elicit either high or low situational stress. Subjects in the high-stress condition were asked to put their name on the second tape and were told that the tape would be replayed, intact with their name, to undergraduate psychology classes. Low-stress subjects were asked not to mention their names and were told that only unrecognizable sections of their tape would be replayed and only to a limited number of people. The resulting design was a 2 × 2 factorial, varying subjects' attributions for their disfluencies (self versus situational) and situational stress while making the second tape (high versus low).

Subjects' tapes were coded for the frequency of three types of speech disfluencies, including stammers (repetitions of the beginning of a word), repetitions of entire words, and the use of paralinguistic sounds such as "uh" or "er." A baseline was computed for each subject by averaging the number of each type of disfluency per minute over 3 minutes of the practice tape. These baselines were then compared to the disfluency rates during the first and second minutes of the experimental recording, thus providing change scores as the chief dependent measure.

It is hypothesized in the exacerbation model that subjects' attributions about their disfluencies on the first tape should interact with the level of situational stress to affect their speech on the second tape. Thus, an interaction was predicted such that self-attribution/high-stress subjects would show the greatest increase in their disfluency rates. Figure 1 presents the pretest baselines and the scores from Minutes 1 and 2 of the posttest for stammers, the type of disfluency most commonly associated with stuttering. It can be seen from Fig. 1 that the stammers of the self-attribution/high-stress group increased more than the other groups. An analysis of the change from the pretest baseline to the first minute of the posttest showed no differences. However, by the second minute of the posttest, the effects were more pronounced and were revealed in a significant

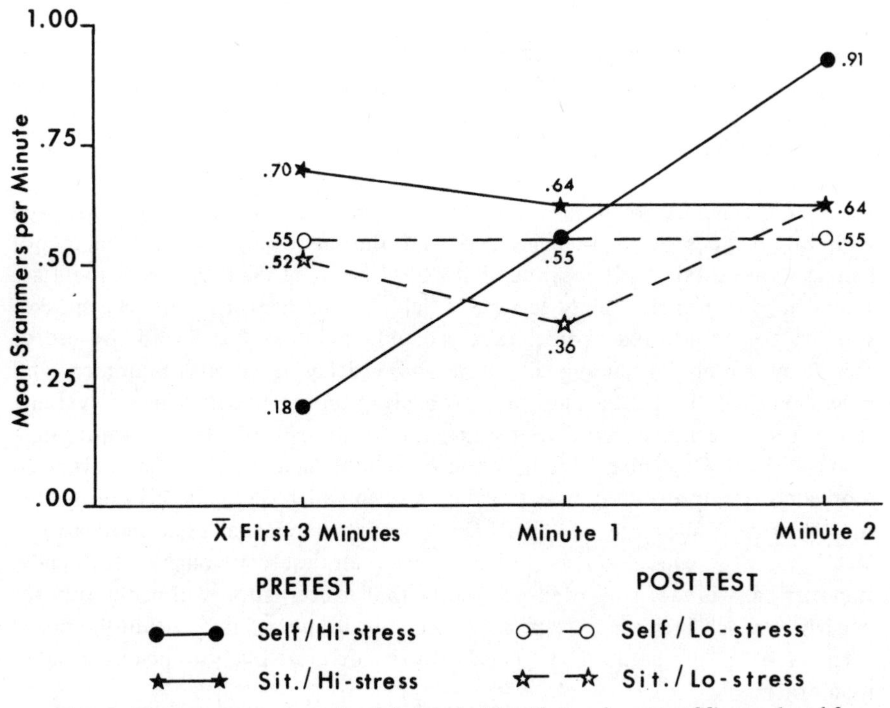

FIG. 1 Mean stammers per minute for pretest baseline and posttest Minutes 1 and 2.

interaction, $F(1,40) = 4.28, p < .05$. This interaction resulted from the fact that the increase in stammering for the self-attribution/high-stress group was significantly greater than for any other group (all individual comparisons yielded $p <$ .05). Moreover, significant exacerbation was shown in the increase in stammers for self-attribution/high-stress subjects alone, from baseline to Minute 2, $t(20) = 22.8, p < .001$.

It should be noted that the self-attribution/high-stress group started from a much lower baseline than the other groups (a difference which was statistically significant). This suggests the possibility that the increase in stammering for this group was simply due to regression to the mean. The use of change scores, while most appropriate for showing true exacerbation, may simply capitalize on this regression. This possibility is ruled out by an analysis of the simple posttest stammers at Minute 2. Once again the self-attribution/high-stress subjects differed significantly from each of the other groups ($p < .05$). Therefore, with regard to stammering, the hypothesis was firmly supported. Self-attributions interacted with greater situational stress to produce exacerbation of speech disfluency.

Although the results for stammers were strongly supportive of our thesis, the results for the other types of disfluencies were somewhat unexpected. For the most part, the cell means for repetitions followed the same pattern as stammers, but yielded no significant differences. The results for the "uhs" were even more surprising. The change scores were exactly the opposite of those for stammering, with self-attribution/high-stress subjects showing the greatest improvement. This contrary pattern of results for "uhs" may reflect important differences between types of disfluencies. Mahl (1958) has noted that "uhs" are negatively correlated with other types of disfluencies and that the frequency of "uhs" does not increase with anxiety. It may be that "uhs" occur at regular intervals and are somewhat functional, usually serving as delaying mechanisms while the speaker searches for appropriate words. It is probably quite easy to avoid the use of "uhs" by simply remaining silent at those delaying points. Stammers and repetitions, on the other hand, are probably more unexpected, more dysfunctional, more accidental, and more difficult to control. If self-attribution/high-stress subjects were most upset by their speech disfluencies, they may have been motivated to improve that aspect of their speech which is most easily controlled, thus showing a decrease in "uhs." On the other hand, that same motivational state evidently caused an exacerbation of their less controllable disfluency, stammering. We have suggested previously that exacerbation will occur only for behaviors which increase in response to anxiety. It appears that within the broad category of speech pathology, some behaviors are more likely to be exacerbated than others.

While the results of this study strongly suggest that stammering is subject to exacerbation, whether the model can be generalized to actual clinical stuttering is still conjectural. We will, therefore, turn our attention to the clinical literature on stuttering, drawing on extensive theoretical and empirical reviews of the speech pathology literature (for example, Van Riper, 1971). Our review of the stuttering literature is admittedly selective; we focus on studies which are most relevant to the two steps in the proposed exacerbation model–the role of self-attributions and the role of anxiety. In each case we have found supportive evidence.

Self-Attributions and Anxiety

The exacerbation model and the Storms and McCaul (1975) study suggest that one causal factor in stuttering is the internalization of a negative self-attribution. This appears to be the case for actual stutterers. Luper (1968) proposed that through listener reaction or by self-imposed penalities, the individual moves from a situation-dependent attribution, "I sometimes speak poorly," to a revised self-concept, "I am a stutterer." Furthermore, Sheehan (1954) found that this self-concept of being a stutterer predominates over all other self-concepts the individual might have. Clark and Murray (1965), reporting on three case studies, noted that a major part of their clients' problem was a "disturbed" concept of themselves as different from others. This self-attribution of incompetence or disability probably solidifies over time. Woods (1974), for example, found that disfluent sixth graders were much more negative in evaluations of their own speaking than were third-grade boys. Williams (1957) has recognized the problem of negative self-attributions and suggests as a part of clinical therapy that stutterers think of their problem more in terms of a temporary, situational difficulty as opposed to the more permanent, dispositional lable of stutterer.

There is additional evidence that debilitating, negative self-attributions result from society's reaction to and labeling of the stutterer's flawed speech. Johnson (1949) describes a dramatic example of this possibility in his description of a patient:

> All through his previous years at school he had been known not merely as a normal speaker, but as a definitely superior speaker . . . Then one day a 'speech correction teacher' examined him in the course of a school survey, and for some reason told him that he was a *stutterer,* and advised him to 'watch' his speech and be careful when he talked. Within a few months he would have been regarded as a stutterer and a fairly severe one . . . . In short, that teacher made him into a stutterer [p. 175].

Society's reaction to stuttering may have two important consequences. First, as already suggested above, it enhances the individual's tendency to make harmful, dispositional attributions for his behavior. Second, it may function even more directly in producing anxiety in the stuttering speaker. Johnson (1946), who was himself a stutterer, added support to this latter contention when he described social reaction to stuttering: "It is likely that if you have never been regarded as a stutterer, you can come nowhere near appreciating the uncanny, crushing power of the social disapproval of whatever is regarded as stuttering. It is probably one of the most frightening, perplexing, and demoralizing influences to be found in our culture [p. 458]."

Perhaps the most obvious source of anxiety for stutterers is the fear of having to speak. Much as an insomniac fears a bout with insomnia, so the stutterer fears emitting "stutters." Umeda (1960, cited in Van Riper, 1971) found changes in galvanic skin response (GSR) as stutterers changed from silent to oral reading, with the latter producing an increase in GSR relative to nonstuttering control group. Similarly, Valyo (1964) alternated periods of silence and speech for stutterers and found that subjects' GSRs doubled when they spoke. Lingwall (1967) found that adult stutterers had a higher GSR just before they were given a signal to speak.

Although the stutterer evidently experiences a good deal of anxiety any time he must speak, certain situations may elicit even higher levels of arousal. As Storms and McCaul (1975) show, self-attributions interact with situational variables. In line with those results, Maxwell (1965, cited in Van Riper, 1971) found that anxiety (measured by the Palmar Sweat Index) was greater in audience situations than when stutterers read to a single listener. Stuttering also increases in response to anxiety-provoking characteristics of the audience, including unfavorable audience reactions (Hansen, 1955, cited in Van Riper, 1971), increased audience size (Siegel & Haugen, 1964), and the presence of authority figures in the audience (Sheehan, Hadley, & Gould, 1967).

It is evident that stutterers adopt negative self-attributions for their speaking problems, that they often suffer from extremely low self-esteem, and that they experience intense anxiety about speaking, especially in stressful audience situations. In terms of self-attributions and anxiety, the exacerbation model would seem to describe the plight of the stutterer quite accurately.

Anxiety and Increased Stuttering

The exacerbation model and the Storms and McCaul (1975) research further suggests that the anxiety experienced by stutterers may cause an even greater occurrence of their dysfunctional behavior. Several investigators have performed standard anxiety manipulations on stutterers and normal speakers and have measured resulting changes in speech. Their research supports the notion that anxiety exacerbates some speech disfluencies.

Van Riper (1937), for instance, found that stutterers who were threatened with shock increased their stuttering. Savoye (1954, cited in Van Riper, 1971), using normal speakers as subjects, sounded a tone and then delivered a random shock. He found a greater number of disfluencies preceding the shock than after it (mirroring the periods when one would expect high or low levels of anxiety). Stassi (1961) punished the disfluencies of normal speakers with the word "wrong" and found greater disfluency than when the word "right" was contingently applied. Schwartz (1974) suggested a direct relationship between anxiety and stutterer's blocks (the hesitation that occurs while stutters attempt to pronounce a word). He presented evidence that anxiety interferes with the breathing necessary to facilitate speech.

Another mechanism by which anxiety may interfere with speech is suggested by recent work with delayed auditory feedback. A number of investigators (cf. Van Riper, 1971) have shown that delivering a normal speaker's voice back to him at delayed intervals produces hesitations and stammers which are remarkably similar to the disfluences of stutterers. Moreover, Goldiamond, Atkinson, and Bilger (1962) found that when subjects were instructed not to pay attention to the delayed feedback, they were able to read more words per minute than a control group that was instructed to listen to the feedback. From this data on normal speakers, it has been suggested that stutterers may create their own delayed auditory feedback by being overly attentive to their own speech. If anxiety causes a speaker to be overly self-conscious of his own speech, auditory feedback may result in increased disfluency. Although little is presently known about the attentional and perceptual processes underlying stuttering, the delayed auditory feedback studies suggest that anxiety may interfere with normal speech at this very basic level.

Research on the effects of anxiety on speech clearly indicates that the etiology of stuttering may be described using the exacerbation model. However, we must proceed cautiously toward implying that studies which demonstrate increased disfluency for normal speakers point to the causes of pathological stuttering. For one thing, the disfluencies of normal speakers and the disfluencies of stutterers are not always qualitatively identical (Van Riper, 1971). By the same token, we must be equally cautious about implying that studies which demonstrate changes in the disfluencies of stutterers are indicative of the original causes of stuttering. Wingate (1972), a consistent detractor of evaluational theories of stuttering, offered criticisms which are directly relevant to our present model: "It seems fairly clear that in many cases the individual's concern about his stuttering has the effect of exacerbating the difficulty, but to acknowledge such an effect is quite different from accepting the position that the concern *caused* the stuttering [p. 409]." Nevertheless, in our analysis we have approached the issue of stuttering from both directions and have found a remarkable convergence of the evidence. Storms and McCaul (1975) indicate that self-attributions of disfluency interact with anxiety-provoking situational factors to produce exacerbated stammering in normal speakers. Our review of research with stutterers reveal that the same factors, self-labeling and anxiety, are important determinants of the degree of speech disorder. At a time when the field of speech pathology lacks a comprehensive cognitive theory of stuttering, the exacerbation model may provide a tenable framework.

SUMMARY AND CONCLUSIONS

In this chapter we have elaborated a model of dysfunctional behaviors which are exacerbated by negative self-attributions and anxiety. The model was derived from experimental studies of attribution processes and emotional behavior, case studies of pathological emotional disorders, and theoretical studies of deviant behavior. Together, they suggest that individuals may make negative self-attributions for their own unwanted, symptomatic behaviors, that these self-attributions may lead to increased emotional states such as anxiety, and that anxiety may exacerbate the very same dysfunctional behaviors.

The contributions of two specific negative self-attributions to increased anxiety were discussed in greater detail. It was suggested that attributions of inadequacy and of loss of control may characterize many exacerbation cycles. Attributions of inadequacy were compared to the state of objective self-awareness. It was noted that many dysfunctional behaviors are likely to initiate an objective self-awareness state, that individuals in this state are likely to arrive at self-attributions of inadequacy, and that this state produces increases in negative self-evaluations and unpleasant emotions. Attributions of loss of control were compared to research in reactance, learned helplessness, and urban stress. It was noted that perceived loss of control increases anxiety, psychological upset, and dysfunctional reactions to stress. Finally, it was proposed that negative self-attributions of inadequacy and loss of control may interact with situational factors.

Research was described which indicates that negative self-attributions create even greater anxiety in relevant situations where more severe outcomes are at stake.

An analysis of the effects of anxiety on dysfunctional behavior suggested three categories of behavior which may be subject to exacerbation. Arousal properties of anxiety may exacerbate some dysfunctional physiological responses, drive properties of anxiety may exacerbate some habitual emotional behaviors, and distraction properties of anxiety may exacerbate cognitive and attentional activities. It was concluded that the exacerbation model may describe any behavior which can be attributed negatively to the self and which is also increased by anxiety.

Finally, the specific dysfunctional behavior of stuttering was examined in light of the exacerbation model. New research was described which demonstrated that the occasional stammering of normal speakers can be exacerbated by self-attributions of disfluency and situational stress. In a review of clinical research on stutterers, we indicated further that negative self-attributions and excessive anxiety contribute to the degree of speech pathology in chronically poor speakers. It was suggested that the exacerbation model may accurately describe the development of this particular dysfunctional behavior.

A clear implication of the exacerbation model is that clinical workers should be sensitive to the inferences their patients make about the causes of their symptomatic behavior. If those inferences produce anxiety, the patient may be trapped in a cycle of exacerbation. Rather than emphasizing a change in behavior, which may only underscore the patient's feelings of inadequacy, the clinician would do better to help the patient reinterpret the severity and implications of his symptoms.

ACKNOWLEDGMENTS

The original research conducted by the first author and his colleagues and reported here was made possible through National Institute of Health Grant No. 1-RO3-MH26344-0 and University of Kansas Bio-Medical Grant No. 4139-5706. We are indebted to students at the University of Kansas, including Karen Blank, Tona Stewart, and Tracy Spellman, for their assistance in reviewing the literature on social labeling theory and stuttering for inclusion in this chapter. Finally, our appreciation is extended to Richard Nisbett, Shelley Taylor, and Sara Kiesler who read earlier drafts of this chapter and provided valuable criticism.

REFERENCES

Becker, H. S. *Outsiders.* New York: Free Press, 1964.

Bloodstein, O., Alper, J., & Zisk, P. Stuttering as an outgrowth of normal disfluency. In D. A. Barbara (Ed.), *New directions in stuttering.* Springfield, Illinois: Charles C Thomas, 1965.

Brehm, J. W. *A theory of psychological reactance.* New York: Academic Press, 1966.
Clark, R. M., & Murray, F. P. Alterations in self-concept: A barometer of progress in individuals undergoing therapy for stuttering. In D. A. Barbara (Ed.), *New directions in stuttering.* Springfield, Illinois: Charles C Thomas, 1965.
Davison, G. C. Differential relaxation and cognitive restructuring in therapy with a "paranoid schizophrenic" or "paranoic state." *Proceedings of the American Psychological Association,* 1966, 177–178.
Duval, S., & Wicklund, R. A. *A theory of objective self awareness.* New York: Academic Press, 1972.
Duval, S., & Wicklund, R. A. Effects of objective self awareness on attribution of causality. *Journal of Experimental Social Psychology,* 1973, **9,** 17–31.
Easterbrook, J. A. The effect of emotion on cue utilization and the organization of behavior. *Psychological Review,* 1959, **66,** 83–201.
Ganzer, V. J. Effects of audience presence and test anxiety on learning and retention in a serial learning situation. *Journal of Personality and Social Psychology,* 1968, **8,** 194–199.
Glass, D. C., & Singer, J. E. Behavior aftereffects of unpredictable and uncontrollable aversive events. *American Scientist,* 1972, **60,** 457–465. (a)
Glass, D. C., & Singer, J. E. *Urban stress.* New York: Academic Press, 1972. (b)
Goldiamond, I., Atkinson, C. J., & Bilger, R. C. Stabilization of behavior and prolonged exposure to delayed auditory feedback. *Science,* 1962, **135,** 437–438.
Ickes, W. J., Wicklund, R. A., & Ferris, C. B. Objective self-awareness and self-esteem. *Journal of Experimental Social Psychology,* 1973, **9,** 202–219.
Johnson, W. Letter to the editor. *Journal of Speech and Hearing Disorders,* 1949, **14,** 175.
Johnson, W. *People in quandaries.* New York: Harper, 1946.
Lemert, E. M. *Social pathology.* New York: McGraw–Hill, 1951.
Lemert, E. M. *Human deviance, social problems, and social control.* Englewood Cliffs, New Jersey: Prentice–Hall, 1967.
Lingwall, J. B. Galvanic skin responses of stutterers and nonstutterers to isolated word stimuli. Convention address, American Speech and Hearing Association, 1967.
Luper, H. L. An appraisal of learning theory concepts in understanding and treating stuttering. In H. H. Gregory (Ed.), *Learning theory and stuttering therapy.* Evanston, Illinois: Northwestern University Press, 1968.
Mahl, G. F. On the use of "ah" in spontaneous speech: Quantitative, developmental, characterlogical, situational, and linguistic aspects. *American Psychologist,* 1958, **73,** 349.
Mandler, G., & Watson, D. L. Anxiety and the interruption of behavior. In C. D. Spielberger (Ed.), *Anxiety and behavior.* New York: Academic Press, 1966.
Masters, W. H., & Johnson, V. E. *Human sexual inadequacy.* Boston: Little & Brown, 1970.
Merton, R. K. *Social theory and social structure.* New York: Free Press, 1959.
Monroe, L. J. Psychological and physiological differences between good and poor sleepers. *Journal of Abnormal and Social Psychology,* 1967, **72,** 255–264.
Nisbett, R. E., & Schachter, S. Cognitive manipulation of pain. *Journal of Experimental Social Psychology,* 1966, **2,** 227–236.
Ross, L., Rodin, J., & Zimbardo, P. G. Toward an attribution therapy: The reduction of fear through induced cognitive–emotional misattribution. *Journal of Personality and Social Psychology,* 1969, **12,** 279–288.
Schachter, S., & Singer, J. E. Cognitive, social and physiological determinants of emotional state. *Psychological Review,* 1962, **69,** 379–399.
Schur, E. M. *Labeling deviant behavior.* New York: Harper & Row, 1971.
Schwartz, M. F. The core of the stuttering block. *Journal of Speech and Hearing Disorders,* 1974, **39,** 169–177.
Seligman, M. E. P., Maier, S. F., & Solomon, R. L. Unpredictable and uncontrollable

aversive events. In F. R. Brush (Ed.), *Aversive conditioning and learning.* New York: Academic Press, 1971.

Sheehan, J. Self-perception in stuttering. *Journal of Clinical Psychology,* 1954, **10,** 70–72.

Sheehan, J. G., Hadley, R. G., & Gould, E. Impact of authority on stuttering. *Journal of Abnormal Psychology,* 1967, **72,** 290–293.

Siegel, G. M., & Haugen, D. Audience size and variations in stuttering behavior. *Journal of Speech and Hearing Research,* 1964, **7,** 381–388.

Spence, J. T., & Spence, K. W. The motivational componants of manifest anxiety: Drive and drive stimuli. In C. D. Spielberger (Ed.), *Anxiety and behavior.* New York: Academic Press, 1966.

Stassi, E. J. Disfluency of normal speakers and reinforcement. *Journal of Speech and Hearing Research,* 1961, **4,** 358–361.

Storms, M. D. Videotape and the attribution process: Reversing actors' and observers' points of view. *Journal of Personality and Social Psychology,* 1973, **27,** 165–175.

Storms, M. D., & Leviton, L. C. Exacerbation of perceptual motor performance following self-attribution and stress. Unpublished manuscript, University of Kansas, 1975.

Storms, M. D., & McCaul, K. D. Stuttering, attribution, and exacerbation. Unpublished manuscript, University of Kansas, 1975.

Storms, M. D., & Nisbett, R. E. Insomnia and the attribution process. *Journal of Personality and Social Psychology,* 1970, **16,** 319–328.

Tannenbaum, F. *Crime and community.* Boston: Ginn, 1938.

Valins, S., & Nisbett, R. E. *Attribution processes in the development and treatment of emotional disorders.* Morristown, New Jersey: General Learning Press, 1971.

Valyo, R. PGSR responses of stutterers and nonstutterers during periods of silence and verbalization. *American Speech and Hearing Association,* 1964, **6,** 422.

Van Riper, C. The influence of empathic response of frequency of stuttering. *Psychologie Monographs,* 1937, **47,** 244–246.

Van Riper, C. *The nature of stuttering.* Englewood Cliffs, New Jersey: Prentice–Hall, 1971.

Wachtel, P. L. Anxiety, attention, and coping with threat. *Journal of Abnormal Psychology,* 1968, **73,** 137–143.

Wicklund, R. A. Objective self–awareness. In L. Berkowitz (Ed.), *Advances in experimental social psychology* (Vol. 8). New York: Academic Press, 1975.

Williams, D. E. A point of view about 'stuttering'. *Journal of Speech and Hearing Disorders,* 1957, **22,** 390–397.

Wine, J. Test anxiety and direction of attention. *Psychological Bulletin,* 1971, **76,** 92–104.

Wingate, M. E. Evaluation and stuttering. In L. L. Emerick & C. E. Hamre (Eds.), *An analysis of stuttering.* Danville, Illinois: Interstate, 1972.

Woods, C. L. Social position and speaking competence of stuttering and normally fluent boys. *Journal of Speech and Hearing Research,* 1974, **17,** 740–747.

Wortman, C. B., & Brehm, J. W. Responses to uncontrollable outcomes: An integration of reactance theory and the learned helplessness model. In L. Berkowitz (Ed.), *Advances in experimental social psychology* (Vol. 8). New York: Academic Press, 1975.

8

Extensions of Objective Self-Awareness Theory: The Focus of Attention-Causal Attribution Hypothesis

Shelley Duval
Virginia Hensley
University of Southern California

That man structures his world in terms of cause and effect relationships and acts on the basis of these constructions evidences the importance of causal attribution as an area of psychological investigation. Unfortunately, contemporary attribution researchers share no comprehensive theoretical base. A starting point for building an integrative theory of attribution is, however, provided by Heider's (1944, 1958) notion of the causal unit formation process: that man's tendency to reduce the diversity of the stimulus manifold leads him to connect individual objects and events in causal units. As this theory now stands, the crucial factors which govern the outcome of the causal unit formation process concern the relationships between objects and events on various phenomenological dimensions. Principal consideration has thus far been given to the phenomenological dimensions of time, space, and form/substance. Our purpose in this chapter is to broaden the base and increase the predictive power of causal unit formation theory by showing the way in which relationships between objects and events on other phenomenological dimensions function in determining the outcome of the causal attribution process. Primary attention will be given to those phenomenological dimensions which are contributed and defined by properties of the system 'consciousness.'

THE FOCUS OF ATTENTION–CAUSAL ATTRIBUTION HYPOTHESIS

In the theory of objective self-awareness (OSA) Duval and Wicklund (1972) define a relationship among states of consciousness and the phenomenon of self-evaluation. Specifically, it is suggested that self-evaluation and the behaviors that attend negative self-evaluation will occur only when consciousness is in a state of objective, as opposed to subjective, self-awareness. Recently, researchers have concentrated on the integrative and experimental ramifications of this particular theoretical formulation. However, the application of OSA theory is not limited to this area of investigation. The self-evaluation/self-awareness hypothesis is derived from a more general model in which consciousness is characterized as a variable, rather than a fixed constant, and is assumed to influence a wide range of psychological activities.

The assumptions which underlie the model are developed in this first section. In the second section, the operations of this model are integrated with the primary causal attribution mechanism proposed by Heider. Experimental tests of the model are detailed in Section 3. In Section 4 we discuss the various implications of the unit formation theory of causal attribution.

Assumptions of the General Model

1. There are mediational processes which correspond to the organism in the stimulus–organism–response paradigm. These mediational processes can be treated as a complex but interrelated set of systems.

2. All stimuli of either internal or external origin are processed by some system within the *mediational complex.* (The processing may, as in information-processing theory, take the simple form of transmission through a particular channel.)

3. The interaction between a stimulus and the particular system of the mediational complex which processes that stimulus culminates in responses which will be referred to as *psychological end-products.* These end-products can be divided into three primary components: (a) percepts (the basic apperception of mediational complex-processed stimulus as it is given in awareness [Heider, 1958] or in the "life space" [Lewin, 1936]. Percepts take the form of either objects or events, (b) relationships between objects or events (percepts which are organized in various associational structures, and (c) behaviors (reactions to objects and events and/or to the relationships between objects and events).

4. All object and event end-products have certain phenomenological characteristics. These consist of values on each and every dimension of being along which an object or event is registered and defined in awareness (for example, a position in time, space, an intensity, quality, etc.).

5. All primary inputs into the mediational complex (that is, stimuli) have definite properties associated solely with those stimuli. These properties contribute to the phenomenological characteristics of the object or event end-product associated with each individual stimulus.

6. The systems which make up the mediational complex also possess properties which are as definite and real as the properties possessed by stimuli. These system properties also contribute to the phenomenological characteristics of the object or the event end-product which represents the outcome of the stimulus-mediational complex interaction.

7. Combining these assumptions, we arrive at the following principle. The phenomenological characteristics of the object or the event end-product which results from the stimulus-mediational complex interaction is a function of the real properties of the stimulus introduced into the mediational complex and the real properties of the systems involved in processing that stimulus.

The above assumptions and the concluding general principle regarding the relationship among stimuli, mediational processes, and end-products represent the minimal structure required to support the general stimulus–organism–response paradigm. Using this as a frame of reference, we can now analyze the role of consciousness.

Consciousness

Let us assume that consciousness is one of the systems which makes up the mediational complex, and that this system, like all other components of the complex, possesses real and definite properties. Thus, to the extent that consciousness is involved in processing a given stimulus, the properties of that system will contribute to the phenomenological characteristics of the end-product of the interaction of that stimulus with the mediational complex.

Not only can the assumption be made that consciousness possesses real properties which affect the end-products of stumuli processed in that system, but evidence also indicates that the properties of the system consciousness are variable. From initial observation, Duval and Wicklund (1972) concluded that consciousness of one's self is not monotypic. Rather, consciousness of self was found to alternate between two quite different modes: subjective and objective. (This conclusion has been verified in numerous experiments which show that processing the object self in the objective mode results in self-evaluation, whereas the object self processed in the subjective mode is not evaluated.) Further work has shown that this division is not peculiar to the consciousness of self, but that consciousness, regardless of the stimulus referent, consists of two discrete but interrelated subsystems. The terms focal and nonfocal consciousness will be used to refer to these two subsystems. (Focal consciousness is a hypothetical construct which roughly corresponds to the notions of attention, the orienting response, and, in the earlier (Duval & Wicklund, 1972) work with self as the stimulus referent, objective self-awareness. Nonfocal consciousness is also a hypothetical construct which corresponds in the earlier work to subjective awareness. Although both subsystems perform essential operations and we do not assign functional primacy to either, focal consciousness appears to take place within the more pervasive fabic of nonfocal consciousness.)

In accordance with the principle concerning the outcome of a stimulus–mediational complex interaction, the assumption that each subsystem possesses a unique set of properties implies that the end-product of a stimulus processed in focal consciousness will differ in phenomenological characteristics from the end-product of the same stimulus processed in nonfocal consciousness. By virtue of this variability, the subsystems of consciousness not only contribute to the final phenomenological characteristics of end-products but also produce differences between the phenomenological characteristics of stimulus–mediational complex end-products.

In summary, this rudimentary model suggests that the subsystems of consciousness differentially contribute to the phenomenological characteristics which are associated with the percepts (object or event end-products) extant in awareness. Suppose that the general association process and, in particular, the choice of a specific object or event as the cause of some "effect" event, is based on and determined by the phenomenological characteristics associated with the objects and events in question. This would require that the contributions of each subsystem of consciousness be taken into account in a complete theory of causality. In the following section, we provide evidence that the attribution of causality is indeed a function of the phenomenological characteristics possessed by objects and events and that the subsystems of consciousness play an integral role in the process which links objects and events in causal relationships.

THE CAUSAL ATTRIBUTION SYSTEM

The literature concerning causality is strongly divided on the issue of the origin of the impression that "X causes Y." Some theorists argue that causality does not exist in the objective world and that causal attribution represents the projection of an internal cognitive structure onto the external environment. In essence, this projection hypothesis asserts that man's experience of the cause–effect relationship is merely an illusion. Were this the case, how could one account for the fact that the external environment can be altered and manipulated on the basis of these causal constructions of the world?

Opponents of the projection hypothesis argue that causality is real but insist that causal relationships are given by the external structure of events and objects themselves (that is, the existence of causal relationships are totally independent of the observer's mediational systems). This objectivist point of view implies that all one has to do is observe the world in order to obtain knowledge as to the causal structure of that world. If this were true, man must judge himself myopic-to-blind in order to account for attributional discord and error.

The difficulties associated with the projection and objectivist hypotheses make it untenable to assume that attribution of causality is exclusively given by internal cognitive organization or by some "essence" displayed by objects and events as they exist in the external world. Thus, in accordance with the general prescription that all end-products, in this case causal associations, are a function of both the person and the environment, attribution of causality will be treated as a function of an interaction between various components of the mediational complex and the properties of stimuli.

Cause and Effect Component of the Mediational Complex

Heider, among others, has pointed out that man has a basic disposition to reduce the diversity present within the stimulus manifold which constitutes the environment. We assume that this tendency reflects the operation of one system of the mediational complex. The primary manifestation of the operation of this system is the organization of raw discrete sensory data into coherent wholes (object and event end-products). A secondary manifestation of the operation of this system is the organization of individual objects or events into higher order units. Heider refers to this second operation as the process of *unit formation.* The end-products of this process are units in which individual objects and/or events become parts or elements of a larger whole.

Within the framework of this theory, the linkage of an object or event with a second effect event in a cause–effect relationship is one type of unit end-product which results from the interaction between the unit formation process and a stimulus manifold made up of discriminant changes occurring successively and referring to the activities of separate and distinct objects. For example, if the stimulus manifold is made up of the discrete and successive motions associated with one billiard ball striking another, the unit formation process (the operation of the mediational system which tends to simplify the environment) will tend to fuse the individual and separate movements of the two billiard balls into elements of a single, and thus simpler, cause–effect unit. (The interaction between the unit formation process and the stimulus manifold does not always result in the formation of cause and effect units. When the stimulus manifold is made up of objects at rest, events which occur simultaneously, or the successive and continuous occurrence of events which refer to the activities of the same object, such as the swinging of a pendulum, the interaction between the unit formation process and the stimulus manifold will result in what could be called grouping units.)

Elaborating upon the nature of unit relationships, Heider suggests that the elements of all unit formations exhibit an invariant relationship to one another.

This relationship is defined as a type of cognitive organization in which the object and/or event elements of the unit (or whole) belong or are connected to one another to a greater extent than to any other objects and/or events. In the case of cause–effect units. the nature of this belonging or connected relationship is easily illustrated.

For example, if a car owner is asked to predict what will happen if the ignition key of his automobile is turned on, he will say that the motor of the car will start. If a homeowner is asked to predict what will happen when a light switch in his house is turned on, he will say that the light bulb in this or that particular lamp will light up. If a TV repairman sees the horizontal hold tube in a TV set removed and is asked to predict what will occur, he will say that the TV picture will roll. On the other hand, if a person who knows nothing about the cause–effect structure of cars, electric lights, or TVs is asked the same questions, his predictions will reflect random guessing at best. These examples demonstrate that a person who has formed causal relationships between events tends to call out the effect event when he observes that the causal event has taken place, whereas a person who has not formed the cause–effect relationships has no definite expectations upon observing the causal events. This, in part, defines the nature of the belonging or connected relationship characteristic of the causal type of unit. When an event is joined with a second event or object in a causal relationship, observing or thinking of the occurrence of one of the elements of the unit tends to call the other element into cognition.

In essence, then, the unit formation approach to causality defines causal attribution as one type of end-product which is generated by the interaction between the stimulus manifold and a mediational system which tends to simplify that manifold. The result of this interaction is the organization of objects and events into belonging or connected causal unit relationships. Obviously from this point of view, the basis and laws of causal attribution will not be those of a logicodeductive methodology (as suggested by most causal attribution theorists), but the basis and law of the general process of unit formation.

The Basis and Law of Unit Formation

In most research designed to investigate the processes of unit formation stimuli have been used which tend to produce grouping, rather than causal, units. However, this situation presents no problem since the basis and principles of one type of unit formation reflect the operation of the general system. Thus, data derived from research on grouping units apply equally to the formation of cause–effect units.

From the research regarding the basis and law of the unit formation process (Koffka, 1935) two clear conclusions can be reached. First, it is evident that a person's tendency to reduce diversity does not result in the indiscriminant coupling of objects and events but, rather, organizes objects and events into unit

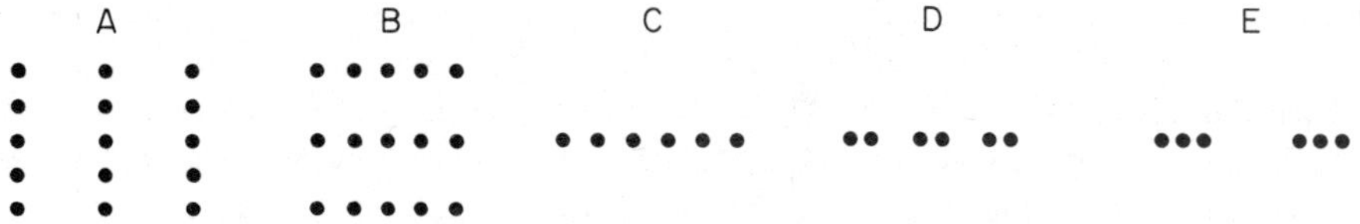

A B C D E

FIG. 1 Aggregates A, B, C, D, and E.

relationships on the basis of the phenomenological characteristics associated with those objects and events. This conclusion is derived from the fact that any variation in the phenomenological characteristics of objects and events along any of the three recognized dimensions of time, space, and form/substance affects the way in which the objects and events are organized into units. Figure 1 illustrates that variations along the spatial dimension affect unit formation within a given set of objects. Figure 2 illustrates the case in which variations along the dimension of form/substance affect which objects will be grouped in unit relationships. Variations in tempo and phrasing in a concerto illustrate how changes along the temporal dimension alter unit formation.

Second, research regarding the law of unit formation indicates that the degree of similarity between the phenomenological characteristics of objects and events determines which objects and events will be connected as elements of a particular unit. To the extent that the phenomenological characteristics of two or more objects or events are more similar to each other than to the phenomenological characteristics of other objects or events, the tendency to connect those objects or events into a higher order unit increases (Koffka, 1935; Wertheimer, 1958). This similarity–unit formation principle is the fundamental law of unit formation. Its operation is illustrated in the perception of the Stimulus Aggregates A and B in FIG. 1 as either columns or as rows. Aggregate A is perceived as columns because the spatial coordinates of the dots are more similar to each other on the vertical than on the horizontal dimension. Thus, the dots are formed into units, that is, lines, from top to bottom rather than from left to right, giving the impression of columns. Conversely, in Aggregate B, the dots exhibit greater spatial similarity to each other on the horizontal plane. Consequently, the unit formations occur from left to right, giving the impression of rows.

If causal attribution represents one manifestation of the tendency to simplify the stimulus manifold by connecting individual objects and events into higher

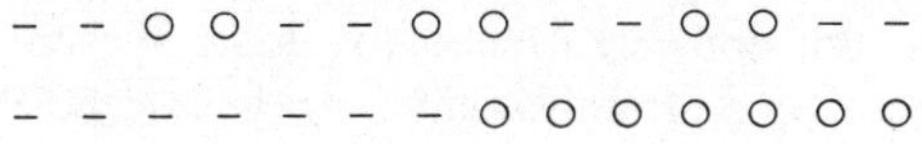

FIG. 2

order units, causal attribution would have the same basis and be governed by the same law which governs all unit formation processes. From the research and examples cited, we know that the phenomenological characteristics of object and event end-products constitute the basis for unit formation. We also know that the general law which determines the outcome of the unit formation process concerns the degree of similarity between phenomenological characteristics of objects and events. To the extent that the characteristics of objects and events are similar to each other on the three dimensions which have thus far been identified (time, space, and form/substance), the tendency to group those objects and events into units will increase. Applied to the formation of cause–effect unit relationships, the similarity–unit formation principle has yielded the following hypotheses. As the similarity among the spatial coordinates, temporal coordinates, and form/substance of an event or object A and an effect event X increases, the tendency to attribute causality for event X to A will increase. In the case of multiple plausible causes and a dichotomous measure of attributed causality, these hypotheses take the following forms. Given an effect event X and multiple plausible causal objects or events, causality for X will tend to be attributed to the object or event which is most similar to X in terms of spatial coordinates, temporal coordinates, and form/substance.

Due to limitations of space, it is impossible to discuss the procedural details of experiments designed to test the spatial and temporal hypotheses. However, research by Michotte (1963) clearly supports the spatial and temporal hypotheses. In a variety of experiments, he found that decreasing the time and space (that is, distance) between the discrete movements of two objects increased the subject's tendency to perceive the movements of the two objects as a causal unit. Research which is clearly related to the form/substance hypothesis is unavailable at present.

Thus far we have applied the similarity-unit formation principle to phenomenological dimensions defined by the properties associated solely with the stimulus referent of object and event end-products. However, our model of the stimulus–mediational complex–end-product interaction sequence suggests that some of the phenomenological characteristics associated with an object or event end-product may be contributed by the properties of the systems which process the stimulus referent of those end-products. Since there is no reason to believe that the basis of unit formation and the similarity principle are limited to those phenomenological dimensions defined by the properties of stimuli, we must conclude that the degree of similarity on any phenomenological dimension associated with an object or event, whether given by the properties of the stimulus alone or added by the properties of the systems which process the stimulus, will affect the tendency to link any object or event with a second event in a causal unit formation. If this conclusion is correct, then a complete unit formation theory of causality would have to take into account not only those dimensions of object and event end-products defined by properties of stimuli, but also those dimensions defined by the properties of the mediating systems.

Many systems may participate in processing a given stimulus and each system may contribute to the phenomenological characteristics of the object or event end-product. However, at the present time, we shall limit ourselves to an analysis of the effects which the subsystems of consciousness have on the phenomenological characteristics of object and event end-products.

Contributions of the Subsystems of Consciousness to the Phenomenological Characteristics of Objects and Events

In the brief introduction to the model of consciousness which was presented earlier, we suggested that consciousness consists of two discrete but interrelated subsystems: focal and nonfocal consciousness. Inferential, experiental, and experimental evidence suggest that these two subsystems possess different properties which contribute qualitatively different phenomenological characteristics to the end-product of any stimulus which is processed in those subsystems.

For example, it has been demonstrated that the operations of focal and nonfocal consciousness differ with regard to evaluation of self (Duval & Wicklund, 1972). This means that either (a) the fundamental properties of the two subsystems are the same but obey different laws or (b) that the fundamental properties of the two subsystems differ but obey the same laws. Since variations in the operations of subsystems usually represent differences in the properties of the subsystems rather than differences in the basic laws governing the operations of those subsystems, the fact that evaluation of self (or any object) occurs when that object is processed in focal consciousness but not when it is processed in nonfocal consciousness implies that the properties of the two subsystems differ. Thus, strictly on the basis of inference from the object-evaluation data, there is good reason to suspect that the properties of focal and nonfocal consciousness differ.

This evidence is suggestive but not conclusive. We must also demonstrate that the properties of the two subsystems do contribute to the phenomenological characteristics of the end-products of stimuli processed in those subsystems and that the contributions of each subsystem are qualitatively different. A technique which could be used to demonstrate the validity of these two hypotheses is to divide the environment so that some stimuli are processed in focal consciousness while others are processed in nonfocal consciousness: for example, expose an individual to an array of stimuli with instructions to direct and restrict focal consciousness (attention) to one and only one of an array of stimuli. This exercise should cause the one stimulus to be processed in focal consciousness while all other stimuli are processed in nonfocal consciousness. If the two subsystems do have distinct properties which contribute qualitatively different phenomenological characteristics to the object end-products of the stimuli processed in each subsystem, then the object end–product of the stimulus processed in focal consciousness should have some critical phenomenological characteristics which object end-products of stimuli processed in nonfocal consciousness do not and vice versa. That this is in fact the case is clear. Quite apart from the phenomenological characteristics contributed by the stimuli's particular positions on the time, space, and some of the form/substance dimensions (which seem to remain constant regardless of the processing system), the person will describe the object end-product of the stimulus processed in focal consciousness as having much sharper boundaries, as being more set-off, distinct, and separate from its surroundings than the object end-products of stimuli processed in nonfocal consciousness. In addition, the person will describe the object end-product of the stimulus processed in focal consciousness as more dense, substantial, compact, three dimensional, and, in some cases, more intense than the object end-products of the stimuli processed in nonfocal consciousness (Koffka, 1935). Of course, in the above test, it is possible that these critical phenomenological characteristics may be due to some unique properties of the stimuli rather than to contributions of the subsystems of consciousness. This possibility can be checked by simply instructing our subject to redirect focal consciousness away from the original stimulus, A, to a second stimulus, B. If those critical phenomenological characteristics of the object end-product of Stimulus A were due to the fact that A was being processed in focal consciousness, then shifting focal consciousness away from A to B should alter the critical phenomenological characteristics of both A and B; A should assume the critical characteristics of a stimulus end-product processed in nonfocal consciousness while B, now being processed in focal consciousness, should take on the critical phenomenological characteristics particular to that subsystem. That this is the case can and has been verified by direct experience (Koffka, 1935). Thus, we conclude that focal and nonfocal consciousness do contribute to the phenomenological characteristics of the end-products of stimuli processed in those subsystems and that the contributions of the two subsystems are qualitatively different.

Besides immediate, direct, and unambiguous experiential verification, data from a substantial amount of formal research lead to the same conclusions; the end-product of a stimulus processed in focal consciousness possesses different phenomenological characteristics than the end-product of the same or an equivalent stimulus processed in nonfocal consciousness, and these critical characteristics are due to the effects of the processing system. For example, in an experiment by Rey and Richelle (cited by Piaget, 1969), subjects were exposed to two parallel lines, both 22 cm in length, .5 cm thick, and 3 cm apart. These lines were drawn in black ink on a sheet of glass. Subjects were asked to close their left eye and to fixate the left-hand line with the right eye. Even though subjects were only instructed to determine if either line appeared closer to them than the other, 26 of the 35 subjects reported in debriefing that the fixated line was longer than the nonfixated line. The remaining 9 subjects indicated that the fixated line fluctuated in length but was never shorter than the nonfixated line. Since both lines were equal in length and the subject's eye was equidistant from both lines, the only reasonable explanation for the effect is that processing a stimulus in focal consciousness tends to exaggerate the magnitude of the object end-products.

In a second experiment designed to investigate the effect of fixating focal consciousness on a stimulus, Piaget (1969) presented subjects with a standard rod, 40 mm in length, with one of a group of variable rods ranging in length from 34 to 46 mm. The standard rod and variable rod were presented successively, and order of presentation was counterbalanced. Each time the standard rod was presented with one of the variable rods, the subject was asked to indicate whether the standard rod was longer than, equal to, or shorter than the variable rod. The experimental manipulation consisted of varying the amount of time subjects were allowed to focus attention on each rod. In Condition 1, the standard rod was presented for 1.5 seconds, while each of the variable rods was presented for 1 second. In Condition 2, both the standard rod and each variable rod were presented for 1 second. An analysis of the results of the experiment indicated that the variable rods which subjects estimated to be equal in length to the standard rod were longer in Condition 1 than in Condition 2. In addition, the mean length of the variable rods which subjects in Condition 1 matched to the standard rod was greater than the actual length of the standard rod. Since the only difference between conditions was length of exposure to the standard rod, Piaget attributed the results of the experiment to the fact that duration of fixation on the standard rod was greater in Condition 1 than in Condition 2.

A clear inference from the above experiments is that the phenomenological characteristics of stimuli processed in focal and nonfocal consciousness differ and that the contributions of focal consciousness increase as the duration of processing in that system increases. In the first experiment the object end-product of the stimulus processed in focal consciousness was represented in awareness as being greater in intensity (represented as overestimation of length) than the object end-products of an equivalent stimulus processed in nonfocal consciousness. In the second experiment, increasing the length of time a stimulus was exposed to focal consciousness tended to increase the intensification effect. In further experimentation, Piaget suggested that the tendency to process some areas of a figure in focal and some in nonfocal consciousness may be responsible for the quantitative distortions which we refer to as illusions (see Piaget, 1969, for a complete discussion).

Thus far we have presented experiential and experimental data which suggest that the end-products of stimuli processed in focal consciousness exhibit some phenomenological characteristics which are different from the end-products of the same or equivalent stimuli processed in nonfocal consciousness. As may have become apparent to the reader, the differences between the characteristics of a stimulus end-product in focal and nonfocal consciousness correspond closely to the different phenomenological characteristics associated with figure and ground. Thus, Gestalt research concerning the nature of figure and ground will provide further evidence concerning the nature and reality of the contributions of focal and nonfocal consciousness to the phenomenological characteristics of object and event end-products.

For example, in our description of the phenomenological characteristics of the end-product of a stimulus processed in focal consciousness (a figural object) and in nonfocal consciousness, (ground) we noted that the object end-products of stimuli processed in focal consciousness appeared to be more dense, substantial, etc., than the object end-products of stimuli processed in nonfocal consciousness. Noting these same differences in the appearance of figure and ground, Geld and Granit (cited by Koffka, 1935) reasoned that the more solid figural object would offer greater resistance to the intrusion of a secondary figure than would ground provided that the different appearances of figure and ground corresponded to functional differences between the two types of end-products. To test this hypothesis, they divised an apparatus whereby a small colored spot could be projected onto the figure and ground component of a visual display. Holding the luminosity of the figure and ground constant, they measured the intensity of light necessary to make the colored spot visible when it was projected onto either the figure or the ground component. It was found that the intensity of light necessary to make the spot visible on the figure was considerably greater than the intensity of light necessary to make the spot visible within the ground area of the display. These results can be interpreted as additional evidence that focal and nonfocal consciousness contribute qualitatively different phenomenological characteristics to the object end-products of stimuli processed in those subsystems and that these contributions (at least those dealing with density, substantiveness, etc.) have the same existential status (are as real) as the phenomenological characteristics contributed by the properties associated with the stimulus referent of an object end-product.

The nature and reality of subsystem contributions have also been demonstrated with regard to the dimensions of intensity and displacement. For example, using an ambiguous figure in which two colored crosses, A and B, easily switch from a figure–ground to a ground–figure organization, Koffka (1935) demonstrated that shifts in the figure–ground relationship (that is, Cross A becomes the figure and Cross B becomes ground, or vice versa) produced changes in the perceived color and intensity of the crosses. The cross which changed from ground to figure gained in color and intensity, while the cross which changed from figure to ground lost color and intensity. Likewise, the textbook examples of figure–ground clearly indicate that the object, usually a black cross on a white background or vice versa, which is processed in focal consciousness (is figural) is displaced upward and away from the area of the field being processed in nonfocal consciousness. This gives the impression that the figure is suspended above the ground.

In general then, inferential, experiential, and experimental data support the various assumptions of our model. Consciousness is made up of two distinct (though interrelated) subsystems which we have called focal and nonfocal consciousness. These subsystems have real but different properties. These system properties contribute to or impart particular and specific phenomenological characteristics to the object and event end-products of stimuli which are processed in those subsystems. Finally, these subsystem contributions appear as functional differences in experimental research, indicating that those contributions are as genuine as the contributions of properties associated solely with stimuli. On the basis of these conclusions, the object and event end-products of the stimulus–consciousness interaction can be divided into two groups. The end-products of stimuli processed in focal consciousness form one group which we shall refer to as focal. All members of this group will have the same critical phenomenological characteristics contributed by the focal subsystem (for example, greater distinctness, density, etc.) in addition to any unique phenomenological characteristics contributed by the properties of the stimuli references themselves (for example, a particular position on the space, time, and form/substance dimensions). The end-products of stimuli processed in nonfocal consciousness form a second group which we shall refer to as nonfocal. All members of this group will have the same critical phenomenological characteristics contributed by the nonfocal subsystem in addition to any unique characteristics contributed by properties of the stimulus referents themselves.

Variability in Phenomenological Characteristics of Focal and Nonfocal Objects and Events

We have sought to establish that consciousness is divided into two basic subsystems which contribute different phenomenological characteristics to the end-products of any stimulus processed in those subsystems. It is evident from further investigation that the properties of nonfocal consciousness, and, thus, the phenomenological characteristics of nonfocal objects are relatively invariant. The properties of focal consciousness, on the other hand, possess at least four dimensions of variability: incidence, directionality, duration, and intensity. We will briefly define each of these dimensions.

Incidence means that the focal subsystem is not continually in operation; at any given point in time, the focal subsystem may or may not be processing any part of the stimulus manifold. By directionality we mean that, given the onset of the operation of the focal subsystem, that focal consciousness is normally concentrated on only one area of the environment at any given moment and that this concentration may and usually does shift rapidily from area to area of the environment. Duration is the length of time which focal consciousness is concentrated on any given stimulus. Given 5 stimuli in the environment, the person may focus on one for a total of 10 seconds, a second for 30 seconds, a third for only 2 seconds, etc. By intensity, we mean that extent to which the total subsystem, focal consciousness, is concentrated on a particular stimulus. Thus, given that a stimulus is being processed in focal consciousness for some amount of time, the intensity of the focus during that period may be very weak, very strong, or may fluctuate.

The four dimensions of variability associated with focal consciousness may generate variability in the *degree of focalization* associated with the objects and events belonging to the focal group. Thus, not only are there focal and nonfocal objects and events, but the degree of focalization associated with a focal object or event may vary. Given that focal consciousness is directed toward a particular stimulus, the degree of focalization associated with the object or event end-product of that stimulus will increase as the duration and intensity of focal consciousness used to process that stimulus increases.

In summary, we have sought to establish that the two subsystems of consciousness have definite properties just as any transmission channel or system of the mediational complex has definite properties. We then attempted (a) to describe the phenomenological characteristics which these properties imparted to any stimulus processed within each subsystem, and (b) to show that the phenomenological characteristics have the same existential status as the properties of the stimulus referents of objects and events. This led to the categorization of object and event end-products into two dichotomous groups, focal and nonfocal, on the basis of the commonalities and differences in the critical "system" phenomenological characteristics associated with each. Finally, we suggested that objects and events vary not only with regard to the dichotomous dimensions of focalization and nonfocalization, but that, if focal, may vary in degree of focalization. This more detailed model of consciousness and its contributions to the phenomenological characteristics of objects and events can now be integrated with the primary causal unit formation process.

Causal Unit Formation and Focalization

Research and direct experience indicate that the end-products of stimuli which are processed by consciousness will have the phenomenological characteristics of being either focal or nonfocal and, if focal, may vary with regard to the degree of focalization. Stimuli which eventually become object and event elements of a causal unit formation are, of course, processed by consciousness prior to being connected in a unit relationship. Thus, prior to the formation of a causal unit between an object or event and a second event, those objects and events will necessarily possess the phenomenological characteristics contributed by the subsystem of consciousness in which they were processed. Since the focal and nonfocal phenomenological characteristics associated with objects and events possess a concrete reality demonstrable in an experimental setting, these characteristics must define dimensions which are as significant to the process of causal unit formation as the phenomenological dimensions defined by the properties associated solely with the stimulus referents of objects and events. When placed within the context of the similarity–unit formation principle, this conclusion implies that the similarity between objects and events on the dimensions of focalization and nonfocalization will affect attribution of causality in the same way as the similarity between objects and events on the spatial, temporal, and form/substance dimensions. Objects and events which are focal or nonfocal will tend to be causally linked with objects and events which have similar system-contributed phenomenological characteristics. Further, given that objects and events are focal, the tendency to connect an object or event with a second event in a causal unit will increase as the degree of focalization associated with those elements becomes more similar.

The postulated relationship between similarity on the dimensions of focalization and nonfocalization and the tendency to link particular objects and events in causal units constitutes the fundamental proposition of the focus of attention–causal attribution hypothesis. Transformation of this basic proposition into hypotheses amenable to experimental tests requires further inquiry into the nature of the elements of cause–effect units.

The postulated tendency to link only focal or nonfocal objects together implies that the cause–effect unit will be made up of elements which are either both focal or both nonfocal. However, further analysis of the nature of the cause–effect unit indicates that the effect element in the relationship must be an event (a discriminant change in the preexisting environment). The proof of this assertion lies in the fact that there is nothing to be explained causally unless a change in the environment has taken place. In other words, some change or movement is inherent to the notion "effect" and is thus, by definition, an event.

Given that the effect element must be an event, it may be the case that an event is always focal or nonfocal. If either were the case, it would be possible to delimit the class of plausible causes to events and/or objects which match the effect event in terms of being either focal or nonfocal. This possibility may be stated in more precise terms as the following question: Does the general quality which defines something as an event (that it must be a stimulus which represents a discriminant change in the preexisting environment) determine which of the two subsystems of consciousness will process the stimulus? The answer to this question involves a theory of the operation of focal consciousness. If the nature of a stimulus which represents a change in the environment is such that it necessarily engages and attracts the focus of attention, the event end-product of that stimulus will always be focal.

The solution to this problem can be found in Sokolov's (1960) theory of attention (also discussed in Lynn, 1966). He suggests that the incidence and direction of focal consciousness are functions of the degree and points of fit between a pattern of neuronal organization which represents the encoded traces of previous stimulation and the pattern of stimulation which is from moment to moment impinging on that organization via the senses. To the extent that the two patterns do not match, focal consciousness as a system is brought into operation. Furthermore, the area of the environment to which focal consciousness will be directed is that area of stimulation which does not fit with the previously established neuronal pattern.

The structure and operation of Sokolov's model can best be illustrated by experiments conducted by Sokolov and reported in Lynn. In one study, subjects were exposed to a 5-second tone. Since the subjects' neuronal models presumably did not include this tone, there should have been a lack of fit between their neuronal models and the pattern of incoming stimulation. In terms of Sokolov's theory, this lack of fit should have produced a strong orienting response (which is roughly synonymous with focal consciousness). This prediction was confirmed. However, as a further test of the theory, subjects were exposed to the tone until presentation of the tone no longer elicited an orienting response (that is, habituation occurred). In terms of the theory, the lack of an orienting response indicated that the subjects' internal neuronal pattern had accommodated to and thus included the presentation of a 5-second tone. Following this procedure, subjects were again exposed to the tone. In this case, however, the tone was interrupted after 2 seconds. The interruption of the tone should have created a lack of fit between the pattern of incoming stimulation and the neuronal pattern adapted to a 5-second tone. This, in turn, should have elicited the operation of the focal system and generated an orienting response precisely at the point at which the pattern of incoming stimulation and the neuronal pattern diverged. The subjects' orienting responses clearly confirmed this prediction. Although subjects no longer oriented to the presentation of the tone itself, strong orienting responses occurred when the tone was interrupted. In further experimentation, subjects were first habituated to a high pitched tone, A. After habituation to Tone A had occurred, subjects were exposed for the same amount of time to a lower pitched tone, B. Again, as predicted by Sokolov's theory, subjects exhibited a strong orienting response when the pattern of incoming stimulation changed from Tone A to Tone B.

As the experimental data indicate, the critical factors in Sokolov's theory are the degree and points of fit between the neuronal pattern established by previous stimulation and the pattern of stimulation which is being introduced into the neuronal system. If a particular region of incoming stimulation is at variance with the established neuronal pattern, focal consciousness is elicited and directed toward the region in which the deviation is occurring. In terms of this theory, it is clear that any event (any discriminant change in the pattern of incoming stimulation) must always be associated with focal as opposed to nonfocal consciousness.

In terms of Sokolov's theory, the nature of any stimulus which represents an event will necessarily result in the processing of that stimulus by focal consciousness. This implies that, in addition to whatever other phenomenological characteristics events may possess, they will invariably possess the phenomenological characteristics associated with focalization. In terms of the similarity principle of causal unit formation, this conclusion implies that the object or event which will be linked to an effect event must also possess the phenomenological characteristics peculiar to being processed in focal consciousness. This conclusion emphasizes the importance of taking the dimension of focalization into account in a theory of causality and allows us to delimit the set of objects and events which may become the causal elements of a cause–effect unit to those which are focal in their phenomenological characteristics. Given this, we can now present the experimental form of the focus of attention–causal attribution hypothesis. Given an effect event X and a set of objects and events which are both focal and nonfocal, (a) a focal object or event will always be selected as the cause over a nonfocal object and (b) the focal object or event which is most similar to the effect event in terms of degree of focalization (duration × intensity of focus) will tend to be seen as the cause of the effect.

As can be seen, application of the similarity–unit formation principle to the dimension of focalization generates several general but testable hypotheses provided that the degree of focal similarity between the effect events and the various plausible causal objects and events is known. Before proceeding to data relevant to tests of these hypotheses, we must briefly discuss the parameters associated with three types of situations in which multiple plausible causes are present.

First, there is the case in which the set of plausible causes is composed entirely of events. In this case, predictions based on the application of the similarity principle to the dimension of focalization are straightforward. Given an effect event, X, and multiple causal events, A, B, etc., the event which will be seen as the cause will be the event which is most similar to the effect event on the focalization dimension. In the case where the multiple plausible causes are events and objects, the form of the hypothesis remains the same. However, in the case in which the plausible causes are all objects, the similarity principle remains the same, but the form of the hypothesis is altered. For example, Sokolov's model of focal consciousness makes it necessary to assume that the degree of focalization associated with *any* observed event will exceed the degree of focalization associated with *any* object not perceived in the act of changing the environment. This is the case since any event is a discriminant change in the environment while objects not observed in a process of change are, by definition, not discriminant changes. On the contrary, objects are characterized by stable dispositional properties. Thus, by virtue of having stable dispositional properties, the maximum degree of focalization associated with any object is less than the minimum degree of focalization associated with any event.

Given that the degree of focalization associated with any event exceeds the degree of focalization associated with any object, the plausible causal object which is attended by the greatest degree of focalization is necessarily more similar to the effect event on the focalization dimension than is any plausible causal object associated with a lesser degree of focalization. Thus, in the case of an event, X, and plausible causal objects, A, B, etc., the focus of attention–causal attribution hypothesis predicts that causality will be attributed to the object which is associated with the most intense degree of focalization. When the measure of causal attribution is not dichotomous, this hypothesis takes the following form. Given an event X, as the degree of focalization associated with any plausible causal object increases, the amount of causality attributed to that object will also increase.

Experimental Tests of the Focus of Attention–Causal Attribution Hypothesis

The unit formation process takes all phenomenological dimensions into account when organizing the stimulus manifold into simpler structures. This implies that the tendency to link any object or event with an effect event in a causal unit is a function of the combined values of those objects and events on the space, time, form/substance, and focalization dimensions. However, in the following research, the values of the various plausible causes are held constant on the time, space, and form/substance dimensions (with one exception). Thus, the research which is discussed is primarily devoted to tests of the hypothesis concerning the role of the dimension of focalization in causal unit formation.

Experiment I

In the original experiment (Duval, 1972) designed to test the focus of attention–causal attribution hypothesis, subjects were presented with a series of five hypothetical situations in which some discriminant change occurred in the environment, that is, an event. The subject was instructed to imagine herself in each particular situation. Each situation also included a second person who performed certain actions. Each subjects was told that either her actions or the actions of the other person could have caused the events in each of the five hypothetical situations. For example, one hypothetical situation read as follows: "Imagine that you pull up behind another car at a stoplight. The light turns green but the car in front of you does not move. As you pull out around the car, the other driver turns right and runs into you. To what degree did your actions cause the accident and to what degree did the actions of the other driver cause the accident?" After exposure to each situation, the subject was asked to estimate the extent to which she and the other person in the situation caused the event in question. Responses were elicited in terms of the percentage of total causality which the subject assigned to herself and to the other person.

Three critical facets of the experiment made a test of the focus of attention–causal attribution hypothesis possible. First, focal event end-products were introduced into the subject's life space by the inclusion of discriminant changes in each hypothetical situation. Second, the subject was told that either she or the other persons in the situations could be the cause of the events. This provided the subject with two objects which could have caused the events in question, objects which were focal by virtue of being pointed out by the experimenter. Third, the degree of focalization associated with the subjects was varied by seating approximately one-half of the subjects in front of the reflecting surface of a large mirror (mirror condition) and the remaining half in front of the nonreflecting back of the mirror (no mirror condition). Since exposure to one's image increases the duration of a person's self-focus (Davis & Brock, 1975), subjects seated in front of the mirror should have been associated with a greater degree of focalization than subjects in the no-mirror condition.

In summary, the stimulus referents of the events in the hypothetical situations were discriminant changes in the preexisting environment and, thus, were focal events. The hypothetical situations represented cases in which the set of plausible causes for the events were two focal objects. The mirror/no-mirror manipulation should have increased the degree of focalization associated with subjects in the mirror condition. Since increasing the degree of focalization associated with an object increases the focal similarity between that object and any event, subjects in the mirror condition should have been more similar to the five events on the focalization dimension than were subjects in the no-mirror condition. According to our earlier hypothesis, as the level of focal similarity between an object and an event increases, the amount of causality attributed to the object should also increase. Thus, subjects in the mirror condition should have attributed more causality for the hypothetical events to self than subjects in the no-mirror condition. This hypothesis received strong confirmation. Mean attribution to self in the mirror condition was 60.10%, while mean attribution to self in the no-mirror condition was 50.86%, $F(1,39) = 14.43, p < .001$. In addition, the desirability of the effect events was varied. For approximately one-half of the subjects in the mirror and no-mirror conditions, the hypothetical events were desirable. The remaining subjects were exposed to situations in which the events were undesirable. Given that the focalization dimension may in many cases outweigh other phenomenological dimensions (for example, good–bad) in the causal unit formation process, subjects in the mirror conditions should have attributed greater causality to self regardless to the affective quality of the event. This expectation was supported by the absence of any mirror/no-mirror × desirable/undesirable event interaction.

Experiment 2

In the initial experiment, it is possible that the subject interpreted the presence of her image in the mirror as a cue to attribute greater causality to self. This is unlikely since precautions were taken to represent the mirror as a prop left over from another experiment. Nevertheless, it seemed reasonable to rule out this possibility by reducing rather than increasing the duration of the subject's self-focus. In order to accomplish a reduction in self-focus, another study was run (Duval & Wicklund, 1973) in which approximately half of the subjects in an experimental situation engaged in motor activity while responding to the causality problems presented by five (in this case, all undesirable) hypothetical events. The other half of the subjects did not participate in any motor activity while responding to the causality problems. Since motor activity reduces the duration (and perhaps the intensity) of self-focus (Duval & Wicklund, 1972), subjects who engaged in motor activity should have been associated with a lesser degree of focalization that were the subjects in the no motor-activity condition. Thus, in this experiment, subjects in the treatment condition should have been *less* similar to the hypothetical events in terms of degree of focalization than were subjects in the no-treatment condition. This leads us to the prediction that subjects in the motor-activity condition would attribute *less* causality to self than would subjects in the no motor–activity condition. This expectation was clearly confirmed. Mean attribution of causality to self in the motor-activity condition was 49.65% compared to a mean of 57.63% in the no motor-activity condition, $F(1,28) = 5.06, p < .05$.

Experiment 3

In interpreting the results of Experiments 1 and 2, we assumed that the experimental manipulations actually affected the subject's duration of focus on self and, thus, the degree of focalization associated with self. A more definitive test of the causal attribution–focus of attention hypothesis would be to present a person with several plausible causes for an event and to control directly the length of time (duration) the person could focus on each of the objects. Under these circumstances, the application of the similarity–causal unit formation principle to the dimensions of focalization (the focus of attention–causal attribution hypothesis) clearly predicts that subjects would attribute causality to the object that was exposed to their focus of attention for the longest period of time. Since this third experiment (Duval, Hensley, & Cook, 1974) has not appeared previously, the procedure will be presented in some detail.

Subjects were ushered into the experimental cubicle and seated before a television monitor. They were separated by barriers which allowed them to see the television screen but not each other, and they were not allowed to communicate after the instructions had begun. The experimenter gave a general description of problematic situations in which any one of several objects could be the cause of some event. After giving examples, the experimenter explained that the study concerned methods and strategies which people use to arrive at conclusions about which of several objects is actually causal in a particular situation.

At this point, the experimenter informed the subjects that the television set located in front of them was hooked into three live television cameras in the next room and that each of the three cameras was focused on one of three motorized disks. She said that at her signal the three disks in the next room would begin rotating and that, shortly thereafter, live pictures of each of the three wheels would be projected onto the TV monitor. In addition, the experimenter indicated that the subjects would see each of the three disks presented separately, that each disk would be presented more than once, and that the order of presentation of the disks would be random. At this point, the experimenter explained that one of the three disks was attached to a sound-producing device that could generate a piercing sound that would be audible over the TV speaker, while the two remaining disks were not attached to anything. She indicated that after the attached disk had rotated for a predetermined number of times, it would cause the sound-producing device to begin generating the piercing sound. The experimenter then went behind a screen, said that she had started the wheels in the other room, and that the experiment had begun.

The subjects saw a videotape which presented a picture of a cardboard disk with a horizontal line drawn across it. This disk was attached to a turntable which rotated at 33.3 rpm. To give the subject the impression that he was seeing three separate disks, a number–1, 2, or 3–was attached to the turntable for each separate presentation of the disk. To enhance the effect that each disk was distinct, we randomly varied the angle at which the disk was videotaped through five distinct positions. This made it appear as though each disk was shot from a camera angle that differed from that of the disk presented immediately before.

In the similar-duration condition, the videotape showed each rotating disk eight times. Each separate presentation lasted for 3 seconds, and each disk was presented for a total of 24 seconds. The order of presentation of the differently labeled disks was randomized except that the disk labeled 2 was never presented immediately before the sound started. The reason for this latter precaution was to eliminate the possibility that subjects might choose Disk 2 as causal because it was presented as temporally contiguous with the presentation of the sound.

In the second condition, the 2 longest condition, the videotape showed a rotating disk labeled 2 eight times, a rotating disk labeled 3 five times, and a rotating disk labeled 1 three times. Each separate presentation lasted 3 seconds. Thus, Disk 2 was presented for a total of 24 seconds, Disk 3 for a total of 15 seconds, and Disk 1 for a total of 9 seconds. (In pretesting, breaking down the total time each disk was presented into 3-second segments and randomizing the order of presentation made it virtually impossible for subjects to discriminate correctly among the three disks in terms of number and total length of presentation.) The order of presentation was again randomized, except that the disk labeled 2 was never presented immediately before the presentation of sound for the reasons previously stated. The sound (which was, in actuality, feedback from a microphone dubbed in over the video) began after the complete presentation of the rotating disks and lasted for a total of 77 seconds.

After the presentation of the disks and sound, subjects were asked to indicate which of the disks was causally related to the sound. Subjects responded by writing down the number of the disk which they thought caused the sound. After these choices were made, the experimenter said, "As you know, we are interested in why people make the choices they do. Please indicate why you chose the disk that you did." This was an attempt to determine whether subjects' choices were in response to any perceived experimenter demand. Responses indicated that no subject was aware of the experimental hypothesis. However, two subjects in the 2 longest condition did indicate an awareness that Disk 2 was presented longest. Data from these subjects were dropped from the analysis. (Inclusion of that data had no effect on the results.)

In the context of the present experimental design, the focus of attention–causal attribution hypothesis suggests that the frequency of subjects' choices of the disk labeled 2 would be greater when Disk 2 was presented for a longer period of time than the disks labeled 3 and 1. Sixty-eight % of the subjects in the 2 longest condition chose Disk 2 as causal while only 38% of the subjects in the similar duration condition chose Disk 2. This difference was reliable by a chi-square analysis, $\chi^2(1) = 4.56, p < .05$ and clearly supports the hypothesis.

Alternative explanations. It still might be argued that the subjects in the 2 longest condition were, at some level, aware that the total length of presentation varied for each disk. However, even granting this assumption, it is not clear that differences in length of presentation imply some clear experimental demands within the context of the cover story. For example, subjects could just as easily have concluded that the experimenter was testing their attentiveness and, thus, wanted them to choose the disk which was presented for the least amount of time. Alternatively, subjects might have concluded that the experiment was designed to show that people tend to choose the moderate values on a continuum anchored by extremes. In this case, subjects would have concluded that the disk presented for the medium length of time was the "correct" disk to choose as causal. Thus, within the context of the experiment, differences in length of presentation, even if noticed by subjects at some level, did not seem to imply a definite choice. This is supported by the fact that no subject indicated that he chose the disk he did because it was presented longer than the other disks.

Experiment 4

To demonstrate further the role which focal consciousness plays in causal attribution, an experiment (Arkin & Duval, 1975) was carried out in the context of actor–observer causal attributions. This context seemed valuable for several reasons. First, the effect event to be explained in the actor–observer situation is a spontaneous action (a person's decision or action) rather than a hypothetical or experimentally contrived effect. Second, dependent measures taken in the actor–observer paradigm clearly represent causal attributions. Third, the causal attributions of actor and observers have been thoroughly documented in the literature. Thus, if the actor–observer attributional phenomenon yields to a unit formation–focus of attention analysis, the experimental generality of this model would be substantially expanded.

In early studies on actors' and observers' attributions, researchers found that actors tend to attribute causality for their actions to external factors while observers tend to attribute causality for the same actions to dispositional characteristics of the actor. A focus of attention–causal attribution analysis of these findings suggests that these biases must be due in part to a tendency for actors to focus attention on their external environment to a greater extent than on themselves while observers tend to focus attention on the actor to a greater extent than on the actor's environment. Temporarily deferring the question concerning the origin of these biases, our analysis suggests that the usual actor–observer attributions could be reversed by introducing a stimulus which tends to reverse actors' and observers' usual loci of attention. Proceeding on this assumption, actors and observers were run in two sets of conditions. One set represented the standard experimental setting for actor–observer attributional research. Actors were exposed to external information, in this case, art objects, and were asked to determine which art object they considered to be best. Observers were instructed simply to watch the events as they transpired. The second set of conditions replicated this procedure but included an additional stimulus–a TV camera pointed at the actor. The expected effect of this stimulus was to reverse the usual actor–observer attentional bias while at the same time making sure that the actor and observer were exposed to the same experimental stimuli. For example, as far as the actor was concerned, the TV camera should have increased his tendency to focus on self rather than on the external environment, the opposite of the postulated actor-attentional bias under normal circumstances. For the observer, the TV camera should have been a novel stimulus in the actor's external environment. Since attention is attracted to novel stimuli (Berlyne, 1958: Caron & Caron, 1968; Sokolov, 1960), the observer's focus of attention should have been pulled toward the camera and away from the actor. This would obviously produce a tendency for the observer to focus attention on the actor's external environment rather than upon the actor, the opposite of the usual observer-attentional bias. Under these conditions and assuming that the focus of attention–causal attribution hypothesis is valid, we predicted that the between-condition differences in the direction and duration of actors' and observers' focal consciousness would produce concomitant differences in actors' and observers' causal attributions. In order to test this hypothesis, we subtracted the degree of causality which actors and observers attributed to the actor's external environment from the degree of causality actors and observers attributed to the actor (Nine-point scales were used to measure attributions and were anchored by "completely" at the ninth point and "none at all" at the first.) The difference scores obtained from this analysis clearly confirmed the interaction hypothesis. Under the usual actor–observer conditions, i.e., (the no-camera conditions) observers attributed more causality to the actor, $\bar{X} = +2.99$, than did actors, $\bar{X} = +2.07$. In the camera conditions, this pattern was reversed. Observers attributed less causality to the actor, $\bar{X} = +1.20$, than did actors, $\bar{X} = +3.75$, the actor–observer × camera–no-camera interaction reached significance, $F(1,104) = 5.53, p < .02$.

The data from the previous experiment show that the normal biases in actor–observer attributions can be reversed by introducing a stimulus which reverses the usual direction and duration of actors' and observers' foci of attention. Storms (1973) also reversed normal actor–observer attributional biases through use of manipulations which we assume reversed actors' and observers' foci of attention. However, Storms interpreted the reversal he obtained as a visual perspective phenomenon rather than an effect of unit formation on the focalization dimension. That the reversal of actors' and observers' usual causal attributions does not depend on visual perspective, which implies an information-processing model of attribution, has been demonstrated by Blanche (1975). He found that actors and observers who were blindfolded during an actor–observer attribution experiment and run in camera and no-camera conditions made causal attributions which exactly paralleled the interactional pattern of attributions made by actor and observer subjects in the Arkin and Duval and Storms studies. (We might add parenthetically that the actors' and observers' attributions in the Blanche (1975) study were primarily based on sounds which occurred during the experiment. This indicates the focal consciousness is not an artifact of the visual sensory apparatus, but is a more central structure which affects stimuli received via any of the senses.)

The original actor–observer attributional biases found by Jones and Nisbett (1971) may be reinterpreted accordingly. In the earlier analysis we suggest that actor–observer attributional biases are due to the tendency for actors to focus on their external environment and for observers to focus on the actor. But what factors would bring about such attentional biases in the first place? On the one hand, it is fairly obvious that actors are forced to negotiate their external environment and, thus, must direct attention toward that environment. As the focus of attention–causal attribution hypothesis predicts, this tendency to focus externally will produce external attributions. However, discounting the fact that subjects who play the role of observers in experiments are usually instructed to pay attention to the actor, the reasons for the observer's tendency to focus on the actor rather than on the actor's environment are more complicated. One possible solution is to assume that the observer's focus of attention is attracted to the actor rather than to the actor's external environment because the actor is normally more novel in the sense of being more dynamic (that is, changing over time) than is the environment in which he appears. Data supporting this hypothesis (Arkin, Duval, & Hensley, 1975) was obtained by varying the dynamism of a confederate who played the part of an actor. In one condition (actor stable), subjects were shown a videotape of a person judging art works and making notes by speaking into a tape recorder. In the second condition (actor dynamic), subjects were shown a videotape of the same person judging the same art works. However, in this case, the actor took notes by writing on a note pad at intervals which averaged approximately 20 seconds. The differences between the conditions were intentionally understated to counteract informational explanations, but the actor who took notes by writing was clearly more dynamic than the actor who took notes by speaking into a tape recorder. Evidence was found in this study that observers attributed more causality to the actor than to the actor's external environment when the actor was dynamic (mean actor minus external environment attribution equals +23.33%). When the actor was more stable, the observers' tendency to favor internal attributions over attributions to the actor's external environment (mean actor minus external environment attribution equals +11.71%) was significantly reduced, $t(66) = 2.08$, $p < .05$; attributions were elicited in terms of percentage of total causality). As the data indicate, the focus of attention–causal attribution hypothesis appears to account for the essential aspects of actor and observer attributions and, more importantly, to predict accurately some of the major variables which alter those attributions. Whether this hypothesis, in combination with the influence of other phenomenological dimensions, can account for all other variables which have been shown to alter actor–observer attributions (for example, violation of expectation, choice, etc.) remains to be seen.

Taken together, the data from the series of experiments discussed above are substantial support for the focus of attention–causal attribution hypothesis in particular and for the similarity–unit formation model of causal attribution in general. At this point, however, we must raise an important issue. Are theorists of causal attribution limited to predicting how individuals will structure the world cognitively, or do their theories also have implications for attitudes and behaviors? In the case of the unit formation theory of causal attribution, it seems plausible that the joining of previously separate objects and events into elements of a cause–effect unit has substantial implications for a person's future attitudes and behaviors. The following is a discussion of these implications.

As noted earlier, the invariant relationship between elements of any unit is one of belonging or being connected. Turning the key of an automobile "belongs" to the starting of the car. Speaking harsh words to people belongs to the anger which appears in their faces. We have illustrated the nature of this relationship by showing that the mention of one of the two elements of a causal unit calls the other element into cognition. This is what is commonly meant by the term causal association. When objects or events are causally associated, the occurrence, mention, or thought of one tends to generate the expectation or cognition of the other. The existence of this phenomenon in itself means that causal attribution has substantial implications for a person's attitudes and behaviors. When a loud roar is heard in the garden, one expects a very large, angry animal to appear and adjusts one's behavior accordingly. We believe, however, that there is more to the unit formation notion of belonging than expectation of occurrence.

As previously stated, the basis of unit formation is simplification of the stimulus manifold. Given this as the basic motive for the causal unit formation process, the end-products of this process might be expected to reflect the establishment of some degree of identity between the elements involved in the unit. Heider, for example, has pointed out that estimates of a person's abilities are directly related to the kind and magnitude of the effects which that person's actions cause within the environment (provided that the effects are attributed solely to the person). A man who receives a $100,000 return on a $1,000 investment is perceived as having more ability to make financial investments than a man who receives $2,000 for the same amount of initial investment. A mass murderer is perceived as more evil or deranged than a man who kills only one person. Since in both cases, the people and the effects merely represent individual objects and events, the tendency for effects to color the impression of the person who caused the effects suggests that causal attribution brings the cause and effect elements together in a relationship of oneness or unity. This conclusion can be stated in a more general hypothesis. To the extent that a person connects an event, A, and a second event or object, B, in a causal unit, A and B will tend to become equivalent elements within the person's subjective life space.

In order to test this general proposition, we constructed experimental conditions to result or not result in the formation of a causal unit between an object (in the two experiments reported, the object is the subject's self) and some event. Subjects' behaviors toward the event were then measured. As will become apparent, subjects who were led to attribute causality for the event to self behaved quite differently than subjects who did not connect self and the event in a causal unit.

If our analysis of the nature of a causal unit is correct, leading a person to attribute any event to self should cause self and the event to come to belong to each other. The behavioral consequences of this linkage seem to depend on the desirability of the event. If an undesirable event is causally linked to self–the object–self will, in essence, acquire the negative characteristics of the event. In all probability, these characteristics would be at variance with the person's standards of correctness. In accordance with the earlier model of self-evaluation (Duval & Wicklund, 1972), this new discrepancy between the real self and standards of correctness will produce negative affect. The person should then attempt to eliminate the negative affect by altering the self in the direction of the standards of correctness. When the aspects of self in violation of the standards of correctness are the aspects of an undesirable event causally linked to self, the person can reduce the intraself discrepancy by eliminating the existence of the event. This hypothesis, tested in the following experiment, can be stated more formally. To the extent that a person attributes causality for an undesirable event to self, the person's desire to act in ways which will eliminate the occurrence of the event will increase.

Experiment 5

In Experiment 5 (Duval, Hensley, & Neely, 1975), the effect event was the epidemic spread of venereal disease (VD). To introduce this as an event into the subject's life space, we exposed subjects to a videotape which detailed the unusual consequences and scope of the VD epidemic. The objects introduced as plausible causes were the subjects themselves as members of the general public and the victims of VD. These seemed to be plausible causes because the VD epidemic could in fact have been due to apathy and ignorance on the part of the general public, including the subjects, or inappropriate behavior on the part of the VD victims themselves. In order to vary the extent to which subjects would attribute causality for the VD epidemic to self, they were first told that the experiment was being submitted for a grant, that the funding agency required videotapes of each experimental session, and that they should not be surprised if their live image appeared on the TV monitor at some point during the experiment. Subjects were then exposed to a live image of themselves on a TV monitor either 4 minutes before, immediately before, immediately after, or 4 minutes after exposure to the videotape on VD. (A control group was also run. Subjects in this condition were exposed to the videotape on VD but were never exposed to their image on the TV monitor.) Since decreasing the time interval between the acts of focusing attention on self and on the event would increase the tendency to link self and the event in a causal unit, subjects in the immediately before and after conditions should attribute greater causality for the VD epidemic to self than subjects in the 4-minute before and after conditions. Our hypothesis concerning attribution of an undesirable event to self and desire to eliminate that event leads to the prediction that subjects in the immediately before and after conditions should express greater willingness to act in ways which would tend to eliminate VD than subjects in the 4-minute before and after conditions. (We have argued that temporal proximity is an important factor in the formation of cause and effect unit relationships. At first glance, it might seem that the significant time interval which affects causal attribution is solely the time interval which separates the actual occurrence of events in the external world. However, the focus of attention–causal attribution hypothesis postulates that the formation of cause–effect relationships depends not only on the objective occurrence and existence of events and objects in the external world but also on the person's acts of focusing attention on those events and objects. Given that these acts of focusing attention are considered crucial in the model, it seems reasonable to assume that the time interval which affects causal attribution is the time interval which separates the acts of focusing attention on separate objects and events. In some cases, this time interval will be determined by and will correspond to the time interval which actually separates occurrences, as in observing and focusing on one billiard ball striking another. In other cases, it will not. The experiment involving the attribution of causality for VD represents the latter case since the time interval which separated subjects' acts of focusing on themselves (exposure to their images on TV) and their acts of focusing on the event (exposure to the videotape on VD) was experimentally manipulated and clearly does not correspond to the time interval which actually separated the subjects and the event. Nevertheless, we would argue that a manipulated time interval is as effective in producing causal attributions as the time interval which actually separates the events.)

To provide a check on the manipulation of attribution to self and to measure willingness to act in ways which would eliminate VD, we asked subjects to fill out a questionnaire 5 minutes after exposure to the tape on VD. This questionnaire included the following items: To what extent would you say that the failure to control the current VD epidemic is due to the public's, including your own, lack of concern rather than to the VD victims themselves? To what extent would you be willing to take a 1-hour course dealing with the prevention and treatment of VD and then conduct two half-hour educational sessions with groups of University of Southern California (USC) freshmen? A VD clinic exists in the USC community but is underfinanced and greatly understaffed. How much time per week would you be willing to spend doing volunteer work at the clinic? How large a financial contribution would you be willing to make to the National Center for the Prevention of VD? All questions were answered on 15-point scales anchored at both ends with the appropriate positive and negative labels.

Our analysis of the data obtained from this questionnaire indicated that subjects in the immediately before and after conditions did attribute greater causality to self than did subjects in the 4-minute before and after conditions, $F(1,41) = 5.01$, $p < .05$; means presented in Table 1. Thus, it appears that the manipulation of attribution to self was successful. In terms of the equivalency hypothesis, the creation of a unit linkage between the subject's self and the experimental event should have resulted in negative self–evaluation. This negative self–evaluation should have then motivated the person to act in ways that would tend to eliminate the occurrence of the event. Thus, given that subjects attributed greater causality for VD to self in the immediately before and after conditions, we expected that subjects in those conditions would also show a greater willingness to contribute time and money toward the prevention of VD. As can be seen from Table 2, this prediction was confirmed, $F(1,41) = 6.41$, $p < .05$. In a conceptual replication of this experiment, we found that the obtained effects were not an artifact of the use of VD as the event or the use of the

TABLE 1
Mean Attribution of Responsibility for VD to Self

Condition	No interval	4-minute interval
Before videotape	9.0	6.9
After videotape	10.27	7.9
Control	6.5	–

Note. The larger the number, the greater the attribution to self. $N = 11$ in each condition.

subject's image to induce self-focus and self-attribution. Subjects who filled out an innocuous biographical questionnaire immediately before being exposed to a videotape on poverty in Latin America attributed greater causality for that event to self and indicated a greater willingness to contribute time and money to eliminate poverty in Latin America than did subjects who filled out a biographical questionnaire 4 minutes prior to exposure to the videotape.

Alternative explanations and issues. It might be argued that putting the subject's image on the TV screen immediately before or after the presentation of the videotape caused the subject to suspect that the experimenter wanted her to feel personally responsible for the VD epidemic. There are several features of the experimental design which can be used in an argument against this interpretation. First, at the beginning of the experiment, the subject was told not to be surprised if her image appeared on the TV monitor during the course of the experiment. She was also given a reason why this would occur. If the subjects believed this cover story (and there was no indication in debriefing that they did not), the experimenter's motives in turning on the camera were accounted for entirely. If the subject was not suspicious about the appearance of her image on the TV screen, why would she assume that the experimenter had some ulterior

TABLE 2
Subjects' Mean Desire to Act to Eliminate VD

Action	No interval before	No interval after	4-minute interval before	4-minute interval after	Control
Take 1-hour course	11.2	11.2	8.5	7.9	7.0
Volunteer work at clinic	7.6	8.5	7.6	6.0	8.2
Personal contribution	6.9	7.7	3.3	6.6	5.2
Combined	25.7	27.8	19.4	20.5	20.4

Note. Numbers represent means for each condition. The higher the number, the greater is the desire to engage in the various acts.

motive? We believe that the subject did not harbor any such suspicions. Second, each subject saw her image on the TV screen at some point during the experiment. If the simple presentation of the subject's image in the context of a videotape presentation about VD was sufficient to induce suspicion about the experimenter's motives, all subjects should have given approximately the same responses, which of course, they did not. Thus, to argue that the subjects' responses in the immediately before and after conditions were a result of perceived experimental demand, it would be necessary to establish that the length of the time interval between presentation of the subject's image and presentation of the event indicated something to the subject about the experimenter's motives. However, to argue this position (that juxtaposition of the subject's image with material related to the VD epidemic produced experimental demand) ignores the results in the 4-minute after condition. Subjects in that condition saw their image immediately before filling out the questionnaire. If subjects were susceptible to the belief, that "the experimenter is trying to get me to feel personal responsibility for VD by putting my face on TV," it seems reasonable that the experimenter's introduction of the subjects' images immediately before the questionnaire was presented would be just as likely to make subjects suspicious about the motives of the experimenter as would introducing their images immediately before or after the VD videotape. Since subjects in the 4-minute after condition did not differ from subjects in the control group, we must assume that juxtaposing the subjects' images with material related to VD did not generate experimental demand. This implies that the effects of juxtaposing the subject's image and the videotape in the immediately before and after conditions were not due to some experimental demand induced by the mere fact of juxtaposition, but reflect the genuine effects of a causal unit formation between the person and the undesriable event. The results in the 4-minute after condition also can be used to argue against the possibility that subjects' responses reflected some type of self-evaluation.

One of the major questions which will probably be raised about these studies is whether people really do see themselves as partial causes of events such as poverty in Latin American and the VD epidemic. To determine if subjects' assumptions of causal responsibility in the experiments is present only in the laboratory, we administered to over 500 undergraduates at USC a measure of perceived personal causality for poverty in Latin America, included in a 22-item questionnaire. (A direct question concerning personal causality for VD was considered inappropriate for a survey questionnaire.) Fully 53% of this sample stated that they felt causally related to poverty in Latin America to some degree. Thus, people apparently do assign causality for events like poverty in Latin America and presumably VD to self outside as well as in the laboratory regardless of the reasonableness of the attribution.

The foregoing analysis and predictions concerning the behavioral consequences which attend attribution of an undesirable event to self were predicated on the assumption that elements of a causal unit become equivalent elements within the person's subjective life space. Applied to the case in which the event attributed to self is neutral–to–positive on the desirability dimension, this assumption implies that the person will attempt to maintain an event causally associated with self. This derivation is based on two factors. Even if the event is neutral on the desirability dimension, that event becomes intrinsically linked to self as a consequence of the causal attribution process. Since we assume that people normally attempt to maintain all aspects of self which do not violate their standards of correctness, we would expect a person to attempt to maintain the existence of any neutral–to–positive event which is causally linked to self. This hypothesis was confirmed by Mayer, Hensley, and Duval (1975) who showed that subjects exposed to their images on a TV monitor immediately before making a decision tended to exhibit less willingness to change their decisions, $\overline{X} = 5.00$, than did subjects not exposed to their image, $\overline{X} = 6.52$, $t(37) = 2.01$, $p < .05$; the higher the mean, the greater the willingness to alter the decision).

This effect was obtained even though subjects in both conditions were told that they were being videotaped during the entire experiment, and a TV camera was pointed at all subjects. Thus, the only difference between the two conditions in terms of experimental stimuli was that subjects in the camera-image condition saw their image immediately before making their decision while subjects in the camera/no-image condition did not. Subjects were also given the same rationale for being videotaped as in the VD study. These facts argue against attributing the effect of the image/no-image manipulation on commitment to some vague manipulation of the public–private dimension. In one of the major theories of commitment, Kiesler (1971) suggests that the degree of perceived choice associated with making a decision is the primary variable which determines the extent of the person's commitment. Greater perceived choice produces greater commitment. To determine whether the camera image/camera no-image manipulation affected commitment by altering perceived choice, we asked subjects in the above experiment to indicate how much choice they felt they had in making their decision. The results of this analysis indicated that subjects in the camera/image condition felt *less* choice in making their decision, $\overline{X} = 7.14$, than subjects in the camera/no-image condition, $\overline{X} = 8.02$. Although this difference was not significant, it was in a direction opposite to a perceived-choice interpretation of the results and apparently rules out that particular alternative explanation for the commitment data. Furthermore, we would suggest that perceived choice may be a variable which significantly affects causal attributions. Under conditions of high-perceived choice, the major plausible cause for the decision is the person himself, since the external environment exerts little or no pressure on the person to decide one way or the other. Under conditions of low-perceived choice, the external environment exerts considerable pressure on the person to decide to do X, and thus becomes a plausible cause for the decision. Thus, it is possible that the effect of choice on commitment

(Kiesler, 1972) is due to increasing attribution for the decision to self in high-choice conditions and decreasing attribution for the decision to self in low-choice conditions.

SUMMARY AND IMPLICATIONS

Our interpretation of the unit formation model of causal attribution suggests that the phenomenological characteristics of events and objects determine the outcome of the causal attribution process. We have concentrated on demonstrating that the subsystems of consciousness contribute phenomenological dimensions to objects and events. Also, we attempted to show that variations along the dimension of focalization can determine which of several plausible causal objects or events is seen as the cause of a given effect event when the positions of those objects and events are held constant on other phenomenological dimensions. Specifically, data were presented which indicate that the greater the degree of focalization associated with an object (and, thus, the greater the focal similarity between the object and event), the greater the tendency to attribute the cause of the effect event to that object. In addition, we suggested that, because the objects and events of a causal unit formation become equivalent elements within a person's subjective life space, causal attributions do have substantial implications for the person's future behavior. The data presented support this hypothesis with regard to the behavioral ramifications which attend attribution of undesirable and neutral–to–desirable events to self.

Experimental work on the general unit formation model of causal attribution, the role of the focal subsystem in the model, and the implications of the equivalency of elements within a causal unit, has just begun. Consequently, many problems remain to be solved. In terms of the general model, we must investigate the manner in which the various phenomenological dimensions of time, space, form/substance, and focalization interact to determine causal attribution. Whether other components of the mediational complex, such as the affective system, contribute to the outcome of the causal attribution process is another problem under investigation (Hensley, 1975). The following are crucial issues which concern the role of focal consciousness in the model: (a) a demonstration that focal similarity affects causal attribution when the set of plausible causes consists of only events or of both objects and events; (b) the development of methods for measuring the degree of focalization associated with objects and events in order to confirm more directly the similarity hypothesis; (c) the manipulation of intensity of focalization, as well as incidence, directionality, and duration of focalization; and, (d) an integration of theory and research pertaining to variables which determine the incidence, directionality, duration, and intensity of the focus of attention. A solution to this latter problem would allow us to predict beforehand the degree of focalization associated with various objects and events, thus providing a basis for predicting the outcome of the causal attribution process in terms of focal similarity.

In conclusion, the significance of the causal attribution phenomenon depends on the overall role which causal attribution plays in human behavior. An increasing amount of experimental data indicate that the energy which moves the human organism to action is, as Lewin (1936) suggested, scalar rather than vectored. This means that motivational energy impels a person to act, but this energy in no way implies the direction or content of that action. Happily, this hypothesis is now receiving serious consideration. Bandura (1973), among others, has mounted a persuasive attack against the notion that a particular type of energy compels a person to carry out acts of aggression. Valenstein (1969, 1973) questions whether behaviors, such as eating, drinking, and sexual activity, which have long been considered manifestations of instinctual and, thus, vectored motivational forces, represent anything other than the arousal of prepotent response centers. Viewing motivational energy as scalar implies that there is some system which provides the direction and the content (that is, form/substance) of behavior. Lewin proposed that this is the system which determines "means–ends" (cause and effect). In other words, it may be that causal attribution is the system which dictates the direction and content of behavior by locating the cause of an increase or decrease in tension (negative or positive affect) in one specific area of the environment. Thus, a person who experiences negative scalar energy may attempt to diminish or eliminate a certain object or event to which he has attributed the negative affect. The experience of positive scalar energy may result in the attempt to maintain the existence of a certain object or event because the causal attribution system locates the positive affect in that specific area of the environment. If motivation and causal attribution are functionally separate systems, understanding and predicting any behavior would require an analysis of both motivational and attributional processes. In consideration of the importance of this possible role of causal attribution, we have attempted to direct attention to a potentially comprehensive and integrative theory of this process and to expand its predictive power.

REFERENCES

Arkin, R., & Duval, S. Focus of attention and causal attribution of actors and observers. *Journal of Experimental Social Psychology,* 1975, **11,** 427–438.

Arkin, R., Duval S., & Hensley, V. Attributions of observers as a function of the dynamicity of the actor. Unpublished manuscript, University of Southern California, 1975.

Bandura, A. *Aggression: A social learning analysis.* Englewood Cliffs, New Jersey: Prentice–Hall, 1973.

Berlyne, D. E. The influence of complexity and novelty in visual figures on orienting responses. *Journal of Experimental Psychology,* 1958, **55,** 287–296.

Blanche, J. Effects of focus of attention, visual and nonvisual perspectives on causal attributions of actors and observers. Unpublished doctoral dissertation, University of Southern California, 1975.

Caron, R. F., & Caron, A. J. The effects of repeated exposure and stimulus complexity on visual fixation in infants. *Psychonomic Science,* 1968, **10,** 207–208.

Davis, D., & Brock, T. C. Use of first person pronouns as a function of increased objective self awareness and prior feedback. *Journal of Experimental Social Psychology,* 1975, in press.

Duval, S. Causal attribution as a function of focus of attention. In S. Duval & R. A. Wicklund, *A theory of objective self awareness.* New York: Academic Press, 1972. Also in Duval, S. & Wicklund, R. A. Effects of objective self awareness on attribution of causality. *Journal of Experimental Social Psychology,* 1973, **9,** 171–31.

Duval, S., Hensley, V., & Neely, R. Attribution of an event to self and helping behavior as a function of contiguous vs. noncontiguous presentation of self and event. Unpublished manuscript, University of Southern California, 1975.

Duval, S., & Wicklund, R. A. *A theory of objective self awareness.* New York: Academic Press, 1972.

Duval, S., & Wicklund, R. A. Effects of objective self awareness on attribution of causality. *Journal of Experimental Social Psychology,* 1973, **9,** 17–31.

Heider, F. *The psychology of interpersonal relations.* New York: John Wiley & Sons, 1958.

Heider, F. Social perception and phenomenal causality. *Psychological Review,* 1944, **51,** 358–374.

Hensley, V. Preliminary tests of the effects of the affective system and sensory modality on causal attribution. Unpublished manuscript, University of Southern California, 1975.

Jones, E. E., & Nisbett, R. E. *The actor and the observer: Divergent perceptions of the causes of behavior.* Morristown, New Jersey: General Learning Press, 1971.

Kiesler, C. A. *Psychology of commitment.* New York: Academic Press, 1971.

Koffka, K. *Principles of gestalt psychology.* New York: Harcourt, Brace, 1935.

Lewin, K. *A dynamic theory of personality.* New York: McGraw Hill, 1936.

Lynn, R. *Attention, arousal, and the orientation response.* New York: Pergamon Press, 1966.

Mayer, S., Hensley, V., & Duval, S. Causality and commitment. Unpublished manuscript, University of Southern California, 1975.

Michotte, A. *The perception of causality.* New York: Basic Books, 1963.

Piaget, J. *The mechanisms of perception.* New York: Basic Books, 1969.

Sokolov, M. D. Neuronal models and the orienting reflex. In M. A. Brazier (Ed.), *The central nervous systems and behavior.* New York: Josiah Macy, 1960.

Storms, M. D. Videotape and the attribution process: Reversing actors' and observers' points of view. *Journal of Personality and Social Psychology,* 1973, **27,** 165–175.

Valenstein, E. S. *Brain stimulation and motivation.* Glenview, Illinois: Scott, Foresman, 1973.

Valenstein, E. S. Behavior elicited by hypothalmic stimulation: A prepotency hypothesis. *Brain, Behavior, and Evolution,* 1969, **2,** 295–316.

Wertheimer, M. *Principles of perceptual organization.* In D. C. Beardslee & M. Wertheimer (Eds.), *Readings in perception.* Princeton, New Jersey: Van Nostrand, 1958.

9

Dissonance and the Attribution Process

Mark P. Zanna
University of Waterloo

Joel Cooper
Princeton University

This is a tale of two theories. On the one hand, there is cognitive dissonance; on the other, there is attribution. Dissonance theory sees man as aroused by inconsistencies among his cognitions. As in the classical arousal theories of experimental psychology (Hull, 1943), man is viewed as motivated to rid himself of the drive-like, uncomfortable tension that accompanies perceived inconsistency. Attribution models, on the other hand, see man as in a constant process of making sense out of his environment. In such models, man is viewed as a scientist, using attributional rules to infer causality in an otherwise chaotic world of social stimuli.

At first, the two approaches had no difficulty in peaceful coexistence. Attributional models (for example, Heider, 1958; Jones & Davis, 1965) concentrated on how we make sense of the actions of others, while dissonance theory was concerned with the somewhat unusual ways that we go about putting our own cognitive houses in order. A virtual state of war, however, broke out between the two theories when it was proposed by theorists who later became recognized as attribution proponents that the rules that had been developed for understanding the environment could be redirected at oneself (Bem, 1965, 1972). If we considered ourselves to be observers of our own actions and applied the rules to ourselves, then many of the effects that had been thought of as deriving from the drive-like process of cognitive dissonance could be explained using the nonmotivational, nonarousal, non-drive-like model of attribution. Proponents of each model set out on fascinating forays to credit and discredit the other point of view (for example, Jones, Linder, Kiesler, Zanna, & Brehm, 1968; Bem, 1968).

In this chapter, we will attempt to shed some light on the controversy while at the same time proposing a truce between the two models. Utilizing recent research, we will suggest that the ways in which we use attributional rules can augment rather than conflict with our understanding of motivational systems such as dissonance. One of the weaknesses of a system which posits arousal as an intervening mechanism is that it is difficult to measure such a construct. Attribution theory can be extremely useful in shedding light on this elusive but crucial construct in the theory of cognitive dissonance.

AROUSAL: AN "AS IF" QUESTION

Festinger's (1957) original assumption that dissonance was arousing was rarely the subject of early investigations. Treating people "as if" they experienced arousal enabled creative researchers to evolve interesting predictions about human behavior. For example, in their classic study of induced compliance, Festinger and Carlsmith (1959) predicted that people who made statements discrepant with their attitudes (that is, engaged in counterattitudinal advocacy) would experience dissonance. Moreover, they hypothesized that people who had less of a reason for making a statement contrary to their beliefs would suffer from dissonance arousal to a greater extent than people who had a better reason. Therefore, they predicted that the greater the suffering, the greater would be the cognitive change needed to reduce the suffering.

In Festinger and Carlsmith's particular study, people who engaged in counterattitudinal advocacy changed their beliefs in order to make them coincide with their public statements. And, such belief change occurred more when they had been offered 1 dollar as an incentive than when the incentive was 20 dollars. Again, the internal mechanism which Festinger and Carlsmith relied upon to make their prediction was the uncomfortable arousal state of the participants. However, no direct measurement of the "arousal" was undertaken.

EARLY MEASUREMENT ATTEMPTS

The first attempts to ascertain, in an indirect way, whether dissonance has arousing properties were published simultaneously by Waterman and Katkin (1967) and by Cottrell and Wack (1967). The paradigm that they established was borrowed from the literature on drive states and learning. Spence, Farber, and McFann (1956) had shown that arousal-producing drive states such as hunger have the effect of energizing dominant, well-learned responses. In general, the correct answer in a complex learning task is not as dominant as the correct answer in a simple task. Energizing the dominant response, therefore, would have the effect of facilitating performance on simple tasks but would interfere with performance on more complex tasks.

Waterman and Katkin (1967) had subjects write an attitude discrepant statement after which they were given either a difficult or a simple learning task to perform. The results were only partially in line with Waterman and Katkin's predictions. Simple learning was facilitated for subjects who had written counterattitudinal essays. However, there was no interference with complex learning. Pallack and Pittman (1972) reviewed the evidence utilizing this procedure (for example, Cottrell & Wack, 1967; Waterman & Katkin, 1967; Waterman, 1969) and pointed out that none of the studies ever provided independent evidence that dissonance was aroused. That is, no attitude change measures were ever taken to demonstrate that the particular attitude discrepant tasks could produce self-justificatory opinion change. One study reported by Pallak and Pittman did establish that dissonance-produced attitude change *and* learning interference occurred. Even in that experiment, however, the conditions necessary to demonstrate the facilitation of simple learning were not included. More recently, Cottrell and his colleagues (Cottrell, Rajecki, & Smith, 1974) have looked for energizing effects in the free-choice paradigm. In this context, subjects did tend to emit more dominant responses on an irrelevant task after having made a dissonance-producing decision $p < .10$.

In summary, while the energizing notion was an ingenious idea for studying the arousal properties of dissonance, the results have been less than conclusive. It seemed desirable to find a different way of providing evidence that dissonance is arousing.

TAKING THE PILL: EVIDENCE THAT DISSONANCE IS AROUSING

In 1974, Zanna and Cooper proposed that an attributional framework could be applied to the question of dissonance arousal. In that study, we suggested that the attributional approach of Schachter and Singer (1962) could be utilized to gain evidence on the problem. In Schachter and Singer's view, an emotion is a function of autonomic arousal and a cognitive label. The arousal is necessary in order to experience the emotion, but the available social cues allow us to deduce which emotion we are experiencing. Several years later, it became clear that emotions could be altered if the arousal which people experienced were mistakenly attributed to an external agent. For instance, Ross, Rodin, and Zimbardo (1969) aroused subjects by making them anticipate undergoing electric shock. Some participants, led to believe that their arousal was due not only to their anticipation of the shocks but also to an extraneous loud noise, reduced their attributions of fear and were able to tolerate more shock.

Storms and Nisbett (1970) similarly reasoned that misattribution could be used to help insomniacs fall asleep more readily. Their subjects believed that they were participating in a "drugs and fantasy" study and were instructed to take a drug which was alleged to have an effect on their fantasy productions. Moreover, the drugs were alleged to have side effects. Some of Storms and Nisbett's subjects believed that the drug would make them tense and aroused while it was taking effect; others thought it would make them relaxed, while still others believed there would be no side effect whatsoever. Storms and Nisbett reasoned that since insomnia is reported to be an uncomfortable state of arousal, then the possibility of attributing the arousal not to insomnia but to a drug should reduce the degree of insomnia. In fact, subjects who expected the pill to have arousing side effects reported that they were able to fall asleep nearly 15 minutes sooner than they customarily did. Conversely, subjects who felt they should be relaxed as a consequence of their pill, but who actually felt aroused, took even longer to fall asleep, presumably because they inferred they were more aroused than usual.

In our 1974 study, we established conditions that typically produce dissonance-related attitude change. Volunteer subjects were asked to write attitude-discrepant statements under choice or no choice conditions. According to the body of research in dissonance theory, attitude-discrepant behavior, when engaged in freely, should produce opinion change in the direction of the behavior but should not lead to opinion change under conditions of no choice (for example, Linder, Cooper, & Jones, 1967). If dissonance is truly a state of arousal and if attitude change following induced compliance is based upon the reduction of that arousal, then it follows that misattribution of the arousal to an external agent should reduce the need for attitude change. Just as Storms and Nisbett were able to convince insomniacs that their arousal was due to a drug they had taken, we proposed to convince people who were "suffering" from dissonance arousal that their tension was due to a little white capsule. People who feel that their arousal is due to a drug should not be motivated to undergo attitude-change processes in order to reduce that arousal.

A similar logic led up to predict that if subjects believed they should feel sedated by a drug, yet still felt arousal after writing the counterattitudinal essay, they should be motivated to alter their attitudes even more than subjects who had no such expectation. This followed from Storms and Nisbett's finding that insomniacs who believed they had taken a tranquilizer took even longer to fall asleep than insomniacs who had no expectation.

In order to test our hypotheses, we established a 3 X 2 factorial experiment. Subjects volunteered for a study on memory and were asked to ingest a drug which was to have an effect on their short-term memory. After receiving assurances that the drugs were perfectly harmless, they were told that the drugs might have certain mild side effects during the 30 minutes which would be needed for total absorption. After taking a standard free-recall test, each subject was placed in an individual cubicle, given a capsule which actually contained milk powder, and a "description" of the drug they were to take. For example, in the arousal condition the subjects were told: "This M.C. 5771 capsule contains chemical elements that are more soluble than other parts of the compound. In this form of the drug these elements may produce a reaction of tenseness prior to the total absorption of the drug, 5 minutes after ingestion. This side effect will disappear within 30 minutes."

In the relaxation condition, the word "tenseness" was replaced with the word "relaxation." In a no side effects condition, subjects were merely told that the total absorption time of the drug was 30 minutes and that there would be no side effect.

While subjects were waiting for the drug to take effect, the experimenter announced that she had "another study going on, not about memory, but about opinion research." Half of the subjects were asked if they would be willing to participate, while the other half received no option but to participate.

In the "other study," the students' task was to write a strong and forceful essay taking a position with which they were known to disagree. In this study, they were asked to take the position that potentially inflammatory speakers should be banned from college campuses. At the end of the study, subjects' attitudes regarding inflammatory speakers were assessed.

The no side effects conditions of the study provided a test of whether attitudes had changed in line with dissonance theory predictions. The uncomfortable tension state should be greatest when attitude-discrepant behavior is engaged in freely. Therefore, subjects in the no side effects/choice condition were expected to show dissonance-reducing attitude change, whereas no side effects/low-choice subjects should have been more adamant in their opposition to bans on inflammatory speakers. The results of the study are presented in Table 1. As can be seen from the table, the typical dissonance effect was alive and well in the study.

The crucial question was, what would happen to the dissonance effect if subjects were permitted to believe that their arousal was due to M.C. 5771? We can see from Table 1 that the attribution of the arousal to the pill effectively eliminated any need for attitude change. Subjects in the high-choice/arousal condition did not change their attitudes at all. Moreover, when subjects were led to believe that the pill was supposed to relax them, they apparently felt more upset than ever by the inconsistency among their cognitions. Subjects in the high-choice/relaxation condition changed their attitudes more than subjects in any other condition of the study.

The results of our first study led us to the conclusion that, as Festinger originally postulated, dissonance does have arousal properties. Any other view of attitude change following attitude-discrepant behavior does not lead to the prediction that misattribution would eliminate the opinion change. Yet our results have provoked at least as many questions as they have answered. First, we

TABLE 1
Mean Attitude Scores toward Banning Speakers on Campus[a]

Decision freedom	Potential side effect of the drug: Arousal	None	Relaxation
High choice	3.40_a	9.10_b	13.40_c
Low choice	3.50_a	4.50_a	4.70_a

Note. N = 10 subjects per cell. Higher means indicate greater agreement with the attitude-discrepant essay. Cell means with different subscripts are different from each other at the 1% level by the Newman–Keuls procedure. The mean in the survey control condition is 2.30.

[a]This table is reprinted from Zanna and Cooper (1974). (Copyright 1974 by the American Psychological Association. Reprinted by permission.)

might delimit our finding by asking whether the arousal state for which we have gathered evidence is an aversive condition, as Festinger postulated, or whether it is a state of general arousal. Second, we might extend our finding by asking whether people who experience dissonance will be motivated to seek out an external attribution for their state. Third, we might ask whether dissonance processes can operate in the absence of arousal. We will now turn to recent evidence pertinent to each of these questions.

ATTRIBUTION AND THE PHENOMENOLOGY OF AROUSAL

Is dissonance a general state of arousal, or is it, as Festinger imagined, a specifically aversive state? Zanna and Cooper (1974) do not specifically answer this question. The results of the several studies using the Waterman and Katkin paradigm could have occurred regardless of whether the arousal were specifically aversive or if the persons were in a state of generally heightened arousal. Zanna, Higgins, and Taves (1975) attempted to answer this question by utilizing the attributional approach.

As in Zanna and Cooper's study, subjects volunteered for an experiment involving short-term memory. They were shown a videotape presented by a doctor of geriatrics of the National Institutes of Health who explained that researchers had recently discovered a compound which improves the short-term memory of aged subjects. The purpose of the present research, the physician stated, was to test its effect on college-aged populations.

After completing a bogus short-term memory pretest, subjects were assigned to one of four drug side-effects conditions. Two conditions were identical to conditions run by Zanna and Cooper. Subjects were told that the experimental capsule would make them tense in the moments before complete absorption; others were told that the pill would have no side effects.

Two additional conditions were employed. (One condition, the no information condition, will be discussed in the following section.) In the one that primarily concerns us here, the subjects were told that the drug might give them the experience of "pleasant excitement." Zanna, Higgins, and Taves reasoned that if dissonance arousal were general in nature, then the tense condition and the pleasant excitement condition should produce identical results. If people who are experiencing dissonance are motivated to explain their arousal with an external attribution and if that arousal is general in nature, then either external label should be seized upon as the explanation. However, if dissonance is experienced as specifically aversive, then only the tense label should mitigate the dissonance-produced attitude change.

Using a now-familiar ruse, the experimenter asked subjects if they would mind participating in another study while they were waiting for the absorption of the drug to be complete. In this study, members of a ficticious organization known as the American Youth Organization were soliciting essays advocating a military draft as opposed to a volunteer army. This was an attitude-discrepant position for the subjects all of whom had been pretested earlier in the context of an unrelated study. Subjects were all given the choice of whether or not to cooperate with the American Youth Organization's request. All participants agreed, engaged in their counterattitudinal advocacy, and then filled out a questionnaire which included a 19-point scale of agreement regarding an all volunteer army.

The results of the study are presented in Table 2. Examining the two conditions that were similar to the original study by Zanna and Cooper, we see the

TABLE 2
Mean Attitude Change Scores toward the Military Draft[a]

Potential side effect of the drug			
No information	Tense	No effect	Pleasant excitement
$.25_a$ (11)	$.12_a$ (12)	3.60_b (10)	4.29_b (10)

*Note. N*s for each cell are in parentheses. Higher means indicate greater agreement with the attitude-discrepant essay. Cell means with different subscripts are different from each other at the 5% level by *t* test. The mean in the survey control condition is .28. All experimental subjects were run under conditions of high choice.

[a]This table is taken from Zanna, Higgins, and Taves (1975).

identical effect. When subjects behave in an attitude-discrepant fashion and have no external agent to which to attribute the arousal, attitude change occurs, mean $\overline{X}$ = 3.60; when subjects can believe it is the pill that is making them tense, attitude change is virtually eliminated, $\overline{X}$ = .12. (This misattribution result has also been replicated by Pittman, 1975.) In contrast, the pleasant excitement condition does *not* suit the subjects' need to misattribute their arousal. Indeed, the pleasant excitement group demonstrated the most attitude change of any group in the experiment, $\overline{X}$ = 4.29. The conclusion that can be reached from these data is that the arousal state of dissonance is not a general one. Rather, as Festinger suggested and the attributional paradigm supports, dissonance is a specifically aversive state of arousal. It should be clear that we are not suggesting that dissonance is a particular physiological state that we might someday identify, but merely that, phenomenologically, dissonance appears to be a state of unpleasant arousal.

THE SEARCH FOR A LABEL

In the Zanna and Cooper study and the Zanna, Higgins, and Taves study, subjects given an explicit alternative explanation for their arousal in the form of a tension-producing pill utilized this explanation in preference to changing their attitudes. The question addressed in the present section is whether dissonance-aroused subjects will actually seek out such an explanation for their state in order to avoid having to reduce their dissonance by changing their attitudes. We will now turn to three recent studies in which this issue was examined, at least indirectly.

Blaming It on the Lights

In many dissonance studies, people engage the behaviors that are at variance with their attitudes. The attitudes are very carefully chosen by the investigators so that they are mildly involving to the participants, but no so involving that attitudes will be impossible to change. A study was conducted by Gonzalez and Cooper (1975) to find out just what would happen to dissonance which could not be reduced by a change of attitude. If dissonance is an arousal state, then people who are experiencing dissonance but who find it difficult to change their attitudes, will be motivated to find some other mechanism for coping with their arousal. Gonzalez and Cooper predicted that such people would find an external "explanation" for their arousal.

The study was conducted at Princeton University which has a unique social institution called the eating club. For years, eating clubs formed the nucleus of campus social activities and had the selective features of most fraternities. In more recent years, the private eating clubs have constituted the social life for less than half of the university students. From time to time, there are rumors to the effect that the university will close the eating clubs and convert the buildings to other purposes. These proposals are not popular with most students. Those students who are members of eating clubs have the most to lose by such proposals, but most students generally feel negative about attempts to close the clubs.

Gonzalez and Cooper's objective was to have members and nonmembers of eating clubs make counterattitudinal statements advocating an end to the clubs. They felt that nonmembers would behave as subjects in dissonance studies typically act: they would reduce their dissonance by changing their attitudes in the direction of their behavior. But club members would find it more difficult to come to believe that eating clubs should be abolished. These individuals were expected to seek another way of coping with their arousal.

In their study, Gonzalez and Cooper contacted 20 club members and 20 nonmembers who had earlier expressed extremely negative attitudes about proposals to close the eating clubs. They came to the session to participate in "an attitude survey." When they arrived, half of all subjects were asked if they would be willing to write a forceful, one-sided essay taking the position that the eating clubs should be abolished from the Princeton campus. They were reminded that they were under no obligation to write the essay. The other half of the subjects were told that it was a requirement of the study that they write the anti-eating-club essay.

The design of the study, then was a simple one. Club members and nonmembers wrote attitude-discrepant essays under choice or no choice conditions. It was expected that dissonance would be aroused for any subject who made a free decision to write the counterattitudinal essay. When the essays were completed, subjects were given a one-item opinion scale assessing their attitudes about the eating clubs remaining on campus. It was predicted that, although both high-choice groups would experience dissonance, only the nonmembers would be able to show self-justificatory attitude change on the questionnaire. It was felt that club members would feel too involved in the issue of the abolition of the clubs to change their opinion. Table 3 shows that this is what occurred. In general, nonmembers changed their opinions more than members, and high-choice subjects changed their opinions more than low-choice subjects. More importantly, the interaction between choice and membership was significant, $F(1, 36) = 8.02$, $p < .01$, indicating that the effect for choice was more important for nonmembers than members. As predicted, the manipulation of decision freedom produced attitude change in accord with dissonance theory predictions for subjects who were not club members but did not produce a differential effect for club members.

How, then, would the high-choice club members deal with their dissonance? How could they explain the arousal they were experiencing, since they did not utilize the avenue of changing their opinions? One possible route was provided

TABLE 3
Mean Attitude Change Scores toward Eating Clubs[a]

Decision freedom	Club membership	
	Members	Nonmembers
High choice	3.00_a	10.90_c
Low choice	3.30_a	6.10_b

Note. N = 10 subjects per cell. Higher means indicate greater agreement with the attitude-discrepant essay. Cell means with different subscripts are different from each other at the 5% level by the Newman–Keuls procedure.
[a]This table is taken from Gonzales & Cooper (1975).

for them by the experimenter. After completing the attitude scales, and just prior to walking out of the door of the experimental room, the experimenter stated to all subjects: "Oh, I just remembered. Before you leave, could I ask a favor of you? In the last month we have installed new fluorescent type lighting in the room. Some of the people using the room have complained that the room made them feel tense or uneasy. We suspect that it might have something to do with these lights, if indeed there is any problem."

He then asked the subject to fill out a brief scale in order to inform the maintenance department if there were any problem with the lights. Subjects were given a 26-point scale on which they could indicate whether they felt the lights made them "not tense or uncomfortable at all" to "very tense or uncomfortable." The lights, of course, were neither new nor discomforting. The suggestion was planted as a way of allowing any subject to make an external attribution of discomfort.

Who blamed their arousal on the lights? Table 4 provides the answer. Again, there was a significant interaction, $F(1, 36) = 9.03, p < .01$. But this time, the

TABLE 4
Mean Lighting Discomfort Scores[a]

Decision freedom	Club membership	
	Members	Nonmembers
High choice	9.50_c	$.80_a$
Low choice	6.40_b	1.00_a

Note. N = 10 subjects per cell. Higher means indicate greater reported discomfort with the lighting. Cell means with different subscripts are different from each other at the 1% level by the Newman–Keuls procedure.
[a]This table is taken from Gonzalez and Cooper (1975).

interaction was produced by a high degree of attribution of discomfort by subjects in the high-choice/club member group. The subjects experiencing a great deal of dissonance but feeling proscribed from altering their proeating-club attitude, attributed their arousal to a thoroughly innocent external source.

The study by Gonzalez and Cooper leaves open an intriguing and unanswered question. It is clear that people with unresolved dissonance were motivated to seek a label for their arousal. But it is not clear whether providing the label was in the service of reducing or merely explaining the arousal. It may be that by viewing the lighting as responsible for the arousal, a person can reduce the tension which in fact was caused by inconsistent cognitions. On the other hand, the attribution of the arousal to the lights merely may have served as an explanation of the arousal. Whether attribution to the lighting is successful at reducing the psychological tension or is merely an explanation is a question that must await better, direct measurement of arousal.

Seeking a Noxious State

The observation that people do seem to be motivated to seek external explanations for their arousal when self-justificatory attitide change is proscribed is supported and extended in a study conducted by Gonzalez, Cooper, and Zanna (1975). Although this study was primarily designed to demonstrate that dissonance-produced arousal would lead to affiliation in the service of social comparison (which it did demonstrate), the results also indicated that people experiencing a great amount of dissonance will actually choose to place themselves in a potentially aversive situation, apparently to "mislabel" their arousal and avoid the less preferable dissonance reduction mode of attitude change.

In the study, under one of three conditions, subjects were first induced to write a counterattitudinal essay supporting the proposition that 7 a.m. classes would improve university class scheduling: low choice to write the essay (low dissonance), high choice (medium dissonance), or high choice and the possibility that as a result of writing the essay they might have to participate in an experimental 7 a.m. class twice a week for a short period of time (high dissonance). Following the counterattitudinal advocacy, subjects were given the opportunity to wait alone or with another person during the interim waiting period before a supposed second study. Under each dissonance level, four groups of subjects were run. In one group, subjects could choose to wait alone or with a similar other in a neutral setting; in a second group, the choice was to wait alone or with a dissimilar other in a neutral setting; a third group of subjects could choose to wait alone or with a similar other in an aversive setting; and the choice for a fourth group was to wait alone or with a dissimilar other in an aversive setting. The similarity or dissimilarity of the target affiliate was manipulated by indicating that he had either just participated in the same study or in another study. The aversiveness of the setting was established by telling subjects that the rooms designated for waiting with someone else had "new fluroescent lighting" that made some subjects "feel very tense or uncomfortable." In the neutral setting condition no reference was made to the waiting rooms.

Table 5 presents mean affiliation scores, assessed by asking subjects to indicate on a 21-point scale (with end-points labeled "very much with(out) someone else," scored −10 and +10, respectively) where they wanted to wait.

The results of interest from the present perspective, as indicated by an analysis of variance, were the main effect for room type, $F(1, 108) = 9.46, p < .01$, and the interaction between dissonance level and room type, $F(2, 108) = 5.18, p < .01$. The main effect is best characterized by noting that, overall, affiliation occurred to a greater extent in neutral, as compared to aversive, rooms. A qualification for the interaction effect is in order. As can be seen in Table 5, the preference for neutral rooms occurred only in the low- and medium-dissonance conditions; in the high-dissonance condition, there was a slight (nonsignificant) preference to affiliate in the aversive context. This sudden desire to place oneself in an aversive setting (as dissonance increased from medium to high levels) was most marked in the dissimilar condition (means of 2.00 versus −5.50, in the dissimilar–aversive context under conditions of high and medium dissonance, respectively). Since social comparison was an unlikely cause of affiliation in the dissimilar condition (cf. Schachter, 1959), the possibility exists that subjects in the high-dissimilar/aversive condition (and, for that matter, the high-similar/aversive condition) chose to place themselves in the unpleasant room in order to mislabel their arousal.

This misattribution interpretation is further supported by the attitude change results. If subjects were mislabeling their arousal in the high-aversive conditions, then attitude change in these conditions should have been attenuated (Zanna &

TABLE 5
Mean Affiliation Scores[a]

Target affiliate type	Dissonance level and room type: Low		Medium		High	
	Aversive	Neutral	Aversive	Neutral	Aversive	Neutral
Similar	−3.10	−1.20	.40	.30	2.60	2.70
Dissimilar	−4.30	−2.30	−5.50	.00	2.00	.70

Note. N = 10 subjects per cell. The more positive the cell mean, the greater the tendency to affiliate.

[a]This table is taken from Gonzalez, Cooper, and Zanna (1975).

TABLE 6
Mean Attitude Scores toward 7 a.m. Classes[a]

Target affiliate type	Dissonance level and room type: Low		Medium		High	
	Aversive	Neutral	Aversive	Neutral	Aversive	Neutral
Similar	4.50_a	5.70_a	9.30_b	11.20_{bc}	9.20_b	15.00_d
Dissimilar	4.60_a	4.30_a	11.00_{bc}	11.60_c	11.20_{bc}	14.40_d

Note. N = 10 subjects per cell. Higher means indicate greater agreement with the attitude-discrepant essay. Cell means with different subscripts are different from each other at the 5% level by the Newman–Keuls procedure.
[a]This table is taken from Gonzalez, Cooper, and Zanna (1975).

Cooper, 1974). After subjects indicated their waiting preferences, they were asked to indicate the extent of their agreement with the proposition that "Holding course meetings at 7 a.m. will improve class scheduling at the university" (on a 26-point scale with endpoints labeled "strongly disagree," scored 1, and "strongly agree," scored 26). The mean attitude scores are presented in Table 6, where it can be seen that, as expected, subjects became more favorable towards 7 a.m. classes as dissonance increased from low to medium levels. As dissonance increased from medium to high levels, however, subjects' attitudes toward 7 a.m. classes became more positive *only* in the neutral room conditions. In both high-aversive conditions attitude change was attenuated–to the amount of change obtained under medium dissonance.

The question remains, however, whether people who experience a medium amount of dissonance or for whom attitude change is not so difficult will be motivated to seek out external attributions to avoid having to deal with their inconsistency. For example, how might we expect subjects in the original Zanna and Cooper paradigm to react if they were told nothing whatsoever about what to expect from the pill? In this situation, subjects could (a) interpret this rather ambiguous situation so as to provide a label for their arousal or (b) simply change their attitudes to reduce their dissonance. Such a no information condition was, in fact, run in the study conducted by Zanna, Higgins, and Taves (1975). As can be seen in Table 2, this no information group apparently seized the opportunity to assume that the drug did have a side effect which caused them to feel aroused. The mere presence of an unidentified pill eliminated dissonance-produced attitude change. Thus, when it is possible to mislabel dissonance-produced arousal, people do seem to prefer misattributing their arousal over changing their attitudes as a means of resolving their dissonance.

AROUSAL: IT IS NECESSARY?

When Bem proposed that attitude change following attitude-discrepant behavior could be accounted for by self-attribution (for example, Bem, 1968), he aptly pointed out that his data did not mean that people necessarily employed this process when confronting inconsistent cognitions. Rather, his data provided one way of interpreting attitude-change phenomena but did not provide evidence that people actually undertook self-attributional processes. To this point, our argument is in somewhat the same epistemological position. For example, we have used a series of attributional analyses to indicate that attitude change will be obviated if cues for misattribution of arousal are available. This does not provide firm evidence that attitude change *must* be accompanied by arousal. A recent study conducted by Cooper, Zanna, and Taves (1975) does provide more direct evidence that arousal is a necessary component of attitude change following induced compliance.

The procedure began in the same way as the original Zanna and Cooper study. Subjects came for a study on short-term memory and were told of experimental work that was being done on various forms of a drug which had an effect on short-term memory. In this study, they were informed that two different drugs and one placebo were being used. The subjects were told that small dosages of amphetamine or phenolbarbitol (a common tranquilizer) were the active ingredients in the experimental drugs. The placebo was mere milk powder. The subjects, whose medical records were carefully checked prior to the study, were asked to sign a consent form agreeing to take any of the three possible drugs.

After taking a brief short-term memory test, subjects were assigned to conditions. They received their capsule and a printed note which informed them of a condition that they were in. In the present study, all subjects were informaed that they were in the placebo condition. That is, they were all led to believe that they were ingesting milk powder and that there would be absolutely no side effect to the pill.

A second difference between this study and previous experiments that we have reported is that most subjects did not take milk powder. Contrary to their expectations, subjects were given (with adequate medical supervision) either 5 mg of amphetamine or 30 mg of phenolbarbitol. An additional group of subjects was given the milk powder that they had been led to expect.

Shortly after taking their capsule, the experimenter reminded himself of "another experiment" which the subjects could participate in while they waited. They were told that the University's Research Institute was interested in attitudes toward the pardoning of former President Richard Nixon. It was indicated that the Institute was going to make a formal presentation to the Executive Branch and that, in this aspect of their investigation, they were attempting to gather all of the possible arguments in support of President Ford's decision to pardon the former president. Therefore, it would be the task of the subjects to write a strong and forceful essay in support of President Ford's decision. (It was known from a pretest that all subjects were against this position.)

Half of the subjects were run under high-choice conditions. After hearing the details of this "other experiment," they were asked if they were willing to participate. The other half of the subjects were simply told that it was required of them while they were waiting for the drug to take effect. The writing of the counterattitudinal essays was so timed that they would be finished at about the time the drug took effect. It is estimated that between 20 and 30 minutes are typically required for complete absorption of the two active drugs. Therefore, the instructions and the writing of the essay were planned to take about 35 minutes to ensure that the drug was active for all subjects.

At the conclusion of the essay writing, the subjects were asked to fill out a questionnaire which included several items about political events, President Ford, and former President Nixon. The crucial item asked subjects to indicate on a 31-point scale the degree to which they agreed or disagreed with the pardoning of Richard Nixon.

The results of the study are presented in Table 7. The placebo condition was a standard replication of an induced compliance study. The data show that the dissonance-produced attitude change was obtained. High-choice/placebo subjects were in greater agreement with the Nixon pardon than their low-choice counterparts. The interesting effects begin to emerge when the results of those who took active drugs are examined. Subjects who took tranquilizers did not show any need for dissonance reduction. They showed no differences in attitude from the survey control group, or from the low-choice/placebo group. More importantly, the high-choice and the low-choice tranquilizer groups did not differ from each other.

TABLE 7
Mean Attitude Scores toward the Pardoning of Richard Nixon[a]

	Drug condition		
Decision freedom	Tranquilizer	Placebo	Amphetamine
High Choice	8.6_a	14.7_b	20.2_c
Low choice	8.0_a	8.3_a	13.9_b

Note. N = 10 subjects per cell. Higher means indicate greater agreement with the attitude-discrepant essay. Cell means with different subscripts are different from each other at the 5% level by the Newman–Keuls procedure. The mean in the survey control condition is 7.9.

[a]This table is taken from Cooper, Zanna, and Taves (1975).

The groups that ingested amphetamines present a different story. Attitude change was found in both the high- and low-choice variations. Whereas eliminating decision freedom usually has the effect of eliminating dissonance, subjects in the present study were in an unusual predicament. They had behaved in a counterattitudinal fashion and, despite their knowledge that they had followed an experimenter's order, they nonetheless felt aroused. The only attribution they could make was that they were upset by the inconsistency between their cognitions. It has been pointed out that the lack of decision freedom renders two otherwise discrepant cognitions in a cognitive relationship irrelevant (for example, Cooper, 1971). The arousal induced by the amphetamine must have convinced subjects that their cognitions were not at all irrelevant and that they were in some way responsible for their attitude-discrepant behavior, despite the experimenter's coercion.

The data on the subjects' perception of their freedom to decline the experimenter's essay writing request provide an insight into the attitude change evidenced by low choice/amphetamine subjects. In all conditions except one, subjects accurately recalled the choice they had been given. However, in the low-choice/amphetamine condition, subjects reported feeling a high degree of choice. In other words, feeling aroused, the subjects apparently deduced that they must have had some role in deciding to engage in attitude-discrepant behavior. In this instance, then, amphetamine arousal appears to have been misattributed to the attitude-discrepant behavior.

Finally, it appears that dissonance arousal combined with the arousal due to amphetamine is additive. Subjects in the high-choice/amphetamine condition changed their attitudes more than in any other condition in the experiment.

DISSONANCE AS AVERSIVE AROUSAL: A SUMMARY AND CONCLUSION

The experiments that we have reported lead to the conclusion that cognitive dissonance is an aversive state of arousal and that the arousal is necessary to motivate cognitive changes. In Zanna and Cooper's study, it was proposed that if dissonance were an arousal state, it should be subject to the same sort of misattributional processes as other arousal states. Once misattributed as something other than dissonance, it was proposed that attitude-discrepant behavior would not lead to attitude change. Our finding that the providing of an external label for arousal eliminated the need for attitude change suggested that dissonance does indeed have arousal properties.

The study by Zanna, Higgins, and Taves (1975) was primarily designed to show that inconsistency among cognitions did not merely arouse an organism in a general way. Instead, the arousal that emanates from cognitive inconsistency is specifically aversive and unpleasant. The Zanna et al. study showed that misattribution affected cognitive change in an induced compliance paradigm only when an external explanation was provided for unpleasant, as opposed to pleasant, arousal.

In the third study presented, Gonzalez and Cooper (1975) found that people who could not easily change their attitudes following counterattitudinal behavior were motivated to seek alternative ways to explain their arousal. When they thought that the room lighting might be causing tension, they misattributed their arousal to the lights. It was reasoned that such an effect could occur only if dissonance did have arousal properties. Their inability to reduce the arousal led to mistaken attributions that could potentially account for the experienced tension. Gonzalez, Cooper, and Zanna (1975) found that people who experienced a great amount of dissonance actually chose to place themselves in a tension-producing room; the study by Zanna, Higgins, and Taves (1975) also indicated that people, given an unlabeled pill, had no need to change their attitudes. This pattern of results adds to the conclusion that people are often motivated to seek out an external label for their dissonance-produced arousal.

Finally, we presented evidence that arousal is a necessary condition for inconsistency to lead to opinion change. When subjects participated in counterattitudinal behavior under free-choice conditions but did not experience arousal (due to a tranquilizer we had given them), they did not show attitude change. Conversely, when they had taken an arousing drug, subjects did change their attitudes, even if they had participated without decision freedom. Arousal appeared to be motivating condition for producing cognitive changes following induced compliance.

DISSONANCE AND ATTRIBUTION

Dissonance theory and self-perception theory have had a long and interesting battle. Their differences of opinion have provided numerous research endeavors which have had a provocative effect on research in social psychology. We may currently be in a position to provide some answers and to recommend a truce.

The essential question is, what shall we take self-attribution in counterattitudinal advocacy situations to mean? If it is taken to mean that people engage in a process of deducing their opinions without the experience of an internal tension state, the data presented here cast serious doubt on that view. On the other hand, if self-attribution is taken to be a method of predicting attitudes from behaviors, then it is still a very viable approach. All that our research has been intended to demonstrate is that aversive arousal accompanies cognitive inconsistency. It does not indicate that people never engage in attributional processes nor that, in most situations, attributional rules will not lead to accurate predictions regarding attitudinal outcomes. Indeed, our reliance on attributional processes to make the predictions that we have presented indicates that self-attributional phenomena are alive and well.

And so, a truce is proposed that will enable us to go beyond the simple conflict between the motivation-free, information-processing perspective of attribution theory and the motivational perspective of dissonance theory. By sharpening the areas of applicability of the two approaches, we may be able to arrive at a more precise understanding of each.

ACKNOWLEDGMENTS

Since the contribution of the two authors was equal, the order of authorship was determined by a coin-flip. The authors wish to thank Susan A. Darley and Russell H. Fazio for their helpful comments on an earlier draft of the manuscript.

REFERENCES

Bem, D. J. An experimental analysis of self-persuasion. *Journal of Experimental Social Psychology,* 1965, **1,** 199–218.

Bem, D. J. The epistemological status of interpersonal simulations: A reply to Jones, Linder, Kiesler, Zanna, and Brehm. *Journal of Experimental Social Psychology,* 1968, **4,** 270–274.

Bem, D. J. Self-perception theory. In L. Berkowitz (Ed.), *Advances in experimental social psychology* (Vol. 6). New York: Academic Press, 1972.

Cooper, J. Personal responsibility and dissonance: The role of foreseen consequences. *Journal of Personality and Social Psychology,* 1971, **18,** 354–363.

Cooper, J., Zanna, M. P., & Taves, P. A. On the necessity of arousal for attitude change in the induced compliance paradigm. Unpublished manuscript. Princeton University, 1975.

Cottrell, N. B., Rajecki, D. W., & Smith, D. U. The energizing effects on postdecision dissonance upon performance of an irrelevant task. *Journal of Social Psychology,* 1974, **93,** 81–92.

Cottrell, N. B., & Wack, D. L. The energizing effect of cognitive dissonance on dominant and subordinate responses. *Journal of Personality and Social Psychology,* 1967, **6,** 132–138.

Festinger, L. *A theory of cognitive dissonance.* Stanford, California: Stanford University Press, 1957.

Festinger, L., & Carlsmith, J. M. Cognitive consequences of forced compliance. *Journal of Abnormal and Social Psychology,* 1959, **58,** 203–210.

Heider, F. *The psychology of interpersonal relations.* New York: John Wiley & Sons, 1958.

Gonzalez, A. E. J., & Cooper, J. What to do with leftover dissonance: Blame it on the lights. Unpublished manuscript, Princeton University, 1975.

Gonzalez, A. E. J., Cooper, J., & Zanna, M. P. Social affiliation and cognitive labeling under differential labels of dissonance-evoked arousal. Unpublished manuscript, Princeton, University, 1975.

Hull, C. L. *Principles of behavior.* New York: Appleton–Century–Crofts, 1943.

Jones, E. E., & Davis, K. E. From acts to dispositions: The attribution process in person perception. In L. Berkowitz (Ed.), *Advances in experimental social psychology* (Vol. 2). New York: Academic Press, 1965.

Jones, R. A., Linder, D. E., Kiesler, C. A., Zanna, M., & Brehm, J. W. Internal states or external stimuli: Observers' judgments and the dissonance–self-perception controversy. *Journal of Experimental Social Psychology,* 1968, **4,** 247–269.

Linder, D. E., Cooper, J., & Jones, E. E. Decision freedom as a determinant of the role of incentive magnitude in attitude change. *Journal of Personality and Social Psychology,* 1967, **6,** 245–254.

Pallak, M. S., & Pittman, T. S. General motivational effects of dissonance arousal. *Journal of Personality and Social Psychology,* 1972, **21,** 349–358.

Pittman, T. S. Attribution of arousal as a mediator in dissonance reduction. *Journal of Experimental Social Psychology,* 1975, **11,** 53–63.

Ross, L., Rodin, J., & Zimbardo, P. G. Toward an attribution therapy: The reduction of fear through induced cognitive–emotional misattribution. *Journal of Personality and Social Psychology,* 1969, **12,** 279–288.

Schachter, S. *The psychology of affiliation.* Stanford, California: Stanford University Press, 1959.

Schachter, S., & Singer, J. E. Cognitive, social, and physiological determinants of emotional state. *Psychological Review,* 1962, **69,** 379–399.

Spence, K. W., Farber, I. E., & McFann, H. H. The relation of anxiety (drive) level to performance in competitional paired-associates learning. *Journal of Experimental Psychology,* 1956, **52,** 296–305.

Storms, M. D., & Nisbett, R. E. Insomnia and the attribution process. *Journal of Personality and Social Psychology,* 1970, **2,** 319–328.

Waterman, C. K. The facilitating and interfering effects of cognitive dissonance on simple and complex paired-associate learning tasks. *Journal of Experimental Social Psychology,* 1967, **5,** 31–42.

Waterman, C. K., & Katkin, E. S. The energizing (dynamogenic) effect of cognitive dissonance on task performance. *Journal of Personality and Social Psychology,* 1967, **6,** 126–131.

Zanna, M. P., & Cooper, J. Dissonance and the pill: An attribution approach to studying the arousal properties of dissonance. *Journal of Personality and Social Psychology,* 1974, **29,** 703–709.

Zanna, M. P., Higgins, E. T., & Taves, P. A. Dissonance and the Phenomenology of arousal: A test of the aversive nature of dissonance arousal. Unpublished manuscript, Princeton University, 1975.

Part III

ATTRIBUTION AT THE INTERPERSONAL LEVEL

Many of the principles governing self-perception are similar to those involved in the perception of others. In this section, attributional processes are examined in more explicitly social contexts. Our reactions to people are determined not only by our own feelings and thoughts but are also guided by our perceptions of the causes of their actions and expressions. In giving meaning to the continuous flow of an ongoing interaction, the perceiver is more than an information processor. At the interpersonal level, the attributions he forms may have direct consequences for his interaction with others. In the following papers, many of the attributional processes examined in the first section are applied at the level of the person's interaction with others.

Investigators in the area of person perception often have addressed problems related to making inferences about the dispositions of other persons rather than those relating to the more basic perception of their ongoing behavior. When we perceive other people, how do we divide and organize the continuous undifferentiated stream of physical stimulation into discrete, meaningful actions? In "Foundations of Attribution: The Perception of Ongoing Behavior," Darren Newtson attempts to answer this question. In doing so, he also discusses how the unitization of perception can be measured, what cues determine when unitization will occur, and how these perceptual processes can be applied to the study of interpersonal behavior.

The act of reciprocating a favor has become so pervasive as to be considered a societal norm. However, what happens when compliance with this norm of reciprocity violates an important behavioral freedom like being able to choose freely whom to invite to a private dinner party? A possible solution to this dilemma is discussed by Stephen Worchel and Virginia Andreoli in their chapter, "Escape to Freedom: The Relationship between Attribution of Causality and Psychological Reactance." The authors present evidence suggesting that the cause of another's behavior may be externally attributed as a means of reducing the threat to freedom posed by the norm of reciprocity.

While inconsistencies among cognitive elements may cause people to change their attitudes to accommodate these discrepancies and reduce the unpleasant feelings that accompany them, such feelings also may influence thoughts and attitudes toward others. In "Attribution of Attitudes from Feelings . . .," Judson Mills, Jerald Jellison, and James Kennedy explore the relationships among balance theory, causal ascription, and attitude change. They present evidence that our attitudes toward others may be inferred from our perception of particular feelings about them. In general, we seem to like another person if we feel good when he is benefited, but dislike the person if we feel good when he is harmed.

Social influence has long been a topic at the core of social psychology. Only recently, however, has attribution research been conducted in this area. Somewhat counter to the many findings which emphasize the importance of attitude similarity in social interaction, George Goethals specifies conditions in which an attitudinally dissimilar other may actually be more influential than a similar other. In his chapter, "An Attributional Analysis of Some Social Influence Phenomena," Goethals introduces the concept of opinion "triangulation" and describes its attributional implications for the validation and modification of a person's opinions.

In "An Attributional Analysis of Helping Behavior," William Ickes and Robert Kidd provide a detailed analysis of a relatively new topic on the social psychological scene, help giving. The past attributional research in this area has been focused primarily on the degree to which the dependency of a person in need is seen as internally versus externally caused. This focus has neglected the effects on helping of a potential helper's inferences about the stability and intentionality of the other's outcome. Similar shortcomings have been observed in previous theorizing about the perceived causes of the potential helper's own outcome as well. In an attempt to correct these problems, the authors discuss attributions of intent and responsibility in the context of a prototypic situation for altruistic behavior.

The role of male/female stereotypes in attribution is discussed by Kay Deaux in "Sex: A Parameter in the Attribution Process." In many western societies, women have traditionally been viewed as less able than men at certain kinds of tasks. As Deaux points out, sex is viewed not only as a characteristic of the person being evaluated but also as a characteristic of the evaluator and of the task which the person has performed. She explains how people of both sexes make attributions about behavior so as to uphold and reinforce cultural sexual stereotypes. For example, women who are successful on a traditionally male task are often seen as having succeeded by luck (an external factor), while men who do well on the same task are seen as being very skillful (an internal factor).

In "Attributional Conflict in Young Couples," Bruce Orvis, Harold Kelley, and Deborah Butler report research which reveals differences in causal attributions between members of young, heterosexual couples. These investigators employed a relatively novel open-format technique where each respondent gave personal explanations and the presumed explanations of his/her partner for examples of behavior which the couples had specified as generating attributional disagreement between them. Orvis et al. interpret their results as providing evidence about a "social context" view of attributional processes. Given their data base, they propose and provide answers to the provocative question, What are the consequences of assuming that the attribution process is originally learned and subsequently maintained primarily in the social context of justification of self and criticism of others?

10

Foundations of Attribution: The Perception of Ongoing Behavior

Darren Newtson

University of Virginia

Jones and Thibaut (1958) once noted that the term "person perception" is a misnomer, since social psychologists investigating the phenomena seem to be interested only in "person cognition." If nothing else, the present discussion should begin to redress that imbalance. It is not my intention to question the fact that persons can and often do reason about each other, and about each other's traits, intentions, abilities, motives. I would like to propose, however, that there are perceptual processes involved in our experience of persons above and beyond the perceptions that others are three-dimensional moving objects with a limited degree of plasticity.

Specifically, I will address the question of how it is that the continuous, undifferentiated stream of physical stimulation that impinges on our senses is rendered into discrete, discriminable, describable actions. In doing so, I will (a) present evidence that this process can be measured, both reliably and validly; (b) present such evidence as I have about the properties of the process as reflected by this measurement procedure; (c) propose a theoretical model for the process of behavior perception; and (d) discuss its implications for current theories of attribution of causality.

A MEASURE OF PERCEPTUAL UNITS

As Heider (1958) and, more recently, From (1971), have pointed out, our experience of the behavior of others is discrete, rather than continuous. That is, we see persons perform a series of discrete actions, rather than seeing continuous, undifferentiated behavior. Persons do this, and then they do that, and so on. These discrete actions constitute a division into segments, or units, of the stream of information presented by ongoing behavior.

The means by which the segmentation of a given action sequence is measured in the laboratory is quite simple. The subject is provided with a button to press and is instructed to press it when, in his judgment, one meaningful action ends and a different one begins, that is, when the person stops doing one thing and begins to do something discriminably different. He is then shown a sequence of action, either via film or videotape. By measuring the points in the sequence where the subject presses the button, his subjective partitioning of the behavior may be identified.

As artificial as it sounds, the procedure has a number of interesting properties. First, persons readily understand the instructions, and, after a minute or two of practice, can perform the task without difficulty. Secondly, after some practice with the procedure, markings can be timed very precisely, as if one were tracking a bouncing ball and tapping every time it hits the floor. Indeed, one of our subjects found that he could track the shifts in the performance of different actions with the marking procedure so well that he insisted we were deceiving him. He was certain that his button operated a signal to an actor on live television (he was told he was watching a videotape) and that the actor was trained to perform a different action whenever he, the subject, pressed the button. In addition, he announced truimphantly that he had tried it, and it worked.

Beyond such anecdotal evidence, moreover, we now have hard experimental data confirming that the measure is reliable, valid, and nonreactive. Unit data are scored by dividing an action sequence into a series of discrete, equal intervals and then tabulating from a group of subjects the number of marks that fall into each interval. Results from an initial series of studies employing this procedure (Newtson, 1973) have shown that the distribution of markings over intervals is not random. Some intervals in a given sequence elicit unanimous, or nearly unanimous, marking while others are consistently left unmarked by subjects. We have termed those points eliciting significant agreement on marking "break points," in that they are the points at which subjects divide or "break up" a sequence into its component actions. Intervals eliciting significant agreement on continuity, in that few or no persons mark in them, we have termed "nonbreak points."

Selection of an interval size for purposes of analysis is somewhat arbitrary. We have employed interval sizes ranging from 1 to 5 seconds. As a rule of thumb, we usually select an interval size such that less than 1% of unit markings from individuals yield multiple markings with that size interval.

Although agreement of break points and nonbreak points far exceeds chance, persons do vary considerably in the number of break points they identify in a given sequence. This difference could be due in part to differing levels of analysis that subjects adopt for our sequences. For example, one might see a person get up from a chair, walk over to a door, close it, turn, and walk back to his chair, and mark off each segment as a discrete, meaningful action, or one might see the whole sequence as just one action—closing the door. Reasoning that persons might be capable of segmenting the same behavior into smaller, or finer, components of action or larger, grosser components, we conducted a study in which level of analysis was varied by instruction (Newtson, 1973). All subjects were given the above instructional example of the man closing the door. In one condition subjects were instructed to mark off "the *smallest* actions that seem natural and meaningful to you." In a second condition, subjects were instructed to segment the sequence into "the *largest* actions that seem natural and meaningful to you."

Results indicated that persons could readily vary their level of analysis by instruction, in that fine-unit instructed subjects employed significantly more units in marking a 5-minute videotaped sequence, $\overline{X} = 54$, than gross-unit instructed subjects, $\overline{X} = 23$. When unitization patterns in the two conditions were compared, moreover, it was clear that unitization was varied over a hierarchical structure. That is, persons marking small units were breaking down the same units employed by large-unit subjects into their component parts. This is consistent with the notion that action units consist of goal-directed behavior; large units, then, would be superordinate goals, while fine units would be the subgoals. The alternative would be that the two levels of analysis are qualitatively different (organized in a completely different way).

At this point, we were still not overly confident in the measure. While considerable and significant agreement in marking was observed, there was also considerable disagreement. We assumed that much of this disagreement was simply measurement error. Serendipity then intervened. We were set up to run two subjects at a time in the procedure and had taken ample precautions to assure that the two subjects could not influence each other's markings: they were separated by a partition, they wore headphones with a low level of white noise coming through, and the boxes upon which the buttons were mounted were cushioned by foam rubber. They could neither see, hear, nor feel each other's marking. A pair of identical twins then signed up for the experiment at the same time and proceeded to mark an 11-minute videotape virtually identically. While marking over 50 units, they disagreed only 7 times. Occasionally, they lagged a half-second or so, but the overall agreement was astounding.

We did not conclude from this event that unit marking is genetically determined. It did suggest, however, that our measure contained far less measurement error and far more individual differences variability than we had expected. That is, as these were two extremely similar individuals, individual differences were relatively slight.

Thus encouraged, we decided to investigate the within-subjects variability of our measure, opting for a 5-week test–retest reliability format (Newtson & Engquist, 1974). By this time we already had evidence that unit size can be influenced by situational factors, so we investigated reliability at the two instructional extremes of fine-unit and gross-unit marking. To anticipate later discussion, we note that this decision reflects our opinion that we are measuring a kind of cognitive-perceptual skill, or ability, rather than a predisposition.

Subjects were asked to segment a 7-minute videotape of a male exploring, inspecting, and unsystematically arranging various objects in a small, cluttered room. For both fine-unit and gross-unit instructed subjects, the number of units marked at the first viewing correlated .87 with the number of units marked at the second viewing 5 weeks later.

When the data were analyzed separately for each of the 7 minutes of the sequence, test–retest reliabilities remained high and significant, despite significant differences in the number of units marked for different minutes of the sequence. These results imply that, in segmenting a sequence, judgments of unit break points are coordinated to something special occurring in particular intervals of the stimulus sequence. That is, whatever the basis for unit judgments, that basis was present in the sequence, as it seems unlikely that subjects could remember their markings over a 5-week interval.

From more detailed analysis of these data, evidence consistent with this implication was found. Comparison of the precise pattern of markings for each subject at test and retest indicated significantly better than chance repeatability at both instructional levels. Subjects could not only repeat the number of units recorded but could also repeat their pattern of marking.

These results are encouraging, both as to the precision of the measurement technique and as to its adequacy as a measure of the subjective unit of behavior perception. Presumably, the subjective organization of behavior, while variable, is not completely arbitrary. Previous evidence that level of analysis varies across hierarchies of goals and subgoals indicated the measure has some sensitivity to the structure of the stimulus sequence and implies that individual intervals in the sequence are not equivalently eligible to be selected as break points. This was further documented in the present study. Correlations were computed over intervals between the number of times each interval was marked at first presentation and the number of times it was marked at second presentation. This value was .85, $p < .001$.

A second reliability study was undertaken (Newtson, Engquist, & Bois, 1975) for several reasons. First, we do not know how unique these reliabilities are to the particular sequence; it would seem reasonable that some sequences could be more reliably segmented over intervals than others. While we have no reason to expect that this sequence is atypical, we have no evidence that it is not. We tried to select a sequence of action that was relatively unstructured. The actor did not follow a systematic program of task performance. It would seem reasonable to expect that, if anything, a highly structured behavior sequence would be more reliably segmented. Secondly, subjects were operating at unitization extremes, which may be an unusual condition, and reliabilities might therefore be abnormally inflated. Reliability at natural or normal rates may differ.

Accordingly, reliability was assessed over a 5-week interval for eight different 3-minute sequences and under fine-, gross-, and natural-unit instructions. The eight sequences were as follows: #1 depicted a man pacing and intermittently answering a phone, #2 showed a man removing stacks of books from a table and shelving them, #3 showed a woman performing an interpretive dance, #4 showed a woman setting a table with plates and food, #5 showed a man clearing a table by knocking everything off of it onto the floor, #6 showed a man systematically building a tower from tinker toys, #7 showed a man taking a test, # showed a woman making a series of identical tinker toy constructions and placing them in a pattern on the floor.

The result was a substantial replication of the previous findings. The number of units employed at first marking correlated significantly with the number of units identified at second marking for the fine-unit (.76), gross-unit (.63), and natural-unit (.85) conditions. No significant differences in these correlations within and between conditions were observed. Comparison of precise marking patterns indicated better than chance repeatability in 21 of 24 conditions (3 conditions X 8 sequences). Correlations over intervals, between the number of times a given interval was marked as a break point for test and retest, were also significant on this measure. The average values were .61, .62, and .63 for fine-, gross-, and natural-unit conditions, respectively.

Two essential properties of an adequate measure of the unit of perception of ongoing behavior were exhibited: it is reliable and reasonably precise. Our next task was to validate it (Newtson & Engquist, 1975). The reliability data indicate that break point judgments are tied to some distinctive feature of the stimulus. That is, there must be something distinctive or important about those intervals that elicit high consensus as break points. Inspection of the series of break points for given sequences revealed an almost comic-strip quality, in that they appeared to summarize the sequences very well; nonbreak points, on the other hand, appeared to be much more ambiguous. Logically, at least, it seems reasonable that any sequence of action could be summarized by a series of still pictures in correct temporal order. It was possible, therefore, that our subjects were discriminating the best summary points, or the highest information points, in our sequences.

If, as proposed, the measurement technique is a means of tapping the subjective unit of perception of organization for behavior, break points should have this property, at least: the points of division or organization should contain more information about the sequence than points of continuity.

Eight 15- to 30-second filmed action sequences were prepared and marked by 20 subjects for purposes of obtaining consensus break points and nonbreak points. Ten of these subjects marked under natural-unit instructions, while 10 marked under fine-unit instructions. Three consecutive break points were selected that were agreed upon by natural-unit subjects, and three consecutive nonbreak points were selected that were agreed upon by fine-unit subjects. This was done to assure that break points were important division points in natural observation and that nonbreak points for naturally observing subjects were not possible fine-unit break points. The three consecutive break points (BP) and nonbreak points (NBP) that were selected were either BP–NBP–BP–NBP–BP–NBP or NBP–BP–NBP–BP–NBP–BP. After viewing each sequence, these standardization subjects wrote a brief description of the action and rated the sequence on a 9-point scale for intelligibility.

Forty-two subjects then participated in an experiment employing the resulting series of slides from each sequence. Two factors were varied: type of slide (break points versus nonbreak points) and order of presentation (correct versus incorrect). Each subject saw two of the sequences in each of the four conditions. Four different groups, with two randomizations of condition order were employed to obtain data for each of the eight sequences in all conditions. No subject saw the same sequence twice.

Subjects viewed a series of three slides from each sequence, and then (a) wrote a description of the action, (b) rated its intelligibility, and (c) judged whether or not the three slides were in the correct or incorrect order. The descriptions were scored for accuracy according to protocols developed from the descriptions of the pretest group.

Results indicated that break points yielded more accurate descriptions than nonbreak points, $p < .001$, confirming the hypothesis that break points contain more information about the sequence. A significant effect was also observed for order of presentation: slides presented in the correct order yielded more accurate descriptions.

In addition, it was found that break points presented in the correct order were as intelligible to our subjects as the actual film sequences; both the film sequences and correctly ordered break points were found to be rated as more intelligible than incorrectly ordered break points, which, in turn, were found more intelligible than nonbreak points regardless of order.

Finally, there was a strong difference in subject's ability to judge the slide ordering of break points and nonbreak points: order was correctly judged 79% of the time for break points; this dropped to 41% for nonbreak points. This implies that break points contain a considerable element of order information.

Strong evidence was provided that subjects are able to pick out critical summarization points in behavior which render the sequence intelligible for alternate observers. In addition, the data confirm that sequences vary considerably within themselves in information value. Two more issues remain to be resolved, however, to establish fully the validity of the marking technique.

First, while we have established that subjects are able to identify high-information points in a behavioral sequence, it has not been established that these are important processing points. This question, in turn, raises an important theoretical issue about the nature of the perceptual process in behavior comprehension. Given that particular points in the sequence are substitutable in an important way for the sequence itself, as indicated from the previous data, it is possible that behavior is perceived by the extraction and processing of sequences of *points* in a sequence, intermittently, rather than as discrete "chunks" or pieces. That is, the process of behavior perception may be itself discountinuous, rather than one of constant comprehension. If this is so, then the break points identified by the procedure would be more than summary points; they would be the points of perceptual organization themselves. Evidence that perceptual processing occurs selectively at break points would constitute powerful validation of the measurement technique as tapping the subjective units of behavior perception.

A second question concerns the reactivity of the measure. That is, does requiring subjects to record break points alter the process it is designed to measure? This issue is particularly important in the attempt to assess the role of break points in processing, as a discrete response–pressing a button–is paired with the particular points the method seeks to identify. Evidence for processing at break points, such as superior recognition, could therefore be artifacts of response-generation processes.

As a means of establishing that processing of behavior sequences does indeed occur at break points, we decided to test recognition for break points against recognition for nonbreak points. Past results using visual recognition memory tasks, however, might seem to work against use here. Shepard (1967), for example, found that visual recognition memory for pictures was not substantially different from perfect. Shepard's stimuli, however, consisted of unrelated pictures varying on many dimensions. Goldstein and Chance (1970) demonstrated that within homogeneous stimulus arrays, recognition is well below perfect. Exact performance was dependent upon stimulus class; recognition for faces was highest at 71% and lowest for snow crystals, where performance fell below chance levels. Further, Bower and Karlin (1974) have employed this measure with good results as an index of depth of processing for pictures of faces.

Six 2- to 6-minute film sequences were prepared portraying an actor taking an exam, dancing, cutting out a dress pattern, fixing a motorcycle, looking for a lost object, and waiting impatiently for a phone call. All of the films were silent. These films were divided in half and presented to 20 pretest subjects who segmented them according to our procedure. Ten subjects segmented each half of each sequence, to guard against the possibility that knowledge of one half would alter unitization of the other. Thirty-six break points and 36 nonbreak points were thus identified and extracted from the sequences.

We called back 12 of these pretest subjects 2 weeks later and asked them to judge whether or not they had seen the resulting slides. All slides for this group were "old," in that they were drawn from sequences these subjects had marked. This was done to make sure that we would not get ceiling effects. These subjects recognized 76% of the break points correctly, while only 45% of the nonbreak points were correctly identified, $p < .001$. Thus encouraged, we proceeded with the full experiment.

Sixty-four subjects participated in the experiment, which was a 2 × 2 × 2 × 2 × 2 mixed factorial design. Between-subjects factors were conditions of viewing (mark versus watch only), the two halves of the sequences, and two randomizations of order of recognition item presentation. Within-subjects factors included prior exposure (old versus new) and slide type (break points versus nonbreak points). Subjects viewed the films, waited for 10 minutes, and then proceeded with the recognition test. Each slide was displayed for 12 seconds; subjects indicated their judgment as "old" or "new," and gave a confidence rating for their judgment.

There were no differences in accuracy of recognition between those subjects who watched the sequences and those who marked them. Results were perfectly in accord with predictions: break points from previously viewed behavior sequences, $\overline{X}$ = 73%, were recognized significantly more accurately than previously viewed nonbreak points, $\overline{X}$ = 67%, and more accurately than new break points, $\overline{X}$ = 62%, and new nonbreak points, $\overline{X}$ = 60%. Signal detection analysis incorporating the confidence ratings confirmed that higher recognition for break points is not due to a positive response bias but reflects real differences in perceptual sensitivity.

Our subjects reported that this was an extremely difficult task. They had just seen the films from which the old items were taken, but the new items were extremely similar to the old ones in employing the same actors, backgrounds, etc., indeed, in being continuations of the same action sequence.

Although this study provides highly significant confirmation of our hypotheses, the absolute magnitude of the results may seem somewhat disappointing. It is important to keep in mind at this point that the study was directed at demonstrating increased processing at break points, not encoding of the break points themselves. This distinction will, it is hoped, become more clear in the theoretical discussion to follow.

As a more persuasive demonstration of the role of break points in ongoing behavior perception, another experiment was performed (Engquist & Newtson, 1975) to demonstrate that observers are perceptually more sensitive to disruption at break points than at nonbreak points. In this study, after a pretest group had identified break points and nonbreak points in a series of sequences, we deleted either 4, 8, or 12 frames at these points (4 frames = 1/6 second). Twenty-one subjects were then asked to detect all occurrences of missing action. Results indicated that deletions at nonbreak points were detected 29% of the time, regardless of the number of frames deleted. Subjects were far more sensitive to deletions at break points, detecting them at a rate of 38, 54, and 70% accuracy for deletions of 4, 8, and 12 frames, respectively. Detection of deletions of break points was significantly superior to detection of nonbreak points at all levels and significantly increased as the duration of the deletion increased.

Taken together, then, these data provide strong evidence that break points are importantly involved in the perceptual processing of behavior sequences. And, while it is difficult to interpret the null hypothesis, there was no evidence that the fact of marking differentially affected accuracy of recognition. Thus far, then, we may conclude that this measure of the unit of perception of ongoing behavior is reliable, valid, and nonreactive.

VARIATIONS IN LEVEL OF ANALYSIS

As is apparent from the studies cited previously, persons, upon request, will readily vary the size of the unit that they employ in behavior perception. The size of the unit, as indicated by the length of the average interval between unit marks, or the total number of units employed by a perceiver for a given segment has been found in our research to vary quite readily in response to situational factors as well. The precise cognitive consequences of such variation are still unresolved and will be discussed later.

It is apparent, however, that each individual has some "range of analysis" within which he may operate. That is, each person can analyze a sequence of behavior into fine units, large units, or at least one level between the two. We have consistently found that, when instructed to analyze a sequence "naturally," the mean number of units for a given sequence will be somewhere in-between that for fine-unit or gross-unit instructed subjects. Where on that continuum the level of analysis falls is very much a function of the particular sequence. In general, natural-unit analysis for sequences portraying highly organized, step-by-step action, with a clear hierarchy of subordinate and superordinate goals, will tend to be closer to gross-unit levels. Irregular, loosely organized action sequences will tend to produce natural unit sizes closer to fine-unit analysis.

One factor we have investigated which affects unit size is that of predictability. In one study (Newtson, 1973), two versions of a videotape were prepared of an actor performing a molecule assembly task. In one of the tapes, a 30-second insertion was made of an unexpected action. The actor removed a shoe, put it on the table, and rolled his other pants leg up to the knee. The primary dependent measure was rate of unitization in the final 3 minutes of the tapes in the two conditions. Tapes were carefully standardized as to the number of problem steps and their elapsed time completed throughout the tapes. The insertion was made after 2 minutes of action.

For the 2 minutes preceding the unexpected action, rate of unitization did not differ. For the last 3 minutes of the two condition tapes, however, subjects who had viewed the unexpected action employed significantly more units per minute than controls. In addition, it was found that unitization rate for the predictable sequence declined over time. This was a very reasonable result, in that as an observer gains understanding and predictive control of a sequence–particularly a regular, highly structured one, as in the present instance–he should be able to organize it over longer intervals. The net result, however, was that the unexpected action prevented this decline in that condition.

Wilder (1974) followed up this experiment, pointing out that it was not clear whether the results on unitization were due to the change in behavior or the change to unpredictable behavior. Wilder prepared stimulus tapes of a person assembling a series of six-page booklets. The larger-unit organization remained constant, in that the stimulus person collated and stapled booklet after booklet. At the finer-unit level of organization, however, the predictability of the behavior was varied. In the predictable tapes, the actor assembled the booklets in a consistent manner (he selected pages in a constant order). In another version, unpredictable tapes, page selection was performed in a random, haphazard, unpredictable manner. For the sequence that was predictable at both levels, Wilder found a constant, low unitization rate over the entire sequence. This seemed to be a floor effect, in that subjects quickly adopted the larger unit organization and remained there. For the sequence that was unpredictable at the lower level of analysis, he found a significantly higher rate of unitization for the first half of the sequence than for the second half.

Wilder also varied predictability at lower levels within sequences. In one condition, lower level organization was predictable for the second half; in a second condition, this was reversed. Results indicated that the change in lower level predictability from unpredictable to predictable simply prevented the decline observed in the consistently unpredictable condition described above. Consistent with the Newtson (1973) data, then, a change per se in behavior may prevent the transition to higher levels of analysis.

Results from the predictable to unpredictable condition, however, indicated a significant increase in unitization rate. This difference was mostly accounted for by a tremendous increase in unitization rate immediately following the change to unpredictable behavior at the finer-unit level. These results demonstrate a powerful effect of unpredictability per se on behavior organization.

Overall, Wilder (1974) suggests that persons begin at fine-unit levels of analysis and work up to higher levels. If behavior changes at the level of analysis the perceiver is employing, he tends to remain at that level. If the behavior changes to unpredictable action, however, it would appear that the perceiver begins again at the finest unit level of analysis (the jump in unitization in Wilder's predictable to unpredictable condition). The subsequent decline in unitization rate observed in Wilder's predictable to unpredictable condition to a unitization rate lower than that in the unpredictable to predictable condition suggests that once a perceiver reaches a higher level of analysis, he may return to it quite readily.

Frey and Newtson (1973) employed a different situational manipulation with implications for predictability, that of social power. In an unequal power dyad, the high-power person has predictability of the low-power person's actions by virtue of his position. The low-power person, however, has less control of the other's actions and, hence, other things held equal, is less able to predict the other's actions. Frey and Newtson introduced subjects to an experiment which ostensibly was designed to study the effects of familiarity on worker–supervisor performance. Each task pair was to consist of a worker and a supervisor. The supervisor's duty was "to make sure the worker understands what to do, see that the worker completes the task, and then to recommend to the experimenter the size of the reward the worker should receive." The worker's duty was "simply to cooperate with his supervisor by performing the work assigned to him in such a way that he will gain the most rewards."

Subjects then watched an 11-minute videotape of a person assembling a block puzzle; they were told, however, that it was a live view. They were asked to keep a continuous record of their partner's activity during the observation period (ostensibly because we could not control for the amount of activity and wanted a measure of it). The rationale for this one-way observation was that it would permit us to separate the effects of workers' familiarity with supervisors from effects of supervisors' familiarity with workers.

Evidence was found for a significantly higher unitization rate when workers watched their supervisors-to-be than vice versa, confirming the hypothesis. We replicated this study with a stronger utility manipulation, thinking that perhaps increased utility of the information would strengthen the effect. The result, however, was that the effect was completely eliminated.

The effects of utility on unitization are problematic. A straightforward prediction would be that as the other's behavior becomes more important to the perceiver, finer unit analysis would result. In several experiments, however, we have found exactly the opposite result. Newtson (1974b) had subjects observe a person via videotape; subjects either expected to meet the person following their observation (interaction expected condition) or were explicitly told they would never meet him (no interaction expected). It was anticipated that information from the other's behavior would have greater utility for persons anticipating interaction than those for whom the person's behavior was less relevant and hence would lead to finer unitization. In fact, the reverse result was obtained.

In another experiment (Newtson, Engquist, & Bois, 1974a), utility was manipulated more directly. In this experiment, subjects were asked to segment a videotape of problem-solving behavior. Subjects were further told that chances were only about one in three that the person would complete his task in the allotted time (indicated by a tone); if he did, however, they would receive either 25¢ (low-utility condition) or $2.00 (high-utility condition) in addition to the money they were already receiving for their services. A third group (no utility) was given the same information, but no money contingent on the stimulus person's performance was mentioned.

Results were consistent with our previous manipulations of utility. Low-utility and no-utility condition subjects averaged significantly more units per minute than high-utility subjects. That our manipulation had succeeded in involving subjects was apparent from condition differences in their estimates of how well the person they observed would perform on subsequent tasks. A dissonance type effect resulted, such that subjects with a small investmant in the stimulus person's performance tended to rate him highly despite his failure; a large investment in his performance, however, produced a significantly lower rating for his failure. Ratings in the no-utility condition fell between the other two conditions.

This puzzle provoked yet another study in an attempt to understand this phenomenon (Newtson, Engquist, & Bois, 1974b). Increased utility, we reasoned, could have its effect upon unitization in one of two ways. First, it could be simply distracting the subject, causing him to think about other things, his own concerns, etc., or at least causing him to ignore behavior that was not relevant to achievement of the goal most personally relevant to him. This assumes that the unitization of behavior is typically under control of higher level processes. The fact that subjects can readily vary level of unitization by instruction confirms that higher level processes are able to affect it; the distraction/selection hypothesis assumes that subjects readily assert their own control of the unitization process.

An alternative hypothesis would be that the effect is due to increased arousal when utility is high. Kahneman (1973) cites evidence that arousal causes an increased tendency to focus on a few relevent cues, that is, it decreases the range of attention while amplifying its intensity. The effect of arousal on task performance is thus quite dependent upon the nature of the task. Where task performance requires close attention to a few, easily discriminable cues, arousal enhances performance; where the task performance requires difficult discriminations between many cues, arousal interferes with task performance. It is possible that high utility causes subjects to focus their attention on fewer, more easily discriminable cues (that is, break points), resulting in larger unit analysis. This would imply that our utility manipulations in previous studies were affecting the unitization process at a lower level of processing.

We decided to explore the effects of two types of interference tasks upon unitization and the recall of behavior. Sixty-six subjects segmented two behavior sequences under one of three conditions. In all conditions, subjects were instructed to segment the behavior into units that seemed natural to them. Two types of interference were employed. In one condition, a cognitive interference task was used; subjects were instructed to count outloud backwards from 100 (cognitive interference condition). In a second condition, subjects were subjected to intermittent white noise (arousal condition); this was selected on the basis of findings that white noise reliably and effectively induces high states of arousal (Glass & Singer, 1972). In a third condition, no interference was introduced (control). Measures were the number of units employed and a recall test of 18 multiple-choice recall items.

Results indicated that only subjects in the arousal condition tended to differ from control subjects. Mean number of units were 44.96 for the control group, 35.14 for the arousal condition, and 42.05 for the cognitive interference condition. There was a marginally significant difference between the arousal consition and the other two conditions. The analysis was complicated due to extreme heterogeneity of variance in several cells of the design. The results should be interpreted cautiously and bear replication.

Results on the recall measure were less ambiguous. The recall test was difficult, as indicated by a mean of only 65% correct in the control condition. Performance in the arousal condition was marginally poorer, while performance in the cognitive interference condition was significantly worse than in both the control and the arousal conditions.

While these results do not resolve the issue of the effects of utility on unitization, they do provide suggestive evidence on two points. First, they suggest that utility manipulations can reduce unitization by arousal; whether this results in disruption or distortion of behavior perception could well depend upon the discriminability of the break points necessary for veridical perception. If the critical break points are easily discriminable, increased arousal could enhance observer accuracy; if ambiguous, however, increased arousal could be highly disruptive. Second, they suggest, congruent with earlier findings on the validity of the measurement of procedure, that unit formation occurs at a very early stage in the perceptual interpretation of behavior. That is, the cognitive interference task apparently disrupted memory encoding, without substantially disrupting unitization. Results from the arousal task could plausibly be interpreted as reflecting interference at the stage of unit formation, in that fewer units were recorded, and a decrement in recall was observed, although less of a decrement than was observed under cognitive interference. This would be consistent with Kahneman's (1973) conclusion that arousal focuses attention more closely on a narrower range of cues. Certainly a considerable number of alternative interpretations are possible. These await further research.

A THEORY OF BEHAVIOR PERCEPTION

Given evidence for the reliability and validity of the measurement technique and some evidence as to the nature of the process of behavior observation, it now becomes possible to formulate a highly specific and testable theory of behavior perception.

The most basic assumption of the theory is that an action is defined by a change in a feature of a stimulus array. Any feature change is necessary and sufficient for an action to be perceived. This can be illustrated quite simply, as follows: consider the four stimulus arrays in Fig. 1.

One can readily comprehend an action if one sees (a) and then (b) (or vice versa, although not all feature changes need have this property), or (c) and then (d). In both sequences a feature has changed: in the first example, the figure raised its arm; in the second, the figure raised its leg. If one is shown (a) and then (c), however, no action is specified, as no common features have changed from (a) to (c). Obviously enough, in an ongoing behavior sequence, there is nearly constant change, and not all changes lead to the perception of action. The perceiver has a great deal of discretion in the feature changes that he uses for purposes of organization.

Behavior perception, then, may be viewed as a feature monitoring process. The perceiver monitors some criterial set of features (a subset of the available features), segmenting the behavior into parts as one or more of the monitored features change their state. Break points, then, are points in the ongoing sequence where a noticeable change in state of one or more of the observer's criterial features has occurred.

It should be noted that this definition of meaningful action specifies that the information gained is from *change* in the stimulus, not from the stringing together of a *series of discrete states* of the stimulus. This definition is critical to the understanding of ongoing action sequences and has important implications for the way one approaches the perception of meaning of behavior. A meaningful action can *only* be portrayed by a minimum of two break points in which a common feature has undergone transformation.

This notion is not an original one in our research. Eisenstein, the great Russian director and film theorist, made this assumption the central principle of his theory of film technique, terming it the principle of montage. The primitive way to think about action sequences, Eisenstein (1949) pointed out, is to see them as composed of a series of building blocks extended linearly in time, with meaning a function of what follows what. That is not the way it works at all, Eisenstein argues. The perception of movement in the film image itself, depends not upon one image *following* another image, but upon one image being *overlaid* over another; the meaning (that is, the movement) is defined by the *change* or

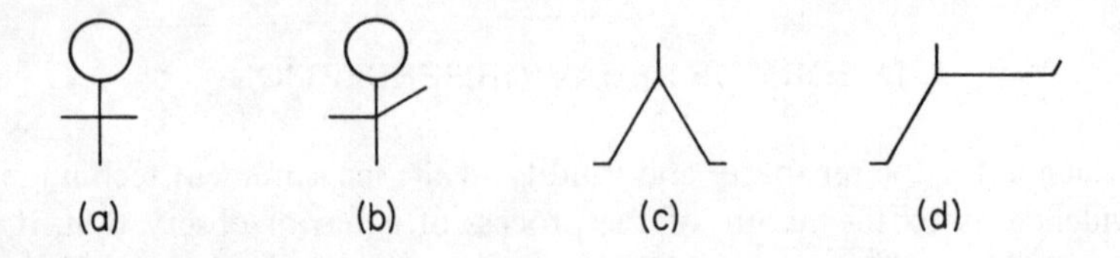

FIG. 1 Four possible action-defining stimulus arrays (see text for explanation).

difference between the two images. The same is true, he suggests, of larger levels of meaning in action. In his view, events should be viewed as having a *depth* dimension in time, consisting of the overlay of successive states, with meaning resulting from the differences between them.

The difference is more than a merely heuristic one. It is similar in its approach to behavior meaning to Gibson's (1966) approach to the perception of optical motion. Gibson argued that motion can be described as a continuous series of perspective transformations. Thus, to choose but one relevant implication for the present discussion, the *same* motion may be perceived in many different objects. The motion is defined by change of a particular type, independent of what is changing.

The same may be true of social meanings, such as intention. Heider and Simmel (1944) produced a geometrical cartoon with which they demonstrated that persons could "see" or organize an event in terms of what they defined as psychological or mental entities. They concluded: "Just as the successive perspective views of a landscape from a moving train can only be 'resolved,' or made to yield a meaningful unit, by reference to distant objects laid out in space, so acts of persons have to viewed in terms of motives in order that the succession of changes becomes a connected sequence [p. 258]." The experience of "motive" or intent is as direct and immediate as the perception of depth, to which Heider (1958) directly compares the process. Such judgments, however, have often been found to be highly variable and hence have been thought to be the product of higher level inference processes, rather than the result of direct apprehension of stimulus information. Given the discretion that perceivers have in feature selection, however, it is possible that this variability is due to differences in whether or not the perceiver "picks up" the existing information, in Gibson's (1966) terms, rather than to problems of construction by higher level processes. At any rate, in their film, and many subsequent geometric cartoons, Heider and Simmel (1944) demonstrated that social meanings such as motive and intent can be conveyed very powerfully and completely by a definite series of changes in a stimulus, regardless of what it is that is changing.

Preliminary results of a detailed analysis of the eight sequences employed in the Newtson, Engquist, and Bois (1975) reliability study provide encouraging evidence for the theory. These sequences were coded with the Eshkol–Wachman movement notation (Eshkol, 1973). This notation system was developed to permit choreographers to write dance scores but has been employed in animal research to render precise descriptions of sequences of movement. It permits the description of behavior sequences complete enough for the actual behavior to be reproducible from the notation itself, by specifying the precise position of the body at successive, predetermined intervals. Sequences were coded at 1-second intervals.

It should be emphasized that this notion system is a rendering of the movement in the sequence, not the meaningful actions. The features employed are not necessarily those used by perceivers in the perceptual organization of behavior. As the coding is exhaustive with respect to movement, however, its elements should correlate closely enough with the functionally operating features employed by perceivers. For example, while the system requires separate codings (treats as different features) the left hand, left forearm, and left upper arm, these different features may be employed as a single entity in the perceptual organization of ongoing action.

Following coding of the sequences at all intervals, we identified four types of interval-to-interval transitions: (a) break point to break point, (b) nonbreak point to nonbreak point, (c) break point to nonbreak point, and (d) nonbreak point to break point. If, as hypothesized, perceptual organization is achieved by means of feature changes in the stimulus, then there should, on the average, be a significantly higher probability of a difference in position of the stimulus person when (a) successive break points are compared and (b) a nonbreak point interval is compared with an immediately following break point interval. Conversely, probability of a difference in position of the stimulus person should be lower when (c) successive nonbreak points are compared and (d) when a break point interval is compared with an immediately following nonbreak point interval. The analysis is tedious, requiring codings for each 1-second interval of the stimulus sequences and then comparison of the codings for the four types of interval transitions. At this writing, analysis has been completed for two of the eight sequences. Consistent with the theory, probabilities of a change in the stimulus are equivalent for break point to break point and nonbreak point to break point transitions, and both are significantly higher than probabilities for nonbreak point to nonbreak point and break point to nonbreak point transitions.

In addition, the sequences differed significantly in which coding features were associated with break point transitions. This suggests that the perceiver may be highly selective in which features are employed for perceptual organization of a given episode of behavior. It may thus be necessary to derive the perceptual features employed in the process empirically, accumulating a large number of instances, before systematic propositions as to the nature of these features are attempted.

Definition of behavior perception as a feature-monitoring process has a number of more concrete implications. First, the low magnitude of recognition differences obtained in the study cited above (Newtson & Engquist, 1975) are more readily understandable. As the definition of action requires information extracted from two break points, it is clear that it is not information from the break point itself that is encoded but, rather, the result of a process of comparison of information from that break point with information from a previous break point. This would imply that a larger magnitude of difference should be obtainable if the recognition items were successive pairs of break points, rather than single break points in isolation. This, in turn, implies that behavior perception must impose a short-term memory load. If an action is defined by a feature change, the state of the feature at Time 1 must be maintained for comparison when the feature changes at Time 2.

Second, the precise time at which the perceiver detects a feature change may not necessarily be critical. That is, perception of an action depends upon detecting its defining change; the action may be thus perceived at any time the feature is changed, whether or not the detection occurs at the precise moment the feature underwent the change. The importance of immediate detection of features changes may depend greatly upon the nature of the behavior sequence. One hypothesis would be that the action is experienced as having occurred at the time the feature change is detected. If this is true, immediate detection of feature change would be essential for a precise or accurate perception of the order of events. A demonstration of this phenomenon is simple to perform, and, incidentally, is the basis of many magic tricks. You need two persons, standing about 10 or 12 feet apart. One must be armed with a throwable object; a wadded-up piece of paper will serve. Let Person A and Person B converse briefly, to get your audience switching their attention back and forth between the two. Have Person B than make a quick gesture with his left hand, and then throw an object at Person A with his right. Person A may duck down as soon as Person B makes the first gesture and still be seen nearly unanimously as having ducked only when B performed the second action, throwing.

The notion that behavior perception is a feature-monitoring process is also consistent with studies of eye-movement during pattern recognition. Noton and Stark (1971) reviewed a number of studies demonstrating that persons viewing a pattern employ fixed "scanpaths" that are characteristic of each subject for each pattern and are, moreover, cyclic, in that the viewer repeats them about seven times. Noton and Stark (1971) noted that such patterns of feature-scanning are very likely habitual but were puzzled to find that the scanpath was recursive, although recognition seemed to be complete after the first scanning cycle. Such repetition is indeed redundant for the recognition of static patterns but would be ideally suited for the detection of feature changes in ongoing behavior. Indeed, the measurement of eye movement during behavior perception may be a powerful method in identifying the features used for perceptual organization of behavior, as it has been for identifying critical stimulus features in pattern recognition.[1]

From the perspective of a feature-monitoring theory, variations in level of behavior analysis could be due to one of two related mechanisms. First, and most simply, variations in level of analysis could reflect differences in the number of criterion features monitored. With increasing numbers of monitored features, the probability of a change in one of those features would be increased, and more frequent break points would be identified. A second possibility is that

[1] I am grateful to Dr. Wayne Shebilske for this suggestion and for bringing this research to my attention.

different levels of analysis reflect the monitoring of features that change at different rates. This is, the features available for perceptual organization of a behavior sequence could be ranked according to how rapidly, or frequently in a given time interval, they change. Some change very rapidly, others slowly. Fine-unit analysis of a behavior sequence could reflect the use of rapidly changing features, while large-unit analysis could reflect reliance on relatively slow changing features. In a hierarchically organized behavior sequence, for example, where there is a clear superordinate/subordinate goal structure, organization employing feature changes defining the achievement of subgoals would be equivalent to reliance on a rapidly changing feature; reliance on feature changes defining the achievement of superordinate goals would be equivalent to relying on a slow-changing feature.

Further, there seems to be reason to expect that a span of apprehension for feature change rates exists, such that features changing too slowly or too rapidly cannot be employed for perceptual organization. Consider, for instance, the experience of someone rapidly dribbling a basketball. The cyclic movement of the ball or perhaps the downward thrusts of the dribbler's hand could, in principle, be employed to organize the sequence bounce by bounce. When the dribbling is rapid, however, we tend to "see," or organize in our mind, the event dribbling from the start of it to the end of it. If, however, the dribbler slows down at a continuous rate, there comes a point at which one begins to see the event bounce by bounce. The relevant features are now slow enough to be within our span of apprehension for feature changes.

A similar example may be given for slow feature changes. Eibl-Eibesfeldt (1970) cites an instance where a newspaper seller was filmed going about his business. At normal speed, nothing unusual was detected about his behavior. When viewed at fast motion, however, Eibel-Eibesfeldt reported that the film revealed that the man patrolled a very precisely defined territory, as if he were tethered on a leash. Eibl-Eibesfeldt noted that fast-motion film techniques are valuable to ethologists because they "make visible certain regularities in behavior which normally escape direct observation [p. 415]." Our interpretation would be that the feature changes defining the organization of the behavior—gradual movement around the perimeter—changed too slowly for an observer to employ them to experience the organization that existed in the behavior. It was outside the span of feature-change apprehension.

Given the discretion of the observer in behavior perception, furthermore, it is likely that expectancies, or sets, could affect behavior perception by the mechanism of altering the set of features monitored. Zadny and Gerard (1974) reported a series of studies demonstrating that information about actors' intentions prior to observation of behavior biases the perception of the behavior, as indicated by increased recall for intent-related information. Evidence that the bias occurred during observation, rather than as biased retrieval, was provided by a control which was provided the intent-relevant information after viewing. This group showed no such effects. This could be tested directly by replicating their study while employing the unitization measure, predicting significant differences in selection of break points between conditions with contrasting intention information.

A final set of issues concerns the perceiver's ability to edit systematically the criterial set of features as observation proceeds. That is, the same features may not be monitored continuously throughout the behavior episode. Given the limits on the number of features monitored, perceivers may adopt monitoring priorities, such that the appearance of a given feathre may cause the observer to cease monitoring another. In addition, shifts in feature-monitoring patterns may reflect the perceiver's dependence upon his causal grammar of the event, such that certain feature changes result in systematic shifts to different features, as the occurrence of one action directs the observer to be vigilant for other actions.

The theory, then, may be summarized as follows. Behavior perception is achieved by monitoring a set of features of the stimulus array; a change in any of the monitored features results in the perception of a meaningful action. Interpretation of a given action sequence for a given observer, then, will be a function of those features monitored. Level of analysis of behavior may be altered by (a) increasing or decreasing the number of features monitored or (b) monitoring features that change at faster or slower rates. Limits on the process are those primarily of the duration of short-term memory.

Finally, it should be noted that the theory may be a more general one of event perception not just behavior perception. While the focus of researchers has been on behavioral events, as these events are both important and frequent in the experience of persons, there is no necessary reason to limit the theory to these instances. We have focused upon perception of behavior of single actors in developing these ideas; the principles derived may be applicable to ongoing events of many other types as well.

BEHAVIOR PERCEPTION AND CAUSAL ATTRIBUTION

Heider (1958) defined the attribution process as the organization into meaningful units of a continuous stream of information from another's behavior. He emphasized that the perceiver responds to the meaning of other's actions, not to overt behavior. That is, "*p* perceives the distal object, the psychological entities that bring consistency and meaning to the behavior; *p*'s reaction is then to this meaning, not to overt behavior directly . . . Generally *p*'s perceptions of *o* are coordinated to psychological dispositional properties of *o*, and not to the cues mediating those properties [pp. 33–34]." These units of meaning, then, constitute the perceptual input to subsequent causal inference processes.

There is an important difference between these units of input in Heider's (1958) formulation and the units of experience postulated in subsequent derivations of attribution theory (Jones & Davis, 1965; Kelley, 1967). Clearly, Heider assumed that the input already includes important causal judgments at some level. He drew an explicit parallel with Michotte's (1946) work on the perception of mechanical causality. Michotte demonstrated that certain physical configurations give rise to immediate, unambiguous experiences of causation. Heider (1958) had something very similar in mind when he argued that animate objects have the potential for patterns of action—what he termed equifinality—that permit a particular set of invariances to be employed in the perceptual organization of action.

Kelley's (1967) definition of what constitutes the unit of input to the process is not the same as Heider's. In Kelley's view, the primary datum for the attribution process is an "entity-effect covariation." Basically, an effect is said to be attributed to that entity which is present when the effect is present and which is absent when the effect is absent. This primary datum is converted to internal or external attribution by examining variation in effects over entities, persons, modalities, and time, with respect to several criteria of validity. It is at this point that Kelley's argument shifts from the singular to the plural, that is, from an attribution to a set of attributions. In the analysis, he thus shifts from the direct perception of causal entities in the stimulus field—Heider's focus—to the combination of successive perceptual entities into stable sets of causal beliefs.

Thus, Heider (1958) focused on attribution as the active construction of meaning in behavior, while Kelley (1967) focused on causal analysis after the perceptual organization of events. To put it another way, Heider was concerned with attribution as the primary perceptual process: causal attribution is the unit of perception. He focused on "the features surrounding the actions of another person which lead us to penetrate the depth dimension and precipitate into reality the meaning of actions" (Heider, 1958, p. 82). In Kelley's (1967) formulation, on the other hand, the perceptual unit is entity-effect covariation, and his concern was with causal analysis following from an analytic combination of these data patterns. Thus, he shifted from analysis of single attributions to the specification of properties of sets of attributions. His central emphasis, the "subjective validity" of causal inferences, is meaningful only in relation to sets of attributions following the accumulation and successive combinations of initial observations.

Jones and Davis (1965) focused on a different aspect of attribution. While Kelley (1967) was most concerned with "the allocation of causality between the environment and self [p. 196]," Jones and Davis (1965) were most involved with the attribution of personal causality to others. They addressed themselves to the problem of specifying the antecedent conditions for the attribution of dispositions to an actor. In particular, they were interested in accounting for the differential salience of particular "effects of actions" in the inference process.

Actions are seen to vary with respect to their information value for the perceiver. When acts have high information value, the perceiver is likely to draw the conclusions thus made possible, that is, attribute a dispositional characteristic to the actor consistent with the revealed information. In contrast to Kelley, then, Jones and Davis (1965) were concerned with the analysis of a single attribution.

Acts are defined as molar responses reflecting some degree of personal choice on the part of the actor, which have one or more distinctive effects. Effects, in turn, are defined by Jones and Davis (1965) as discriminable changes in the "pre-existing state of affairs [p. 225]" that are brought about by action. The specification of choice of actions, and hence choice of effects of actions, is important to this theory. Effects of an act provide information because they are assumed to have been chosen by an actor who had alternative actions available to him.

The unit of input to Jones and Davis' (1965) theory, then, is the effect of actions. Like Kelley (1967), Jones and Davis (1965) were clearly concerned with inference processes subsequent to the perceptual organization of action. And, as in Kelley's theory, the causal information is treated as a deduction from the input–in this case the effects of actions–rather than being part and parcel of the process of perceptual organization itself, as Heider (1958) would have it.

The question at issue here is the degree to which the units of perceptual input to the process include causal judgments and the degree to which they are the result of later stages of inference. In a sense, Kelley's (1967) was concerned with the analytic combination of these units, while Jones and Davis (1965) emphasized the differential weights of individual inputs.

In the research reviewed above, I have been concerned with the formulation of these inputs. The relevance of that research to attribution theories hinges on three questions: (a) Does the unit of input contain some level of causal judgment? (b) Does it vary? and (c) Does that variation have implications for later causal inference processes?

With respect to the first issue, we now have clear evidence that each unit contains some level of causal information. In several studies, we have provided subjects with two additional buttons and asked them to record, as they marked each unit, whether the action they recorded was "chosen by the person" or "produced by the situation." Both reliability studies described earlier included this additional measure (Newtson & Engquist, 1974; Newtson et al., 1975).

In the first reliability study (Newtson & Engquist, 1974), this unit-by-unit attribution judgment was highly reliable over a 5-week test–retest interval for both fine- and large-unit judgments. Correlations of number of person chosen units employed at first test with number marked at retest averaged .86, with all of the within-conditions correlations significant beyond the .01 level. This value was .70 for situation produced units. Test–retest correlations of the number of each type of judgment elicited by each interval of the stimulus was .80 for person caused units and .67 for situation produced units, $p < .001$. The sequence employed in this study averaged 76% person caused units.

In the second reliability study (Newtson et al., 1975), employing the eight sequences described earlier, we observed differences in both the reliability of subjects and intervals on these judgments between sequences. In general, subjects were less reliable than intervals, but substantial and significant reliability was obtained on both measures. These data demonstrate that reliable causal information is available at discrete points in the action sequence itself and that this information is importantly involved in its perceptual organization.

That the unit of perception varies is well-established; that variation in level of analysis may have implications for later causal inference processes is also established, but the nature and direction of those effects are as yet ambiguous. Newtson (1973) found, initially, that fine-unit analysis resulted in increased dispositional attribution, as compared to large-unit analysis, and that more confident and differentiated impressions of the stimulus person resulted. Frey and Newtson (1973) reversed that result on attribution, however, finding that a manipulation leading to finer unit analysis led to increased situational attribution. Results of the second of the two studies in the Newtson (1973) paper also produced a reversal of the result, although it was of marginal significance. In the first study we varied level of analysis by institutional set, while in the latter two we employed situational manipulations. It should not be concluded that manipulations producing finer unit analysis lead to increased situational attribution while instructional variations with similar effects lead to increased dispositional attribution. In one study reported by Newtson (1974b), it was found that a manipulation producing fine-unit analysis replicated the increased dispositional attribution found with fine-unit instruction on the same sequence in an earlier study. The effect seems to be dependent upon the sequence involved. We have found several sequences upon which we can readily replicate the original result (Newtson, 1973). That is, fine-unit analysis will lead to more dispositional attribution, with incresased confidence in impressions of the stimulus person. We have also found sequences where increased situational attribution is produced by fine-unit analysis, replicating the Frey and Newtson (1973) result. In addition, we have found several sequences where attribution is not affected one way or the other by level of analysis.

The difference seems to follow from the nature of the sequence itself. All sequences producing high personal attribution as a result of fine-unit analysis have been "free" behavior sequences, where the actor moves about more or less of his own volition. Sequences producing more situational attribution, by contrast, all involve a stimulus person working at a task, such as assembling a molecule model according to instructions (Experiment 2, Newtson, 1973) or completing a complex block puzzle (Frey & Newtson, 1973). As a preliminary check, we assessed the proportion of situation produced units from the tape employed in the Frey and Newtson (1973) sequence and found it to contain 80% situation produced units.

This result is an intriguing one. We have delayed following it up, pending evidence on the reliability and validity of the marking procedure and are only now beginning to pursue it. One implication is that the units of perception constitute the entity-effect covariations postulated in Kelley's (1967) theory. This would suggest that sequences could be carefully composed of varying combinations of number and types of behavior units, permitting a direct test of Kelley's model of attributional analysis. An encouraging corollary result of all of these experiments investigating attribution is that large-unit analysis yields neutral attribution on the personal-situation dimension; fine-unit analysis, depending on the sequence, produces a displacement toward either the personal or situational end of the scale. If one assumes that fine-unit analysis yields more of Kelley's input units, it would be reasonable to expect that, given more data points, persons are able to make more differentiated attributions as to cause.

In any event, our data show that (a) the units identified by our procedure do contain reliable information for causal judgments and (b) that variation in level of analysis may, under conditions not well understood, alter the output of the attribution process. This result has important implications for research and theory in attribution. Research on attribution processes has, in general, followed two approaches. One approach has been to vary the configuration of input information, in order to verify that the rules of inference postulated in the theory are indeed correct (cf. Ajzen, 1971; Newtson, 1974b). In the second line of research, investigators have focused upon "bias" in attribution, particularly motivational biases. When personal or situational variables are found to alter attribution, the prime explanation is sought in terms of motivated distortions of the inference process. The finding that the input to the process, the unit of perception, readily varies in response to situational variables raises the possibility of an alternative set of explanations for motivational variables, that of input bias. That is, both input processes and inference processes may lawfully respond to motivational variables. Research on these factors must take input variation into account.

The observation that behavior has a temporal dimension is so obvious as to be a truism, yet the process of the perception of behavior has attracted little theory and still less empirical investigation. The process is of particular importance to social psychology, because, as Heider (1985) noted, "It is probably fair to say that the stimulus fields basic for person perception are usually *more extended in time* than those relevant to thing perception [p. 39, italics in original]."

ACKNOWLEDGMENTS

The research reported herein was supported by U.S. Public Health Service Grant No. MH2400-01 and by Grant No. DAHC19-74-G-0016 from the U.S. Army Research Institute for the Social and Behavioral Sciences.

REFERENCES

Ajzen, I. Attribution of dispositions to an actor: Effects of perceived decision freedom and behavioral utilities. *Journal of Personality and Social Psychology,* 1971, **18,** 144–156.

Bower, G. H., & Karlin, M. B. Depth of processing pictures of faces and recognition memory. *Journal of Experimental Psychology,* 1974, **103,** 751–757.

Eibl-Eibesfeldt, I. *Ethology: The biology of behavior* (Transl. by E. Klinghammer). New York: Holt, Rinehart, & Winston, 1970.

Eisenstein, S. *The film form.* New York: Harcourt, Brace, & Co., 1949.

Engquist, G., & Newtson, D. Critical information points in the perception of ongoing behavior. Unpublished manuscript, University of Virginia, 1975.

Eshkol, N. *Moving writing reading.* Tel Aviv: The Movement Notation Society, 1973.

Frey, J., & Newtson, D. Differential attribution in an unequal power situation: Biased inference or biased input? *Proceedings,* 81st Annual Convention of the American Psychological Association, 1973.

From, F. *Perception of other people* (B. A. Maher & E. Kvan, translators). New York: Columbia University Press, 1971.

Gibson, J. *The senses considered as perceptual systems.* Boston: Houghton Mifflin, 1966.

Glass, D. C., & Singer, J. E. *Urban stress: Experiments on noise and social stressors.* New York: Academic Press, 1972.

Goldstein, A. G., & Chance, J. E. Visual recognition memory for complex configurations. *Perception and Psychophysics,* 1970, **9,** 237–241.

Heider, F. *The psychology of interpersonal relations.* New York: John Wiley & Sons, 1958.

Heider, F., & Simmel, M. An experimental study of apparent behavior. *American Journal of Psychology,* 1944, **57,** 243–259.

Jones, E. E., & Davis, K. E. From acts to dispositions: The attribution process in person perception. In L. Berkowitz (Ed.), *Advances in experimental social psychology* (Vol. 2). New York: Academic Press, 1965.

Jones, E. E., & Thibaut, J. W. Interaction goals as bases of inference in interpersonal perception. In R. Tagiuri & L. Petrullo (Eds.), *Person perception and interpersonal behavior.* Stanford, California: Stanford University Press, 1958.

Kahneman, D. *Attention and effort.* Englewood Cliffs, New Jersey: Prentice-Hall, 1973.

Kelley, H. H. Attribution theory in social psychology. In D. Levine (Ed.), *Nebraska Symposium on Motivation.* Lincoln: University of Nebraska Press, 1967.

Michotte, A. La perception de la causalité. Louvain Institut Supérieur de philosophie, 1946.

Newtson, D. Attribution and the unit of perception of ongoing behavior. *Journal of Personality and Social Psychology,* 1973, **28,** 28–38.

Newtson, D. Dispositional inference from effects of actions: Effects chosen and effects forgone. *Journal of Experimental Social Psychology,* 1974, **10,** 480–496. (a)

Newtson, D. Selective variation in the unit of behavior perception. Unpublished manuscript, University of Virginia, 1974. (b)

Newtson, D., & Engquist, G. The reliability of the unit marking procedure. Unpublished manuscript, University of Virginia, 1974.

Newtson, D., & Engquist, G. Unit break points as summary and processing points. Unpublished manuscript, University of Virginia, 1975.

Newtson, D., Engquist, G., & Bois, J. The effects of utility on unitization. Unpublished manuscript, University of Virginia, 1974. (a)

Newtson, D., Engquist, G., & Bois, J. The effects of arousal and cognitive interference on unitization and behavior recall. Unpublished manuscript, University of Virginia, 1974. (b)

Newtson, D., Engquist, G., & Bois, J. Unit reliability for eight sequences. Unpublished manuscript, University of Virginia, 1975.

Noton, D., & Stark, L. Scanpaths in saccadic eye movements while viewing the recognizing patterns. *Vision Research,* 1971, **11,** 929–942.

Shepard, R. N. Recognition memory for words, sentences, and pictures. *Journal of Verbal Learning and Verbal Behavior,* 1967, **6,** 156–163.

Wilder, D. Units of perception and predictability of behavior. Unpublished manuscript, University of Wisconsin, 1974.

Zadny, J., & Gerard, H. B. Attributed intention and informational selectivity. *Journal of Experimental Social Psychology,* 1974, **10,** 34–52.

11

Escape to Freedom: The Relationship between Attribution of Causality and Psychological Reactance

Stephen Worchel
University of Virginia

Virginia Andreoli
Madison College

The concept of "freedom" plays an important role in much of the research and theorizing in social psychology. Researchers have demonstrated that the freedom of choice (or at least the illusion of freedom) is a necessary condition for the arousal of cognitive dissonance (for example, Brehm & Cohen, 1962). Jones and his colleagues (Jones & Davis, 1965; Jones, Worchel, Goethals, & Grumet, 1971) have found that an observer is reluctant to make a dispositional attribution unless he perceives that the actor is freely choosing his course of behavior.

Brehm (1966) has made freedom the central focus of his theory of psychological reactance. According to Brehm, much of an individual's behavior is guided by the motivation to maintain or restore freedom. Reactance theory is based on the simple premise that each individual has a set of behaviors that he considers to be his "free behaviors." A free behavior is an act which the individual feels that he has the necessary physical and psychological tools to perform, and he knows by experience, custom, or formal agreement that he can engage in the particular act either at the moment or at some time in the future. Free behaviors vary in importance, the more important free behaviors being those that satisfy important needs (physical or psychological). It is imperative that the individual maintain his freedom to engage in these behaviors because it is the access to these behaviors that ensures his ability to survive and thrive.

According to Brehm, a threat to or elimination of one or more free behaviors will motivate the individual to reestablish his freedom. The motivational arousal is termed psychological reactance. The strength of the reactance will vary as a positive function of the importance of the freedoms threatened or eliminated and the magnitude of the threat.

Research has demonstrated three types of reactions to threats to freedom. First, the individual may attempt to reestablish his freedom by performing the threatened free behavior. For example, the experimenter in the Hammock and Brehm (1966) study threatened children's freedom to choose a particular candy bar by insisting that the children have a different candy bar. The reaction of many of the children was to choose the "denied" candy. This act restored their freedom to have that candy. A second reaction to threats is that the threatened free behavior increases in attractiveness. This is supposedly a result of the increased motivation to perform that behavior. Finally, aggression aimed at the threatening agent has been found to follow the arousal of reactance. Worchel (1974) argued that aggressing against an individual who threatens freedom can restore behavioral freedom at the moment and ensure that the individual will not threaten freedom in the future.

While research has demonstrated that these three behaviors often follow the arousal of reactance, responding in such a fashion may result in an uncomfortable dilemma for the individual. For example, take the case in which a neighbor unexpectedly invites an individual and his family to have dinner and drinks. In order to be polite, the individual accepts the invitation and partakes of his neighbor's wine and food. On leaving, the neighbor remarks. "We must do this again sometime." The rules of etiquette applying to this situation require that the individual reciprocate by inviting his neighbor to dine at his home. Seen in this light, the neighbor has threatened the individual's freedom by placing him in a situation that dictates that he must reciprocate the dinner invitation. If the freedom not to invite the neighbor is of importance, reactance should be aroused. Here, now, is the dilemma for the individual. According to reactance theory, he can restore his behavioral freedom either by not inviting his neighbor for dinner or by aggressing against his neighbor for threatening his freedom. If he behaves in either manner, he must suffer as his neighbor will think him rude and uncouth. If the individual does not want to run the risk of alienating his neighbor, he can comply with the rules of etiquette and reciprocate the dinner invitation. This course of action, however, is also likely to have uncomfortable consequences since it forces the individual to acknowledge that he has relinquished his freedom to choose whom he will invite to dinner. Thus, the individual is caught; he is damned if he does and damned if he doesn't. He can either alienate his neighbor and restore his behavioral freedom, or he can keep his neighbor as a friend but relinquish an important behavioral freedom.

It was specifically this kind of dilemma that motivated the program of research to be outlined in this chapter. The first question we asked was: In what situations is an individual likely to find himself in the dilemma of either having to perform an undesirable act to restore freedom or having to relinquish his freedom by not performing the act? One such situation seemed to be the case where an individual's behavior was guided by the norm of reciprocity. The norm of reciprocity is a strong one in our society, as evidenced by the numerous adages that instruct us to reciprocate in kind those behaviors to which we are recipients: "an eye for an eye" and "do unto others as we would have them do unto us."

The norm of reciprocity seems to apply chiefly to two classes of behavior: favors and harm doing. Gouldner (1960) argued that there is a universal norm prescribing that the recipient of a favor is obligated to return the favor. There have been numerous examples demonstrating that people do indeed abide by this rule for living: the recipient of a favor does repay the favor doer (Berkowitz & Daniels, 1964; Goranson & Berkowitz, 1966; Pruitt, 1968). Harm doing or aggression is the other class of behaviors guided by the norm of reciprocity. Buss (1961) argued that the most potent instigator of aggression is attack. Helm, Bonoma, and Tedeschi (1971) found that an individual's aggressive response was positively correlated with the strength of the aggression he had initially received from the victim of his present attack. Thus, with respect to favors and harmful acts, the research indicates that an individual's behavior is often predetermined by the actions of another.

It is not difficult to see how reactance could be aroused by the norm of reciprocity. In the case of kindly or aggressive behavior, the norm of reciprocity demands that the individual respond with similar behaviors. The norm reduces the individual's freedom to respond in another manner. Utilizing this line of reasoning, Brehm and Cole (1966) predicted that reactance would be aroused in the recipient of a favor if his freedom not to reciprocate with a favor was important. In their study a confederate either brought the subjects a coke (favor) or did not (no favor). The importance of the subjects' freedom not to reciprocate in the experiment was also varied (high versus low). The results of the study indicated that following the confederate's favor, subjects were less likely to return the favor when their freedom not to reciprocate was important than when it was not. Also, when the freedom not to reciprocate was important, subjects were less likely to perform a favor for the confederate if he had previously done the subjects a favor than if he had not. Brehm and Cole argued that the confederate's favor pressured the subjects to reciprocate and aroused reactance when the freedom not to reciprocate was important. Subjects restored this freedom by refusing to reciprocate the favor.

It is precisely the situation faced by subjects in the Brehm and Cole study that should create a great deal of discomfort in an individual. The subject in their study could restore freedom by refusing to reciprocate the favor. However, by doing this he must act counternormatively and run the risk of being seen as socially insensitive. Or, he could return the favor and relinquish his control over his own behavior in the situation. A similar predicament faces the individual who is the recipient of a harmful act. He can fail to reciprocate and maintain his freedom not to counterattack, or he can counterattack. This latter course of action forces the individual to perceive that his behavior is being determined by the other attacker. Either action (reciprocating or restoring freedom by failing to reciprocate) will have negative consequences for the individual.

This dilemma is an interesting and rather common one. The individual is in a no-win situation and will feel badly regardless of the response he makes. This seems psychologically maladaptive, and it motivated us to question: Is there another manner by which an individual facing such a dilemma could retain his freedom but still respond in a normative fashion? That is, how could he "have his cake and eat it, too"?

In casting about for a solution to this question, we focused on the attribution process. It is a common assumption of most of the researchers in perception and in social psychology that man reacts to the world as he perceives it–not necessarily as it really is. The behavior of one individual toward another is often guided by the attributions he makes about the other person's intentions. This is clearly demonstrated in the judicial process, as the sentence a murderer receives is based on the attributions the jury makes about his intentions and premeditations; involuntary manslaughter involves no premeditation or intention to do harm, while first degree murder comprises both intent and premeditation.

Thibaut and Kelley (personal communication, May 1972) and Schopler (1970) have suggested that the attribution about the locus of causality for an actor's behavior is also important in determining the responses to that behavior. Thibaut and Kelley hypothesized that in order for a dyadic relationship to endure, each actor must perceive the other as committed to the dyad. The perception of commitment to the dyad is enhanced to the extent that the actor's behavior is seen as being internally caused rather than being elicited by environmental demands. According to Thibaut and Kelley, the pressure on an individual to reciprocate the acts of the other party in the dyad increases as a function of the perception of the other's commitment to the dyad. In other words, the pressure to reciprocate increases when the observer attributes the causality for the actor's behavior to a locus internal to the actor rather than environmental constraints. Thus, the norm of reciprocity should be strongest when the locus of causality for an act (favor or harmful act) is attributed to the actor rather than to the environment. In line with this reasoning, there have been numerous demonstrations that favors are not reciprocated when the recipient attributes the locus of causality of the favor to the environment. Goranson and Berkowitz (1966) found that favors were rarely reciprocated when the recipient felt the favor he received was performed involuntarily. Greenberg and Frisch (1972) found reduced reciprocation when the recipient felt the favor he received was performed unintentionally.

Since attributions about locus of causality for behavior determine the strength of the norm of reciprocity, it seems reasonable to assume that an individual can use such attributions to restore freedom and reduce reactance. If the recipient makes the attribution that the norm-evoking behavior of the actor is being environmentally determined, the strength of the norm of reciprocity should be reduced, and, consequently, the threat to the recipient's freedom should be removed. Once this occurs, the recipient should not experience reactance and should feel free to determine his own behavioral response. He can, if he likes, decide to reciprocate the other's behavior. If he does this, he can do it without feeling that he is being forced to reciprocate since he has negated the norm of reciprocity through attribution. Thus, the individual can escape to freedom through attribution.

The aim of the first study in our research program was to investgate whether or not individuals do, in fact, utilize attribution as a means of restoring behavioral freedom that has been threatened by the norm of reciprocity. Our first task was to demonstrate that the behavior of another that evokes the norm of reciprocity could arouse reactance. We wished to demonstrate that both favors and harmful acts that invoke the reciprocity norm would arouse reactance. In our study a confederate either helped the subject (favor) or attacked him (harm doing). The subject believed she would either be competing, cooperating, or not interacting with the confederate in the experiment. We felt that the freedom not to return a favor should be of high importance when an individual expects to be competing with the favor doer. Feeling obligated to do one's opponent a favor is detrimental to performing well in the competition. If the recipient of a favor expects to be interacting cooperatevely with the favor doer, the freedom not to perform a favor should be of less importance since one would like to be able to help out his partner in a cooperative interaction. If the individual does not expect to have future interaction with the favor doer, the importance of choosing how he will behave towards him should be of no importance. Thus, we expected that reactance would be aroused to the greatest degree when the confederate favors the subject, and the subject believes he will be competing with the confederate in the future. The attack by the confederate should evoke the norm of reciprocity and threaten the subject's freedom not to aggress against the confederate. The freedom not to aggress should be of greatest importance when the subject expects a future cooperative interaction with the confederate. Aggressing against one's partner in a cooperative interaction could hinder the effectiveness of the partnership. The freedom not to aggress should be of little importance if the subject expects to be competing with the confederate in future interactions and of no importance if the subject feels she would not be interacting with her in the future. Thus, greatest reactance should be aroused following an aggressive attack by the confederate when the subject expects to be acting cooperatively with the confederate in the future.

Numerous studies have shown that derogation of the threatener accompanies the arousal of reactance, and we thus utilized the subject's rating of the confederate as indication of the arousal of reactance. Since we believed that norm-evoking behaviors arouse reactance only when the locus of causality for these behaviors is attributed to the performer, in one-half of the conditions, we manipulated the situation so that the subject would have to believe the confederate's behaviors were internally caused (forced attribution). Thus, our prediction was that in the conditions where the confederate's behaviors are seen as internally caused, a favor will elicit the greatest derogation when the subject expects to be interacting competitively with the confederate. A harmful act by the confederate should elicit the greatest hostility when the subjects expects to be having a cooperative interaction with the confederate. If these predictions were supported, we felt that it would demonstrate that behaviors evoking the norm of reciprocity could, indeed, threaten the recipient's freedom and arouse reactance.

The second and major purpose of the first study was to investigate the question of whether individuals do utilize the attribution process to restore behavioral freedom. In order to demonstrate this, two results must occur. First, greatest attribution to environmental loci must occur when the norm-evoking behavior is most likely to arouse reactance, that is, when it threatens an important behavioral freedom. Second, the consequence of this attribution must be a reduction in the amount of reactance experienced by the individual. In order to test these hypotheses, we had a set of conditions similar to those described above, except that there was no attempt to indicate to the subject the locus of causality for the confederate's behavior. Instead, the subject was asked to determine why the confederate had acted as she did by attributing her behavior to internal or environmental causes. After the attribution the subject was given the opportunity to derogate the confederate. If individuals do use attribution to the environment to restore behavioral freedom, the greatest tendency to attribute the confederate's behavior to environmental causality should occur in the favor expect competitive interaction and harm/expect cooperative interaction conditions. Further, following the attribution, there should be a reduction in the tendency to derogate the confederate in these conditions as compared to the respective conditions in the forced attribution situation where the behavior was determined to be internally caused. If these predictions were confirmed, then we would demonstrate that individuals do use attribution as a means of restoring behavioral freedom threatened by norm-evoking acts and that this attribution process is successful in reducing reactance.

EXPERIMENT 1

Method

Subjects in the study (Worchel & Andreoli, 1974) were 148 female introductory psychology students. When the subject arrived at the experimental room, she was met by the experimenter who informed her that the other subject who had signed up for the study had not yet arrived. The experimenter suggested that he tell the subject something about the study while they were waiting for the other subject to arrive and then stated that the experiment would involve working on a series of tasks. In the expect cooperation condition, the experimenter told the subject that she would be working cooperatively with the soon-to-arrive subject and that their partnership could win money if they performed well. In the expect competition condition, the subject was led to believe that she would be competing with the other subject throughout the study. And in the expect no-interaction condition, the experimenter stated that the two subjects would have little interaction during the study, but it was just easier to run two subjects at a time. The interaction manipulation was executed at this point so that the informed subject would know that the soon-to-arrive subject would be unaware as to the type of interaction that would exist in the study. The experimenter then had the subject begin completing a personality questionnaire.

At this point the other subject, actually an experimental confederate, entered the room and was given a personality questionnaire to complete. After the questionnaires were turned in, the experimenter explained the experiment. He said that it would involve working on a series of tasks that involved luck and skill. (The experimenter avoided saying anything about the interaction that would exist between the subject and confederate.) He said that after each task, the subjects would be asked a number of questions about the task and about the role that fate or skill played in determining the outcome of the task. At this point he said that he had to leave the room to get some materials for the first task. Before leaving, he handed the subject a large stack of IBM answer sheets and asked her to alphabetize them while he was gone. He said they would be used in the first task.

After the experimenter left, the confederate's behavior varied. In the friendly or favor condition, the confederate offered to help the subject with the task, and she took half the answer sheets. In the hostile condition, the confederate glanced at the subject and stated in a degrading tone, "Boy, that's a really stupid thing to do. They can get some people to do anything in these experiments."

Soon after this, the experimenter reentered the room. He stated that he had been watching the subjects behind a one-way mirror, and they had just completed the first task. The locus of causality variable was then manipulated. In the forced-attribution condition, the experimenter said that he had examined the confederate's personality test and that it revealed that she had the "type of personality" that would motivate her to help (or not to help, in the hostile condition). This statement was designed to indicate that the confederate's behavior had been internally motivated and not under environmental control. In the free-attribution condition, the experimenter said nothing about the personality test or what might have motivated the confederate's behavior.

Next, the subjects were asked to complete a questionnaire. The first question asked them to place the locus of causality for the confederate's behavior. Additional questions asked them to rate the confederate on a series of adjective pairs. The former question allowed us to examine the attributions subjects made about the causality of the confederate's behavior. The second set of questions were the indications of reactance with greater disparagement of the confederate suggesting greater reactance. Thus, the study involved a 2 (forced or free attribution) × 2 (friendly or hostile confederate) × 3 (cooperative, competitive, or no future interaction) design.

Results

Attribution of causality. In order to examine the attributions of causality made by subjects, they were asked: "Why did the other subject in the room help or not help you?" with the endpoints labeled (1) The other subject helped or did not help me because of the nature of the situation (i.e., anyone would have acted the same way in this situation) and (31) The other subject helped or did not help me because of some trait or traits which she possesses (i.e., her behavior was due solely to her). We expected that subjects in the forced-attribution condition would attribute the causality for the confederate's action significantly more to internal causes than would subjects in the free-attribution conditions. The results presented in Table 1 indicated that this was indeed the case, $p < .001$. This finding demonstrates that our forced-attribution manipulation was successful in making subjects attribute internal causality for the confederate's behavior.

The second prediction involved the attribution that subjects in the free-attribution conditions would make. If subjects were utilizing attribution to the environment to restore freedom, the greatest tendency to make environmental attributions should occur in the high-reactance conditions. These were the favor/expect competition and harm expect cooperation conditions, since important freedoms were threatened by the norm of reciprocity. The results in Table 1 confirm these predictions. The double interaction within the free-attribution conditions was significant, $p < .001$, and the attributions in the two high-reactance conditions were significantly different from the attributions in each of the other conditions.

Derogation of confederate. Derogation of the confederate was used to measure the amount of reactance the individual was experiencing. Worchel (1974) and others have shown that hostility accompanies the arousal of reactance. It was predicted in the present study that the friendly actions of the confederate should arouse greatest reactance and, consequently, the greatest derogation in the forced-attribution expect competition conditions. It is here where the subject's freedom not to respond with a favor is of highest importance, and she cannot escape the pressure to do so through attribution. This prediction was supported, as derogation was significantly greater in this cell than in any of the other friendly behavior conditions (see Table 2). The confederate's hostile actions should arouse greatest reactance and derogation in the forced-attribu-

TABLE 1
Attribution Ratings

Locus of causality/ Confederate's behavior	Type of future interaction		
	Cooperative	No future interaction	Competitive
Free attribution			
Friendly	11.75	16.42	6.58
Hostile	7.17	18.25	14.08
Forced attribution			
Friendly	25.25	25.42	25.75
Hostile	25.25	23.25	24.42

Note. $N = 12$ in each condition. Scores are responses on a 31-point scale to the question, "Why did the other subject in the room help or not help you?" with the endpoints labeled (1) The other subject helped or did not help me because of the nature of the situation (i.e., anyone would have acted the same way in the situation) and (31) The other subject helped or did not help me because of some trait or traits which she possesses (i.e., her behavior was due solely to her).

TABLE 2
Derogation of the Confederate

Locus of causality/ Confederate's behavior	Type of future interaction		
	Cooperative	No future interaction	Competitive
Free attribution			
Friendly	43.00	41.33	50.92
Hostile	84.42	109.33	97.75
Forced attribution			
Friendly	63.17	37.67	76.00
Hostile	124.25	102.33	108.67

Note. $N = 12$ in each condition. Scores are the sum of responses on 31-point scales to the statement, "I would enjoy working with the other subject," with the endpoints labeled (1) strongly agree and (31) strongly disagree, and to the confederate rating question, with the endpoints labeled (1) friendly and (31) unfriendly, (1) likeable and (31) not likeable, (1) not hostile and (31) hostile, (1) trustworthy and (31) untrustworthy, (1) honest and (31) dishonest, (1) intelligent and (31) unintelligent. Thus, the most accepting score would be 7, and the most rejecting score would be 217.

tion/expect cooperation condition, since the freedom not to respond with aggression in the future interaction is of greatest importance in this condition. As can be seen from Table 2, this prediction was supported. These results suggest that behaviors evoking the norm of reciprocity can arouse reactance when the norm threatens important behavioral freedoms of the recipient.

The final prediction was that the attribution that the norm-eliciting behavior was caused by environmental forces would reduce reactance. If this were the case, derogation of the confederate should decrease when subjects attributed environmental causes to the confederate's behavior. It was predicted that the greatest decrease in hostility should occur between the forced- and free-attribution conditions of friendly/expect competition and hostile/expect cooperation. These were the conditions where greatest reactance occurred in the forced-attribution conditions, and the strongest tendency to attribute environmental causes to the confederate's behavior occurred in the free-attribution conditions. The results presented in Table 2 show that this prediction was supported. Further, the decrease in hostility in these conditions was of such magnitude as to bring the ratings of the confederate in the free-attribution conditions to the level found in the other conditions where little reactance was aroused.

Discussion

The results of the present study suggest that reactance may be aroused in recipients of behaviors that evoke the norm of reciprocity. Reactance will be aroused when the freedom of the recipient not to reciprocate the behavior is of high importance. These results extend those presented by Brehm and Cole (1966) as they indicate that behaviors evoking the norm of reciprocity other than favors will arouse reactance.

The results also show that individuals can and do use attribution to restore behavioral freedom. In the present study, recipients attributed the causality of norm-evoking behaviors to environmental loci in conditions where important free behaviors were threatened by the norm of reciprocity. Attribution to the environmental loci reduced reactance as indicated by the significant decrease in derogation of the confederate.

We have cast the study in terms of reactance theory and suggested that attributions were used to restore threatened behavioral freedom. A question can be raised as to whether we can disregard the reactance theory interpretation and view the results in a cognitive consistency framework. That is, by performing a favor, the confederate may have created a situation of inconsistency and made it difficult for the subject to compete against her. It may be dissonant (Festinger, 1957) for the subject to act competitively with someone who has just aided her. One possible way to reduce this dissonance may be for the subject to convince herself that the confederate was simply forced to do the favor (situational attribution) and that she is not such a nice individual after all. Similarly, it may be dissonance arousing to anticipate cooperating with a hostile individual, and situational attribution may allow the subject to see the confederate in a better light. Thus, it may be possible that subjects in the present situation made situational attributions in order to reduce dissonance rather than restore behavioral freedom.

It is an interesting possibility that situational attributions may be utilized to reduce dissonance, but it does not adequately explain the present results. Specifically, if subjects were utilizing the situational attribution to convince themselves that the favor doer with whom they would be competing was not such a nice individual, we would expect their ratings of the confederate to become more derogatory as compared to when a personal attribution was made. The results, instead, showed that the ratings of the friendly confederate in the competitive condition were more favorable under situational attributions than under actor-instigated attributions. This is the result that would be expected if the situational attribution were aimed at restoring behavioral freedom.

Despite the convincing support which these results lend to our hypotheses, several questions still remain unanswered by the first study. The first question which aroused our interest dealt with the clarification of the precise attributional conditions which were necessary for the arousal of reactance by a norm-evoking behavior. As with most research on attribution, our first study provided the subjects with only two attributional alternatives: actor's disposition and environment. However, research by McArthur (1972) suggests that there are, in fact, several additional loci to which individuals can and do attribute causality for actions. McArthur (1972) presented subjects with 16 different responses made by other people. Each description was accompanied by information pertaining to the consistency, distinctiveness, and consensus dimensions (Kelley, 1967) of the behavior. She then aksed subjects to indicate what they perceived to be the cause of the behavior. Subjects could attribute causality to either the actor, the stimulus, the circumstance, or some combination of the three. If the subject chose the latter category, he was asked to state the specific combination of factors he felt caused the behavior. While the information variable did significantly influence the type of attribution that was make, the finding of interest here is that, overall, more than 50% of the attributions were to categories other than the personal or situational ones. Of the additional categories which were utilized, the most frequent was the person-stimulus conbination, Thus, the subjects did consider these additional loci of attribution to be valid ones. McArthur's findings suggest that other attributional atlternatives must be made available to subjects if we wish to provide a more accurate and precise description of the recipient's perception of the norm-evoking act.

One aim of the second study in our program of research was to provide subjects with four loci of attribution instead of the traditional two and to examine how use of each loci affected the arousal of reactance. In addition to situational and dispositional attributions, the two additional loci which we employed are referred to as the unique-recipient and unique-interaction attributions. Let us examine more closely each of these attributions and the predictions regarding their effect on the arousal of reactance. We will use the context of a favor situation, since it has been repeatedly shown that favors evoke the norm of reciprocity.

As we suggested earlier in the chapter, if someone does a favor under conditions where he would perform a favor irregardless of the recipient, the behavior would be attributed to a situational cause. For example, if the host gives a drink to a guest at a cocktail party, the recipient would no doubt view the favor as an appropriate act given the circumstances and attribute the cause of the behavior to the situation. As we have already indicated, there should be very little pressure to reciprocate in this instance.

A second possible locus of attribution is the actor himself. In some instances we have information about the unique disposition of the favor doer which suggests that he acts differently than others would in the same situation while behaving in this manner across several situations. For example, we may observe that this individual goes out of his way to perform favors for others regardless of the situation. Typically, this information results in a dispositional attribution or what we will refer to here as a unique-actor attribution. This individual is labeled as someone who likes to perform favors for others; that is, the behavior is perceived as being donor instigated and not dependent in any way on the recipient or the circumstances. The recipient, in this case, recognizes that the actual performance of the favor is, in itself, rewarding for the actor. Whether the reward emanates from an immediate increase in pleasure or a gradual increase in self-esteem, the favor doer has benefited by his actions. Because of this, the favor doer does not need to be rewarded by reciprocation, and the pressure on the recipient is therefore reduced.

While the unique-actor attribution focuses on the characteristics of the actor, the third locus of attribution relies on the characteristics of the recipient (unique-recipient attribution). To illustrate this type of attribution, consider the situation in which an individual performs a favor for a very attractive person but does not do so for someone who is not as well-endowed with good looks. Furthermore, imagine that several people behave in the same way toward this attractive recipient. While the actors treat the recipient as a unique individual by performing favors only for him, everyone responds in a constant manner to this one individual. The recipient may conclude, therefore, that the cause of the favors is some unique characteristic of himself, namely, his good looks. This is not to say that this attribution of uniqueness is confined to positive traits of the recipient. It is possible that people perform favors for another out of pity for the person. Because he receives favors from everyone, the individual may come to believe that he deserves these favors. This feeling should reduce the pressure to reciprocate, since he is simply receiving his "just desserts."

Each of the causal attributions discussed so far has involved a constant factor as the cause of the favor—either the situation, the favor doer, or the recipient. In all of these cases, the norm of reciprocity should be weak as the performance of the behavior either was not caused by the actor, or he already received his reward from the performance of the favor itself. There are times, however, when none of these attributions are sufficient to account for the observed behavior. Instead, the recipient perceives the favor to be a result of the unique interaction between himself and the actor. We have termed this a unique-interaction attribution. This attribution is likely to occur between close friends or lovers who know that neither the other person nor himself generally gives or receives favors with regularity. However, when the two are together, favors are exchanged because of the importance of their interaction. As Thibaut and Kelley have suggested, in order to maintain this relationship, evidence of commitment is necessary on the part of both members. Consequently, a great deal of pressure is exerted on the recipient to reciprocate the favor as a way of demonstrating his commitment. Following this logic, we predict that the norm of reciprocity should be strong under such attributional circumstances and if important freedoms are threatened by the norm, reactance should be aroused in the recipient. Based on this reasoning of the four loci of attribution we have discussed, a unique-interaction attribution should produce the greatest pressure to reciprocate and, consequently, provide the greatest potential for the arousal of reactance.

One aim in the second study was to investigate the arousal of reactance under these four different attributions and, thus, to obtain a broader picture of the reactance–attribution relationship. A second concern of this study was with the relationship between the attribution made and the recipient's behavior. While the first study demonstrated that one consequence of reactance aroused by a dispositionally attributed favor was derogation of the favor doer, it did not provide us with information regarding the effects of attribution on the free behavior that was being threatened, that is, reciprocation. In the second study we were interested in investigating the relationship between loci of attribution and actual reciprocation of a favor. We reasoned that when no reactance was aroused by the favorer's action, the greater the pressure on the recipient to return the favor the greater the likelihood of reciprocation. Reactance should not be aroused when the freedom not to reciprocate is of low importance. Based on the previous discussion, an attribution that the favorer's behavior was caused by the unique interaction with the recipient should create the greatest pressure to reciprocate. Thus, it was predicted that when the reciprocation behavior was of low importance, the greatest incidence of reciprocation would occur when the unique-interaction attribution was made. A situational attribution should create the least pressure to reciprocate, and, thus, when behavioral freedom is of low importance, least reciprocation should occur when a situational attribution is made.

Looking at the instance when the reciprocation behavior is of high importance, we expected that the greater the pressure on the recipient to reciprocate, the greater would be the arousal of reactance and the less likelihood for a return of the favor. Thus, when behavioral freedom is important, the unique interaction attribution of causality should lead to the least reciprocation, and the situational attribution should result in the greatest reciprocation. We were not entirely sure about how much pressure would result from the unique-actor or unique-recipient attributions, and, therefore, we could not make precise predictions regarding the frequency of reciprocation in these conditions. It was presumed, however, that they would fall between the unique-interaction and situational-attribution conditions. The second experiment was designed to test these hypotheses. In addition to measuring the frequency of reciprocation, we also took measures of the derogation of the favorer. We expected that this measure would reflect the amount of reactance generated in the various conditions and that it would vary negatively with the reciprocation measure: more reciprocation leading to less derogation.

EXPERIMENT II

Method

One hundred and twenty female introductory psychology students served as subjects in the experiment (Worchel, Andreoli, & Archer, in press). Upon arriving at the experimental room, the subject was asked to complete a questionnaire entitled "Personality Orientation Scale." When the subject had completed the form, the experimenter asked her to read what was allegedly a description of the study while she was waiting for the second subject. In fact, this sheet provided a cover story for the experiment, as well as the manipulation of the importance of the subject's freedom to make unbiased estimations of her partner's personality. The subject was told that the study was aimed at perfecting the "Whitman Scale" personality test. She was informed that the Personality Orientation Scale was the first part of the scale. In the second part, she would evaluate her partner after listening to her answers on three questions. It was emphasized that this evaluation should be made *only* on the basis of her partner's answers to these questions. In the low-importance condition, the subject was told that little was known about this personality test and that the present study was simply a pretest. In the high-importance condition, the subject was informed that the personality test could be used to assess accurately her own sensitivity and intelligence by showing the accuracy of her evaluations.

Thus, it was very important to the subject that she accurately evaluate her partner. To enhance further the importance of the ratings, the description stated that if the subject's rating of her partner came within one standard deviation of a group of expert's ratings of the recorded answers, she would be paid $10.00.

While the first subject was reading the description, the second subject (actually an experimental confederate) entered, apologized for being late, and completed the Personality Orientation Scale. Unlike the real subject, however, she was not given an instruction sheet. Consequently, the subject believed that the confederate was unaware of their future interaction. The two subjects were then told that while the experimenter was analyzing their scales, they would perform a task for another professor. The task involved choosing psychologically disturbed persons from sets of pictures. The subjects were told that 50¢ would be paid for each correct answer but that only one subject was needed. The confederate was then "randomly" chosen, and she worked on this task. When the task was scored, the confederate won $1.00, and she gave the subject 50¢ saying, "Here, you take half. After all, it was just luck that I got to do the test." This act of sharing constituted the "favor," as it was unsolicited and benefited the subject.

Following the favor, the experimenter manipulated the attribution of causality by providing the subject with a locus of causality for the partner's behavior. This was accomplished by stating that the real purpose of the first task had been to determine how accurately their behavior could be predicted by the first part of the Whitman Scale. In particular, the experimenter continued, he was interested in their sharing behavior. In the situational condition, the experimenter stated that almost everyone so far had offered to share their money with the other person, and, consequently, they did not even need the scale to help make predictions. Prior to the next three attribution conditions, the experimenter commented that few people had offered to share in this situation, and hence there did not appear to be anything about the situation itself that motivated people to share their money. In the unique-actor condition, the experimenter said that the confederate's test indicated that she was the type of person that would be motivated to share with anyone in any situation. The experimenter noted that she had behaved just as her personality test had predicted. In the unique-recipient condition, the experimenter stated that, according to the subject's posttest score, she was the type of person with whom most people would want to share in any situation. Subjects in the unique-interaction condition were informed that neither the subject's nor the confederate's personality test would have predicted the sharing behavior. However, the experimenter said that there was something unique about the interaction of the two personalities, and this unique relationship motivated the sharing behavior.

In addition to these forced-attribution conditions, a control condition was employed to determine the baseline attribution that the subjects would make. Therefore, in this condition, the predictions from the personality test were not mentioned nor was there any attempt to explain the behavior of the test-taker.

After the manipulation of the attribution variable, the subjects completed a questionnaire containing the attribution-manipulation checks. Following this, the experimenter began to explain the second part of the Whitman Scale. At this time, he "noticed" that the confederate had not received an instruction sheet and promptly gave her one. In reality, the confederate received a sheet of irrelevant information so that she would be unaware of the importance condition.

When the confederate had read the information, the experimenter began to administer the second part of the scale. Recall that this evaluation was supposed to be made strictly on the basis of the partner's answers to three questions. However, the confederate's favor could potentially influence this set of ratings and thus threaten the subject's freedom to make an unbiased assessment as indicated by the instructions. Furthermore, in the high-importance condition, this freedom was very important since the evaluation reflected the subject's own ability and intelligence. The subject and the confederate then alternated answering three questions asking them to describe their ideal marriage partner, the place to which they would like to travel, and the occupational field they would like to choose (cf. Brehm & Cole, 1966). A "First Impression Rating Form" was then handed to the subjects, and they were told to rate the other person carefully. This form contained 12 semantic differential scales upon which the confederate was evaluated. Although these ratings provided some indication of the extent to which reactance had been aroused, we also examined the behavioral effects of the attribution–reactance relationship. In order to obtain this measure, a final task was assigned to the subjects which provided an opportunity for the subject to reciprocate the confederate's favor. This task involved rating four more individuals on the basis of their written responses to the three questions on the Whitman Scale. While preparing for the task, the experimenter "realized" that he required additional forms. Before leaving the room to obtain the forms, he handed each subject a bundle of IBM forms. They were told that the bundle handed to the confederate was the data from a university population, while the subject was provided with data from a high-school population. They were asked to sort each population separately into the age groups represented. The confederate received four times as many sheets as did the real subject.

The experimenter left the room, and the subjects sorted the sheets. The confederate deliberately worked slowly so that the subject completed her part of the task before the confederate. In this way, the subject had the opportunity to help the confederate and thus reciprocate the favor. The confederate noted whether or not the subject offered to help, and this constituted the measure of reciprocation.

Results

Manipulation check. To ensure that the importance manipulation was successful, we asked the subjects to indicate how important they personally thought the ratings were. As expected, subjects in the high-importance conditions considered the ratings more important than the subjects in the low-importance conditions.

It was also necessary that the subjects accept our explanation of the cause of the favor. Therefore, we asked subjects to place a percentage score beside each of the following four factors to represent the extent to which it contributed to the cause of the confederate's behavior: situational factor, the other subject's personality, the subject's own personality, and the unique relationship between the two subjects. A brief definition of each factor was included. The results indicated that our manipulation of the attribution was successful. In each condition, the attributional factor which we contended was the cause of the favor received the highest percentage of the four possible factors within a given subject's estimation. For example, in the situational condition, the subjects assigned a greater percentage to the situational factor than to the remaining three factors. Furthermore, each locus of attribution received its highest percentage in the condition in which it was purportedly the cause of the favor. Thus, our manipulations were successful in creating the desired conditions.

Reciprocation. The main dependent variable was the behavioral measure of reciprocation–whether or not the confederate was helped on the paper-sorting task. The frequency of help and no-help responses is presented in Table 3. The data supported our predictions. When the behavioral freedom was of low importance to the recipient, greatest reciprocation occurred in the unique-interaction condition. However, when the freedom was of high importance, the least amount of reciprocation was found when subjects made the unique-interaction attribution. There was no difference in the reciprocation measures in any of the other conditions.

Derogation. The subject's rating of the confederate made on the second part of the Whitman Scale was also examined. We felt that this evaluation of the favor doer might provide some additional evidence of the arousal of reactance. As Table 4 shows, the prediction that the derogation scores would vary nega-

TABLE 3
Frequencies of Help and No Help Responses

	Attribution									
	Unique interaction		Unique recipient		Unique actor		Situational		Control	
Importance	Help	No help	Help	No help	Help	No help	Help	No help	Help	No help
High	2	10	6	6	5	7	6	6	7	5
Low	9	3	7	5	8	4	7	5	7	5

TABLE 4
Derogation of the Confederate

	Attribution				
Importance	Unique interaction	Unique recipient	Unique actor	Situational	Control
High	129.36[a]	86.67	97.50	80.00	95.25
Low	50.00	82.17	56.42	93.58	81.25

[a]The lower the score, the more favorable the evaluation. Scores summed for evaluation of the confederate on the following measures: friendly–unfriendly, mature–immature, intelligent–unintelligent, considerate–inconsiderate, deep–shallow, straightforward–devious, interesting–uninteresting, kind–unkind, genuine–affected, unannoying–annoying, and socially competent–socially incompetent. In addition, they were asked directly, "How much do you like the other person?"

tively with the reciprocation measure was supported. This was indicated by the significant importance × attribution interaction. In the high-importance/unique-interaction condition, the favor doer received a significantly less favorable rating than in the remaining four high-importance conditions. Contrary to this, of the five low-importance conditions, the low-importance/unique-interaction condition yielded the most favorable evaluation of the confederate.

It is interesting to note that the unique-actor condition provided the second most unfavorable evaluation when an important freedom was involved, while it yielded the second most favorable rating when an umimportant freedom was involved. These results are congruent with the data from the first study in which reactance was aroused by the dispositional attribution.

Discussion

The results of the second study support previous research findings that norm-evoking acts can arouse reactance in the recipients when the pressure to reciprocate the behavior threatens an important behavioral freedom of the recipient (Brehm & Cole, 1966; Worchel & Andreoli, 1974). In the high-importance/unique-interaction condition, the recipients of the favor not only strongly derogated the favor doer but also reciprocated less frequently than subjects in the remaining conditions. Both of these results were interpreted as evidence of the arousal of reactance in the situation. While demonstrating that reactance is aroused, this result also indicated that the attribution made concerning the favorer's behavior and the importance of the freedom which is threatened interact to determine the extent to which reactance is aroused. With respect to the actual reciprocation of the favor, we found that neither situational, unique-actor, nor unique-recipient attributions decreased the frequency of reciprocation below a chance level, regardless of the importance of the freedom which may have been threatened. However, whether the unique-interaction attribution led to reciprocation of reactance depended on the importance of the freedom which was threatened by the increased pressure to reciprocate. If the threatened freedom was of little importance to the recipient, then the unique-interaction attribution increased the frequency of reciprocation. However, when an important freedom was threatened, the pressure to reciprocate resulted in the arousal of reactance and a consequent decrease in reciprocation. The evaluation of the favor doer followed a similar pattern.

IMPLICATIONS OF RESEARCH AND QUESTIONS FOR FUTURE RESEARCH

In the research discussed in this chapter, we have also addressed some broad questions in the area of attribution. By providing a wider range of loci of causality, we were able to procure more precise assessments of the recipient's perceptions of the cause of the favor. Furthermore, it appears that subjects did consider the additional alternatives as valid conceptualizations of causality. This is evidenced by the fact that subjects in the control groups allocated approximately one-quarter of the responsibility for behavior to the unique-recipient and unique-interaction attribution alternatives.

Taken together, the evidence from both studies reported in this chapter suggests that the attribution of the locus of causality for the norm-evoking behavior can influence the extent to which reactance is aroused by the pressure to reciprocate. In general, situational attributions tend to reduce or inhibit reactance. Unique-interaction or unique-actor attributions, on the other hand, have the potential to arouse the greatest amount of reactance.

The present studies also deal with the preference that individuals have for making certain types of attributions. Jones and Nisbett (1971) suggested that observers tend to attribute environmental causes to their own behavior and internal or personal causes to the actor's behavior. The present line of research, however, suggests that this statement needs to be qualified. When an actor's behavior threatens an observer's behavioral freedom, there will be a tendency for that behavior to be attributed to environmental factors.

One question that cannot be answered from either study deals with causal linkage between the attribution and the arousal of reactance. Do individuals experience reactance and then make situational attributions in an effort to reduce the reactance? Or does the attribution occur before the experience of reactance and serve to ensure that the individual will experience no reactance? It is somewhat of a "chicken and the egg" dilemma, and the answer cannot be garnered from the present results. The answer will have to await future research.

However, the solution is an important one because it would provide some insight into the attribution process.

A second question on which future research can focus is whether or not attributions can be used in other situations with other behaviors to restore behavioral freedom. Our research has dealt only with actions that threaten freedom by arousing the norm of reciprocity. This may be seen as a more or less indirect threat to freedom since the threat is not the motive behind the actor's behavior. Will attribution be utilized to restore freedom if the threat to freedom is more direct, such as a simple command or demand from the actor? The answer to this question, too, awaits further results.

Another question can be raised about this process, as with most psychological processes that we accuse individuals of engaging in. Does the individual consciously engage in the attributional processing, or does he simply act reflexively and unaware? If, in fact, the individual is well-aware of his actions, does he also make a conscious decision as to whether he will use attribution, aggression, or direct behavior to restore his freedom? This degree of processing represents a lot of cognitive work, but it is possible that the particular situation is taken into account by the individual in deciding exactly how freedom will be restored.

While there are a number of unanswered questions, we feel that the two studies presented in this chapter suggest that attribution of the causality for a potentially reactance-producing behavior will help determine if reactance is actually aroused. Specifically, reactance will not occur when the behavior is attributed to environmental causes. On the other hand, reactance is most likely to result from attributions to unique interaction loci of causality when the behavior threatens important behavioral freedom. We also feel that the present research shows that individuals can and do utilize attributions of causality to minimize reactance and restore behavioral freedom. Attributing reactance-arousing behavior to environmental causes reduces the degree of reactance.

REFERENCES

Berkowitz, L., & Daniels, L. R. Affecting the salience of the social responsibility norm: Effects of past help in the response to dependency relationships. *Journal of Abnormal and Social Psychology,* 1964, **68,** 245.

Brehm, J. W. *A theory of psychological reactance.* New York: Academic Press, 1966.

Brehm, J. W., & Cohen, A. R. *Explorations in cognitive dissonance.* New York: John Wiley & Sons, 1962.

Brehm, J. W., & Cole, A. H. Effect of a favor which reduces freedom. *Journal of Personality and Social Psychology,* 1966, **3,** 420–426.

Buss, A. H. *The psychology of aggression.* New York: John Wiley & Sons, 1961.

Festinger, L. *A theory of cognitive consistency.* Stanford, California: Stanford University Press, 1957.

Goranson, R. E., & Berkowitz, L. Reciprocity and responsibility reactions to prior help. *Journal of Personality and Social Psychology,* 1966, **3,** 227–232.

Gouldner, A. W. The norm of reciprocity: A preliminary statement. *American Sociological Review,* 1960, **24,** 546–554.

Greenberg, M. S., & Frisch, D. M. Effect of intentionality on willingness to reciprocate a favor. *Journal of Experimental Social Psychology,* 1972, **8,** 99–111.

Hammock, T., & Brehm, J. W. The attractiveness of choice alternatives when freedom to choose is eliminated by a social agent. *Journal of Personality,* 1966, **34,** 546–554.

Helm, R., Bonoma, T. V., & Tedeschi, J. R. Counter-aggression as a function of physical aggression: Reciprocity for harm done. Paper presented at the meeting of the American Psychological Association, Washington, D.C., September 1971.

Jones, E. E., & Davis, K. E. From acts to dispositions. In L. Berkowitz (Ed.), *Advances in experimental social psychology* (Vol. 2). New York: Academic Press, 1965.

Jones, E. E., & Nisbett, R. E. *The actor and the observer: Divergent perceptions of the causes in behavior.* Morristown, New Jersey: General Learning Press, 1971.

Jones, E. E., Worchel, S., Goethals, G. R., & Grumet, J. Prior expectancy and behavioral extremity as determinants of attitude attribution. *Journal of Experimental Social Psychology,* 1971, **7,** 59–80.

Kelley, H. H. Attribution theory in social psychology. In D. Levine (Ed.), *Nebraska Symposium on Motivation.* Lincoln: University of Nebraska Press, 1967.

McArthur, L. A. The how and what of why: Some determinants and consequences of causal attributions. *Journal of Personality and Social Psychology,* 1972, **22,** 171–193.

Pruitt, D. G. Reciprocity and credit building in a laboratory dyad. *Journal of Personality and Social Psychology,* 1968, **8,** 143–147.

Schopler, J. An attribution analysis of some determinants of reciprocating a benefit. In J. Macaulay & L. Berkowitz (Eds.), *Altruism and helping behavior.* New York: Academic Press, 1970.

Worchel, S. The effect of three types of arbitrary thwarting on the instigation to aggression. *Journal of Personality,* 1974, **42,** 300–318.

Worchel, S., & Andreoli, V. A. Attribution as a means of restoring behavioral freedom. *Journal of Personality and Social Psychology,* 1974, **20,** 237–245.

Worchel, S., Andreoli, V. A., & Archer, R. When a favor is not a favor: Effects of attribution of a favor and importance of behavioral freedom on reciprocity. *Journal of Personality,* in press.

12

Attribution of Attitudes from Feelings: Effect of Positive or Negative Feelings When the Attitude Object Is Benefited or Harmed

Judson Mills
University of Maryland

Jerald M. Jellison
University of Southern California

James Kennedy
Cathage College

The question of how people come to "know" their own attitudes has recently been a topic of interest. In his theory of self-perception, Bem (1967, 1972) proposed that people infer their attitudes from their overt behavior and the circumstances in which the behavior occurs. This proposition has been discussed by Kelley (1967, 1971, 1972, 1973) in the framework of attribution theory, which deals generally with the processes of perceiving the dispositional properties of entities.

Just as people may use their actions to make inferences about their attitudes, they may also infer their attitudes from their feelings and the circumstances in which the feelings occur. In this chapter we will describe research which is concerned with the effect on attitudes of whether the person's feelings are positive or negative and whether these feelings occur when the object of the attitude is benefited or is harmed. It employs an attributional analysis which involves a new application of Heider's (1958) balance principle.

The attribution of attitudes from feelings can be thought of in terms of the operation of a particular type of causal schema, the term used by Kelley (1972,

1973) to refer to the way in which the person thinks about plausible causes in relation to a given effect which provides him with a means of making causal attributions given only limited information. In this specific causal schema the assumption is that good or bad feelings are caused by what happens to things which are liked or disliked. It can be represented as follows: "I will feel good when good things happen to something which I like. I will feel good when bad things happen to something which I dislike. I will feel bad when good things happen to something which I dislike. I will feel bad when bad things happen to something which I like."

The triad involved in this causal schema (feelings, effect upon the attitude object, and attitude) can be viewed as an instance of Heider's balance principle. According to Heider (1958): "A triad is balanced when all three of the relations are positive or when two of the relations are negative and one is positive. Imbalance occurs when two of the relations are positive and one is negative [p. 202]."

In terms of the balance principle, a positive feeling fits with a positive effect on something toward which there is a positive attitude. A positive feeling fits with a negative effect on something toward which there is a negative attitude. A negative feeling fits with a positive effect on something toward which there is a negative attitude. A negative feeling fits with a negative effect on something toward which there is a positive attitude.

If people use a causal scheme of the type described in trying to account for the way they feel, then assuming that there is no other apparent cause for their feelings, they should infer that if they feel good when an attitude object is benefited, it is because they like it; and if they feel good when an attitude object is harmed, it is because they dislike it. Also, they should infer that if they feel bad when an attitude object is benefited, it is because they dislike it; and if they feel bad when an attitude object is harmed, it is because they like it.

The use of such a causal schema in making inferences about attitudes means that, in effect, attribution of attitudes from feelings follows the balance principle. Positive feelings associated with a positive effect on an attitude object will tend to produce a positive attitude towards it. Positive feelings associated with a negative effect on an attitude object will tend to produce a negative attitude toward it. Negative feelings associated with a positive effect on an attitude object will tend to produce a negative attitude toward it. Negative feelings associated with a negative effect on an attitude object will tend to produce a positive attitude toward it.

EXPERIMENT 1

From the balance analysis of the effect of feelings on attitudes, it follows that persons who experience positive feelings when one attitude object is benefited and another attitude object is harmed will tend to like the one benefited more than the one harmed and that persons who experience negative feelings will tend to like the one harmed more than the one benefited. An experiment was designed to test these hypotheses. In the experiment, false meter feedback was used in order to vary whether positive or negative feelings were experienced.

The experiment was deliberately designed to investigate differences in attitudes toward two different attitude objects which were simultaneously either benefited or harmed. One reason was that such a within-subject comparison minimizes the effect of extraneous factors affecting liking and provides a more sensitive test of the effects of feelings on attitudes. A second reason, even more important, had to do with the choice of previously unknown persons as the attitude objects.

Since most people share the attitude that it is bad to feel happy when another person suffers or to feel unhappy when another person has good fortune, these two situations would probably produce additional feelings which would interfere with the effect of the manipulation of positive or negative feelings. However, this problem should not occur if one person is harmed at the same time another person is benefited or vice versa. There is nothing improper in feeling good when a person suffers if at the same time another person is benefited or in feeling bad when a person has good fortune if another person is harmed at the same time.

If the person has another way to account for his feelings, the positive or negative feelings which are experienced when an attitude object is benefited or harmed should have relatively little influence on the attitude. For example, if a person feels bad on election night, he is not likely to infer that he likes the losing candidate more than the winner if he knows that the reason he feels bad is something he ate for dinner. Such an instance can be understood in terms of the operation of the discounting principle (Kelley, 1971, 1973), which states that the role of a given cause in producing a given effect is discounted if other plausible causes are also present. In the present sphere, the attitude toward the thing which is benefited or harmed should be discounted as a cause of the positive or negative feelings if other plausible causes for the feelings are present.

In the experiment an attempt was made to vary the presence of another plausible cause for the positive or negative feelings by giving the subjects a capsule containing a placebo and telling some that it was a stimulant and others that it was a tranquilizer. It was assumed that subjects who were told the capsule was a stimulant would be more likely to attribute their positive or negative feelings to the capsule and that therefore the feelings would have less effect on their attitudes.

Method

Under the guise of a study of drugs and memory, college women monitored a meter supposedly measuring their positive or negative feelings while they listened to a taped description of an election campaign. For some of the subjects the meter shifted to positive when the winner and loser were announced and for others the meter shifted to negative when the winner and loser were announced. All subjects had been given a capsule containing a placebo; some were told they had received a stimulant and others that they had received a tranquilizer. After the tape, attitudes toward the winner and loser were measured by means of ratings of their positive and negative characteristics.

The subjects were 56 female students in introductory psychology and sociology who were recruited for a study of drugs and memory and were promised $1.50 for participating. Half of the subjects were randomly assigned to the positive feeling condition and half to the negative feeling condition. Half were randomly assigned to the stimulant label condition and half to the tranquilizer label condition. For half, one candidate was announced as the winner, and for half the other candidate was announced as the winner.

The subjects were usually run in groups of four, although some sessions contained less than four. For all subjects at a particular session, the same candidates were announced as the winner and loser. Otherwise, the subjects at a particular session were in different conditions.

When the subjects arrived at the experimental room, they were paid the $1.50 and seated so that partitions shielded their view of one another. The experimenter began by explaining that the purpose of the study was to test the effect of certain drugs on memory. They would be given either a stimulant or a tranquilizer, listen to some tape recorded material, and then be given a memory test on the material. The experimenter emphasized that taking the drugs was completely voluntary and that if they did not want to take one they could leave with full payment. He also stressed that if they had a medical history of heart disease, stomach disorders, kidney disorders, diabetes, or epilepsy, they were not allowed to participate but could keep the full payment. One student was not included as a subject because she had a medical condition.

Each subject was given a sheet which described two drugs, Drug A, classified as a "stimulant" which came in a pink capsule; and Drug B, classified as a "tranquilizer" which had a yellow capsule. The dosage of both drugs was given as 5 mg. The effects of Drug A were described as: "Slight increase in heart rate and metabolism. Increase in sensitivity of central nervous system to emotional arousal. May heighten feelings and sensations. May intensify ongoing mood. Duration approximately 1–1½ hr." The effects of Drug B were described as: "Slight decrease in heart rate and metabolism. Decrease in sensitivity of central nervous system to emotional arousal. May decrease feelings and sensations. May lower the intensity of ongoing mood. Duration approximately 1–1½ hr."

After reading the description of the drugs aloud, the experimenter asked the subjects to read and sign a release certifying that they had carefully read the drug descriptions and that their partitipation, including the taking of one of the capsules, was a voluntary act. All of the subjects agreed. Each was given a capsule containing a placebo with some water. Those assigned to the stimulant label condition were given a pink capsule and told that they had received the stimulant; those assigned to the tranquilizer label condition were given a yellow capsule and told that they had received the tranquilizer.

The experimenter went on to explain that it was necessary to control for factors that could influence how much they might remember. One such factor was whether they were experiencing positive or negative feelings while listening to the tape recorded material. In order to control for that factor, their positive and negative feelings would be measured so that they could be taken into account when analyzing the memory data. For this purpose a physiograph machine was going to be used. This machine was said to be "the most sensitive and accurate instrument that is available in psychology to measure a person's positive and negative feelings because it detects very minute changes in the amount of electrical activity in the central nervous system around the area known as the reticular formation. Increases and decreases in the amount of electrical activity in this area are a very accurate indicator of a person's true positive and negative feelings."

The experimenter directed the subjects' attention to the meters in front of each of them and said that their meters were connected to the physiograph machine to which they would be attached. He noted that there were five zones on the meters: a neutral zone; a + zone "which stands for positive feelings"; a ++ zone "which stands for very positive feelings"; a – zone "which stands for negative feelings"; and a –– zone "which stands for very negative feelings." He told the subjects that their help was needed in keeping track of their own meter readings while the taped material was being played. He gave each of them a data sheet on which they were to indicate how they were feeling at different times during the tape.

The experimenter explained that, rather than using materials such as nonsense syllables to test memory, some material which was more meaningful and interesting would be used, much like something that they might read in a newspaper. The tape that they were to hear was a description of a run-off election which was held for president of the student government at a state university in the midwest, comprised of material from newspaper clippings covering the election.

After attaching two electrodes to the back of each subject's neck, the experimenter explained that since the physiograph is very sensitive and each person reacts differently when attached to it, it was necessary to calibrate it for each person. In order to do this, there were two sounds at the beginning of the tape. The first sound was a very unpleasant noise which gives an indication of negative reactions and the second sound, some soft music which is used to generate positive reactions. Then the tape would continue with the material on which they would be tested.

When the meters were "turned on," the needle on each subjects' meter waivered within the neutral zone. After the first sound, a very loud 5 second blast of noise, the needle moved to the –– zone. Following the second sound, a 15-second segment of soft guitar music, the needle moved to the + zone.

The description of the election campaign included information about the biographies of the two candidates and their views about the issues. A number of attractive qualities were attributed to each candidate, as well as a very strong desire to win the election. The election was described as very close prior to the balloting. Throughout this section of the tape, the needles of all subjects waivered within the neutral zone.

After 10 minutes the winner and loser of the election were announced. It was emphasized that the outcome had been decided by only a few votes in a record turnout. After the election result was announced there was a 90-second segment describing the jubilant scene at the winning candidate's campaign headquarters.

For subjects assigned to the positive feeling condition, the meter needle moved to the ++ zone immediately after the election result was announced and then went to the + zone during the scene at the winner's headquarters. For subjects assigned to negative feeling condition, the needle moved to the −− zone immediately after the election result was announced and then to the − zone during the scene at the winner's headquarters.

After the tape ended, the meters were "turned off," and the electrodes removed. Each subject was then given a questionnaire. The first part of the questionnaire contained items which asked the subject to indicate how well different personality characteristics applied to each of the two candidates on a scale from 0 (extremely inappropriate) to 20 (extremely appropriate). The characteristics were: attractive, boring, competent, friendly, happy, insincere, intelligent, irritating, kind, likable, and unpleasant. The experimenter explained that it was necessary to control for the impressions that they might have of the candidates, in order to eliminate the effect of their impressions on the memory data.

The second part of the questionnaire contained 22 multiple choice items, with three alternatives each, covering the content of the tape recording. At the end of the questionnaire, the subjects were told the experiment was over and were asked to give their comments on it. They were instructed to feel free to write anything they wanted and to be as critical as they wanted. The responses of two students indicated that they suspected that the drug was a placebo; they were not included as subjects. In addition, one student was not included as a subject because of an equipment failure. Finally, the true purpose of the experiment was explained, the deceptions revealed, and the subjects asked to promise not to discuss the experiment with anyone.

Results

A check on the perception of the benefit or harm to the two candidates from the election result was provided by the ratings for the adjective "happy." For both candidates, the ratings for happy were higher when the candidate rated was the winner than when he was the loser. Separate analyses of variance were conducted for the ratings for happy for each candidate and for each candidate the main effect of which candidate was announced as the winner was significant beyond the .01 level. (Unless otherwise noted, significance was tested by means of a three-way analysis of variance of the particular measure, with the three factors being direction of feelings, capsule label, and the candidate announced as the winner.)

No significant effects of the experimental conditions were revealed in an analysis of variance of the number of correct memory items on the questionnaire. The average number of correct items was 19.4 out of 22.

It was hypothesized that persons who experience positive feelings when one attitude object is benefited and another harmed will tend to like the one benefited more than the one harmed, and persons who experience negative feelings will tend to like the one harmed more than the one benefited. From these hypotheses, it would be expected that liking for the winner would be greater than liking for the loser in the positive feeling condition and liking for the loser would be greater than liking for the winner in the negative feeling condition.

Liking for each of the candidates was measured by summing the ratings given the candidate on the positive personality characteristics in the questionnaire (excluding happy) and subtracting the sum of the ratings given him on the negative characteristics. A measure of the difference between liking for the winner and liking for the loser was calculated by subtracting the score for liking for the loser from the score for liking for the winner. A positive score for this measure indicates greater liking for the winner than for the loser and a negative score, greater liking for the loser than for the winner.

The means for the measure of liking for the winner minus liking for the loser are presented in Table 1. As can be seen from Table 1, the results were in line with the predictions. The difference between liking for the winner and liking for the loser was +12.9 for the positive feeling condition and was –5.5 for the negative feeling condition.

An analysis of variance of the measure of liking for the winner minus liking for the loser revealed that the main effect of positive versus negative feelings fell just

TABLE 1
Means for the Measure of Liking for the Winner Minus Liking for the Loser for Experiment 1

	Capsule label		
Feelings	Stimulant	Tranquilizer	Combined
Positive	+5.4(14)	+20.4(14)	+12.9(28)
Negative	–.3(14)	–10.6(14)	–5.5(28)

Note. A positive score indicates greater liking for the winner than the loser and a negative score, greater liking for the loser than the winner. *N*s given in parentheses.

short of significance at the .05 level. The difference between the mean for the positive feeling condition of +12.9 and 0 was tested with the use of t and was not significant. The difference between the mean for the negative feeling condition of −5.5 and 0 was also not significant.

It was also hypothesized that when other plausible causes of the feelings are present, experiencing positive or negative feelings when one attitude object is benefited and another harmed will produce a weaker tendency to like the one benefited more than the one harmed and experiencing negative feelings will produce a weaker tendency to like the one harmed more than the one benefited. From these hypotheses it would be expected that the tendency to like the winner more than the loser in the positive feeling condition would be weaker in the stimulant label condition than in the tranquilizer label condition. The tendency to like the loser more than the winner in the negative feeling condition should also be weaker in the stimulant label condition than in the tranquilizer label condition.

As can be seen from Table 1, the difference between liking for the winner and liking for the loser in the positive feeling condition was less positive in the stimulant label condition than in the tranquilizer label condition. In the negative feeling condition, the difference was less negative in the stimulant label condition than in the tranquilizer label condition. However, the interaction between stimulant versus tranquilizer label and positive versus negative feelings was not significant.

As noted earlier, the experiment was deliberately designed to investigate the relative liking for an attitude object which is benefited and one which is harmed. However, the balance interpretation of the effect of feelings on attitudes also makes predictions concerning the effect of the positive and negative feeling conditions separately on liking for the winner and on liking for the loser. It would be expected that liking for the winner would be greater in the positive feeling condition than in the negative feeling condition and that liking for the loser would be greater in the negative feeling condition than in the positive feeling condition.

For both the measure of liking for the winner and the measure of liking for the loser, the results were in line with the predictions. For the measure of liking for the winner, the main effect of feelings was significant ($p < .05$). For the measure of liking for the loser, it was not.

EXPERIMENT 2

In view of the fact that the results of the first experiment were in line with the predictions from the hypotheses but, with a few exceptions, not strong enough to be statistically significant, a second experiment was conducted in order to test the hypotheses further. In an attempt to strengthen the manipulation of the presence of other plausible causes for the feelings, the description of the capsule was changed. The "dosage" was increased, and the capsule was labeled as either a stimulant or an analgesic, the labels used successfully in studies by Mintz and Mills (1971) and Mills and Mintz (1972). In addition, some subjects were told that they were in a control condition and not given any capsule. Otherwise, the procedure was essentially the same as that of the first experiment.

Method

The subjects were 120 female students in introductory psychology who volunteered in order to fulfill a course requirement concerning participation in research. They were not paid as in the first experiment. One third of the subjects were randomly assigned to the stimulant label condition, one third to the analgesic label condition, and one third to the control condition. Half were randomly assigned to the positive feeling condition and half to the negative feeling condition. For half, one candidate was announced as the winner, and for half the other candidate was announced as the winner.

The procedure was the same as in the first experiment except that the drug descriptions were altered. Drug A was described as a "stimulant" which came in a pink capsule and Drug B as an "analgesic" which had a yellow capsule. The dosage of both was given as 100 mg. The effects of Drug A were described as: "Increase in responsiveness of central nervous system, particularly to emotional stimuli. Duration approximately 1–1½ hr." The effects of Drug B were described as: "Decrease in responsiveness of central nervous system. Duration approximately 1 hr."

Subjects assigned to the stimulant label condition were given a pink capsule and told that they had received a stimulant. Those assigned to the analgesic label condition were given a yellow capsule and told they had received the analgesic. Those assigned to the control condition were not given any capsule and were told that they were in a control group. Six students chose not to take one of the drugs and were not included as subjects. The comments of 4 students indicated that they suspected the drug was a placebo, and the comments of 1 student indicated suspicion both about the placebo and the meter; these 5 were not included as subjects. In addition, 2 students were not included as subjects because of equipment failures.

Results

As in Experiment 1, for both candidates the ratings for happy were higher when the candidate rated was the winner than when he was the loser. For each candidate, the main effect of which candidate was announced as the winner was significant beyond the .001 level.

There were no significant effects for the number of correct memory items, although the number correct tended to be somewhat higher for the positive feeling condition than for the negative feeling condition, $p = .06$. The average number of correct items was 18.1.

The means for the measure of liking for the winner minus liking for the loser for Experiment 2 are presented in Table 2. As can be seen from Table 2, the results of the second experiment were similar to those of the first experiment. The difference between liking for the winner and liking for the loser was +10.9 in the positive feeling condition and was −13.2 in the negative feeling condition.

An analysis of variance of the measure of liking for the winner minus liking for the loser revealed that the main effect of positive versus negative feelings was significant beyond the .001 level. The difference between the mean for the positive feeling condition of +10.9 and 0 was tested by t and was significant beyond the .05 level. The difference between the mean for the negative feeling condition of −13.2 and 0 was significant beyond the .01 level.

As can be seen from Table 2, the difference between liking for the winner and the loser in the positive feeling condition was most positive in the analgesic label condition and least positive in stimulant label condition, with the control condition in-between. In the negative feeling condition, the difference between liking for the winner and the loser was most negative in the control condition and least negative in the stimulant label condition, with the analgesic label condition in-between. However, the interaction between the label conditions and positive versus negative feelings was not significant.

As in the first experiment, the measure of liking for the winner was greater in the positive feeling condition than in the negative feeling condition, and the measure of liking for the loser was greater in the negative feeling condition than in the positive feeling condition. For the measure of liking for the winner, the main effect of feelings was significant at the .01 level. For the measure of liking for the loser it was not.

TABLE 2
Means for the Measure of Liking for the Winner Minus Liking for the Loser for Experiment 2

	Capsule label			
Feelings	Stimulant	Analgesic	Control	Combined
Positive	+6.7(20)	+16.5(20)	+9.7(20)	+10.9(60)
Negative	−9.4(20)	−12.0(20)	−18.4(20)	−13.2(60)

Note. A positive score indicates greater liking for the winner than the loser and a negative score, greater liking for the loser than the winner. *N*s given in parentheses.

EXPERIMENT 3

In spite of the fact that the comments of the subjects in Experiment 1 and 2 did not indicate any suspicion of the true purpose of the study, it might be thought that the subjects did suspect the experimental hypotheses and rated the candidates as they did because of "demand characteristics." In order to determine whether the results of the first two experiments could be attributed to "demand characteristics," a third experiment was carried out. In this experiment, at the end of the procedure, the subjects were asked a direct question concerning the experimenter's expectations about the relation between their ratings of the candidates and their meter readings, in order to determine whether they were aware of the experimental hypotheses. It was assumed that if the subjects were unable to state the experimental hypotheses when asked to do so, the results could not be explained by "demand characteristics". Except for the additional question, the procedure was essentially the same as in the control condition of Experiment 2.

Method

The subjects were 40 female students in introductory psychology who volunteered in order to earn extra course credit. Half were randomly assigned to the positive feeling condition and half to the negative feeling condition. For half, one candidate was announced as the winner, and for half, the other candidate was announced as the winner. All subjects were told that they were in a control group and were not given any capsule or drug descriptions.

The general comments of three students indicated that they suspected that the meter did not actually indicate their feelings; they were not included as subjects. After the subjects had written their general comments about the experiment, the experimenter said the following: "I haven't mentioned it so far, but we were also interested in the relation between your ratings of the two candidates and your meter ratings. If you have any idea what we thought that relationship might be, please write it on the back of your test booklet." Every student wrote a comment, but only 1 wrote something which was close to the actual experimental hypotheses.[1] She wrote the following: "The reaction, according to the meter, when the announcement was made of the winner should coincide with the favoritism exhibited on the rating sheet. I reacted negatively. My rating of Colman is less than that of Rodgers." (Colman was announced as the winner and Rodgers as the loser). Although it is not really clear from what she wrote that

[1] This judgment of the comments of the subjects in Experiment 3 was independently verified by Harold Sigall, who has conducted research on "demand characteristics" (Sigall, Aronson, & Van Hoose, 1970) and also has participated in the development of a new technique (Jones & Sigall, 1971) that involves convincing people that a machine is capable of detecting their true feelings and which can be used to minimize "demand characteristics."

this student actually suspected the experimental hypotheses, she was not included as a subject.

Results

As in Experiments 1 and 2, for both candidates the ratings for happy were significantly higher when the candidate rated was the winner than when he was the loser. (Statistical significance was tested in Experiment 3 by means of a two-way analysis of variance of the particular measure, with the two factors being direction of feelings and the candidate announced as the winner.) There were no significant effects for the number of correct memory items. The average number of correct items was 17.3.

The results of the third experiment concerning the difference in liking for the winner and the loser were similar to those of the first two experiments. The mean for the measure of liking for the winner minus liking for the loser was +13.8 for the positive feeling condition and was −16.6 for the negative feeling condition. An analysis of variance of the measure of liking for the winner minus liking for the loser revealed that the main effect of positive versus negative feelings was significant beyond the .001 level. The difference between the mean for the positive feeling condition of +13.8 and 0 was tested by t and was significant at the .05 level. The difference between the mean for the negative feeling condition of −16.6 and 0 was significant at the .01 level.

When the probability values for the difference between the mean for the positive feeling condition and 0 from Experiments 1, 2, and 3 were combined by Stouffer's method (Mosteller & Bush, 1954), the resulting p value was significant beyond the .001 level. The combined p value for the three experiments for the difference between the mean for the negative feeling condition and 0 was significant beyond the .001 level.

As in the first two experiments, the measure of liking for the winner was greater in the positive feeling condition than in the negative feeling condition, and the measure of liking for the loser was greater in the negative than in the positive feeling condition. For the measure of liking for the winner, the main effect of feelings was not significant. For the measure of liking for the loser, it was significant at the .02 level. The combined p value for the three experiments for the difference in the measure of liking for the winner between the positive feeling condition and negative feeling condition was less then .001. For the difference in the measure of the liking for the loser between the positive and negative feeling conditions, the combined p equals .052.

DISCUSSION

The results of the three experiments together provide strong evidence for the hypotheses that persons who experience positive feelings when one attitude object is benefited and another harmed will tend to like the one benefited more than the one harmed, and persons who experience negative feelings will tend to like the one harmed more than the one benefited. If the meter supposedly indicating their positive or negative feelings shifted to positive when the winner and loser of the election were announced, the subjects liked the winner significantly more than the loser. If the meter shifted to negative when the election result was announced, the subjects liked the loser significantly more than the winner. Liking for the winner was greater when the meter shifted to positive than when it shifted to negative and liking for the loser was greater when the meter shifted to negative than when it shifted to positive.

The hypotheses were based on an attributional analysis which assumes that the effect of feelings on attitudes depends on the perception of benefit or harm to the attitude object and that it follows Heider's balance principle. According to this balance interpretation, positive feelings associated with a positive effect on an attitude object will tend to produce a positive attitude toward it and positive feelings associated with a negative effect on an attitude object will tend to produce a negative attitude toward it. Negative feelings associated with a positive effect on an attitude object will tend to produce a negative attitude toward it and negative feelings associated with a negative effect on an attitude object will tend to produce a positive attitude toward it.

Since the use of false meter feedback is a novel procedure for manipulating positive and negative feelings, a question might be raised about whether the meter readings actually affected the feelings which the subjects experienced. What this question really amounts to is whether or not it is reasonable to assume that the subjects believed that the meter indicated their true feelings.

If one assumes that the subjects thought that they were feeling good when the meter shifted to positive and feeling bad when the meter shifted to negative, then it does not make sense to say that they were not actually feeling good or bad. The statement, "I think I feel good but I don't actually feel good" is comparable with the statements, "I think I feel ill but I don't actually feel ill" or "I think it feels cold (to me) but it doesn't actually feel cold (to me)" or "I think it looks red (to me) but it doesn't actually look red (to me)." None of these statements make any sense.[2]

It might be thought that a Freudian would take the position that a person can be mistaken about his own feelings and think he is feeling good or bad when he is actually not feeling that way. However, this would not be consistent with the writings of Freud. In his 1915 paper "The Unconscious,"[3] Freud (reprinted,

[2] Self-contradictory is the more technical term that philosophers would use to describe these statements, according to a personal communication from John R. Searle, a philosopher who has made major contributions to the philosophy of language (Searle, 1969).

[3] In the introduction to this paper, the editor said, "If the series of 'Papers on Metapsychology' may perhaps be regarded as the most important of all Freud's theoretical writings, there can be no doubt that the present essay on 'The Unconscious' is the culmination of that series [p. 161]."

1957) wrote, "It is surely of the essence of an emotion that we should be aware of it, i.e. that it should become known to consciousness. Thus the possibility of the attribute of unconsciousness would be completely excluded as far as emotions, feelings and effects are concerned [p. 177]." After discussing terms such as unconscious anxiety, Freud concluded:

> Thus it cannot be denied that the use of the terms in question is consistent; but in comparison with unconscious ideas there is the important difference that unconscious ideas continue to exist after repression as actual structures in the system *Ucs.*, whereas all that corresponds in that system to unconscious affects is a potential beginning which is prevented from developing. Strictly speaking, then, and although no fault can be found with the linguistic usage, there are no unconscious affects as there are unconscious ideas [p. 178]."

The comments of those included as subjects, in response to the request to feel free to write anything they wanted and to be as critical as they wanted, did not indicate any doubts about the authenticity of the meter. In fact, some subjects spontaneously wrote comments indicating that they definitely accepted the meter as an indication of their true feelings. The most dramatic example was the following comment of a subject in the positive meter condition of Experiment 3: "I thought it was fun, except the first noise scared me to death. Other than that it was quite interesting. Especially how the physiograph works according to your real emotion. You can't influence it by how you think. It knows the truth." Of course, it is not necessarily assumed that every subject in the positive feeling condition thought they felt positive when the meter shifted to positive and that every subject in the negative feeling condition thought they felt negative when the meter shifted to negative, but only that, in general, there was a difference between the conditions in how positive or negative the subjects thought their feelings were.

It is true that the experimental procedure did not include a direct check on the success of the manipulation or positive or negative feelings. The subjects were deliberately not asked to rate how positive or negative they felt at the time the election result was announced. Since the rationale given for using the physiograph machine to which the meters were supposedly connected was that it was the most sensitive and accurate instrument that is available in psychology to measure a person's positive and negative feelings, it might have raised doubts about the accuracy of the meter feedback if the experimenter had asked the subjects to rate how positive or negative they felt.

In addition, since the experimenter had presented the subjects with an elaborate explanation of the physiograph machine's ability to provide an accurate assessment of positive and negative feelings, it seemed unlikely that they would be willing to give the experimenter responses which did not agree with the meter readings which they had recorded. If a direct check on the manipulation of feelings had been included and has shown differences between the positive and negative feelings conditions, it would not have been clear whether the manipulation really had been effective or whether the subjects were merely succumbing to "demand characteristics."

If one does not assume that the subjects believed the meter readings and thought that they were experiencing positive or negative feelings, one is left with the problem of how to explain the differences which were obtained. Since in the third experiment the same results were found for subjects who were not able to state the experimental hypotheses in response to a direct question by the experimenter, the differences cannot be attributed to "demand characteristics." If after the experimenter told the subjects that he was interested in the relation between their ratings of the candidates and their meter readings and asked them to tell him if they had any idea what he thought the relationship might be, they were not able to state the experimental hypotheses, one can be confident that they did not respond as they did in order to conform to the hypotheses.

The fact that the subjects were not able to state the experimental hypotheses can be attributed to the success of the cover story in disguising the purpose of the meter readings and the ratings of the candidates. The meter readings and candidate ratings were justified in the cover story as necessary in order to control for factors which might affect memory. The meter and the questionnaire were not related to one another insofar as the ostensible purpose of the experiment was concerned.

Another aspect of the procedure which helped to conceal the experimental hypotheses was that the subjects were not directly asked to indicate how much they liked the candidates. Instead, attitudes toward the two candidates were measured indirectly by means of ratings of how well specific characteristics described each of them. This was done on the assumption that the attribution of a positive or negative attitude would influence beliefs about the attitude object. If the person infers that he likes the attitude object, he should tend to believe that it has characteristics which he values positively and does not have those he values negatively. The opposite should tend to occur if the person infers that he dislikes the attitude object.

No assumption was made concerning whether the subjects were directly aware of their attitudes toward the two candidates nor does the interpretation of the results require such an assumption. Unlike feelings, people may have attitudes of which they are unaware. Heider's (1958) warning against expecting the attribution process to be represented in experience has been repeated by Kelley (1967), who quotes Heider's (1958) statement that, "Attributions may not be experienced as interpretations at all, but rather as intrinsic to the original stimuli [p. 256]."

The first two experiments failed to provide evidence for the hypotheses that when other plausible causes for the feelings are present, experiencing positive feelings when one attitude object is benifited and another harmed will produce a weaker tendency to like the one benefited more than the one harmed and experiencing negative feelings will produce a weaker tendency to like the one harmed more than the one benefited. The results were in the direction predicted by these hypotheses but the effects were not strong enough to be statistically significant. It is possible that the attempt to manipulate the presence of another plausible cause for the positive or negative feelings by assigning different labels to the placebo capsule did not produce large enough differences in this factor to permit its effects upon attitudes to be demonstrated.

A factor not investigated in the present research which should influence the strength of the effect upon an attitude of positive or negative feelings when the attitude object is benefited or harmed is the amount of the other information about the attitude. In discussing Bem's proposition that a person judges his attitude from his behavior and the conditions under which it occurs, Kelley (1967) points out that, "A distinction should probably be made in regard to the above issue, a distinction between frequently or recently tested attributions and less well 'rehearsed' ones. The former can probably be recovered directly in memory; i.e., I 'know' in the sense of immediate recall what my attitude is toward the thing or what I believe its properties to be [p. 214]."

If the person previously held a very strong attitude toward the attitude object, it is likely that he will have a good deal of information concerning his attitude and thus experiencing positive or negative feelings when the attitude object is benefited or harmed should have little effect. For example, if a person has a very favorable attitude toward President Kennedy and happens to feel good at the time that he sees a film of Kennedy's assassination on television, it is unlikely that this will have much effect on his attitude toward Kennedy. The effect of feelings on attitudes should be greatest when the prior attitude is relatively neutral. It was for this reason that attitudes toward previously unknown persons were used as the attitude topic in the present research.

The results of this research call into question the common assumption that positive feelings generalize to positive attitudes and negative feelings to negative attitudes (for example, Byrne, 1969). The results indicate that this is true if the feelings occur when the attitude object is benefited but that the opposite occurs if the feelings occur when the attitude object is harmed. In that case, positive feelings lead to negative attitudes and negative feelings to positive attitudes.

The importance of the perception of benefit or harm to the attitude object may have gone unrecognized because previous studies of the effect of feelings on attitudes have simply had subjects think about the attitude object while positive or negative feelings were induced, for example, the excellent studies by Gouaux (1971) and by Zanna, Kiesler, and Pilkonis (1970). That people who feel good when they are thinking about the existence of an attitude object tend to have a positive attitude toward it and people who feel bad when thinking about the existence of an attitude object tend to have a negative attitude toward it is quite consistent with the balance interpretation of the effect of feelings on attitudes. Thinking about the existence of an attitude object can be assumed to be equivalent to perceiving a positive effect on the attitude object.

It should be clear that in the present view feelings are not a component of attitudes as is sometimes assumed. Rather it is assumed that feelings, while different from attitudes, are closely related to them and that the relationship depends on the perception of the consequences for the attitude object. A similar view was expressed by Asch (1952) in his discussion of the cognitive basis of emotions. He stated that, "Emotions are our ways of representing to ourselves the fate of our goals [p. 110]."

As noted in the introduction, the present interpretation of the effect of feelings on attitudes is a new application of Heider's principle that a triad is balanced if all three relations are positive or if two are negative and one is positive. Previously, formulations similar to the balance principle have been used to account for the effect of beliefs on attitudes (Jones & Gerard, 1967, chapter 5; Rosenberg & Abelson, 1960).

According to a balance analysis of the effect of beliefs on attitudes, beliefs that an attitude object is positively associated with something which is valued positively or negatively associated with something which is valued negatively will tend to produce a positive attitude toward it. Beliefs that an attitude object is negatively associated with something which is valued positively or positively associated with something which is valued negatively will tend to produce a negative attitude toward it. For example, if one wanted to get people to like a political figure, one might try to convince them that he has qualities which would insure peace or reduce crime. If one wanted people to dislike him, one might try to make them believe that he has characteristics which would decrease freedom or increase inequality.

The balance principle has also been applied to the effect of behavior on attitudes (Insko & Schopler, 1967; Insko, Worchel, Folger, & Kutkus, 1975). According to a balance analysis of the effect of actions on attitudes, behavior which associates the person positively with a positive effect on an attitude object or which associates the person negatively with a negative effect on the attitude object should tend to produce a positive attitude toward it. Behavior which associates the person positively with a negative effect on the attitude object or negatively with a positive effect should tend to produce a negative attitude toward it. For example, to make people like a political figure, one might try to get them to choose to distribute materials supporting him or to refuse to sign a statement which censured him. To make them dislike him, one might try to get them to choose to applaud a speech which criticized him or to refuse to give him a political contribution.

The extension of the balance principle to the relation of feelings and attitudes, together with its prior application to the relation of beliefs and attitudes and to the relation of behavior and attitudes, means that all three major psychological dimensions related to attitudes can be dealt with in the same theoretical framework. It makes it possible to describe the various aspects of attitudes in terms of one general principle. The balance principle appears to have great usefulness for a comprehensive theoretical understanding of attitudes.

ACKNOWLEDGMENTS

The research reported in this chapter was supported by a grant from the National Science Foundation. The assistance of Steve Brake, Kyle Britt, Joe Priour, and George Sullivant in carrying it out is gratefully acknowledged. Duncan Forest deserves special gratitude for acting as the experimenter in Experiment 3.

REFERENCES

Asch, S. *Social psychology.* Englewood Cliffs, New Jersey: Prentice-Hall, 1952.

Bem, D. J. Self perception: An alternative interpretation of cognitive dissonance phenomena. *Psychological Review,* 1967, **74,** 183–200.

Bem, D. J. Self perception theory. In L. Berkowitz (Ed.), *Advances in experimental social psychology* (Vol. 6). New York: Academic Press, 1972.

Byrne, D. Attitudes and attraction. In L. Berkowitz (Ed.), *Advances in experimental social psychology* (Vol. 4). New York: Academic Press, 1969.

Freud, S. The unconscious. In J. Strachey (Ed.), *The standard edition of the complete psychological works of Sigmund Freud* (Vol. 14). London: Hogarth Press, 1957.

Gouaux, C. Induced affective states and interpersonal attraction. *Journal of Personality and Social Psychology,* 1971, **20,** 37–43.

Heider, F. *The psychology of interpersonal relations.* New York: John Wiley & Sons, 1958.

Insko, C. A., & Schopler, J. Triadic consistency: A statement of affective–cognitive–conative consistency. *Psychological Review,* 1967, **74,** 361–376.

Insko, C. A., Worchel, S., Folger, R., & Kutkus, A. A balance theory interpretation of dissonance. *Psychological Review,* 1975, **82,** 169–183.

Jones, E. E., & Gerard, H. B. *Foundations of social psychology.* New York: John Wiley & Sons, 1967.

Jones, E. E., & Sigall, H. The bogus pipeline: A new paradigm for measuring affect and attitude. *Psychological Bulletin,* 1971, **76,** 349–364.

Kelley, H. H. Attribution theory in social psychology. In D. Levine (Ed.), *Nebraska Symposium on Motivation.* Lincoln: University of Nehraska Press, 1967.

Kelley, H. H. *Attribution in social interaction.* Morristown, New Jersey: General Learning Press, 1971.

Kelley, H. H. *Causal schemata and the attribution process.* Morristown, New Jersey: General Learning Press, 1972.

Kelley, H. H. The processes of causal attribution. *Americal Psychologist,* 1973, **28,** 107–128.

Mills, J., & Mintz, P. M. Effect of unexplained arousal on affiliation. *Journal of Personality and Social Psychology,* 1972, **24,** 11–13.

Mintz, P. M., & Mills, J. Effects of arousal and information about its source upon attitude change. *Journal of Experimental Social Psychology,* 1971, **7,** 561–570.

Mosteller, F., & Bush, R. R. Selected quantitative techniques. In G. Lindzey (Ed.), *Handbook of social psychology* (Vol. 1). Cambridge, Massachusetts: Addison–Wesley, 1954.

Rosenberg, M. J., & Abelson, R. P. An analysis of cognitive balancing. In M. J. Rosenberg & C. I. Hovland (Eds.), *Attitude organization and change.* New Haven: Yale University Press, 1960.

Searle, J. R. *Speech acts: An essay in the philosophy of language.* Cambridge, England: Cambridge University Press, 1969.

Sigall, H., Aronson, E., & Van Hoose, T. The cooperative subject: Myth or reality? *Journal of Experimental Social Psychology,* 1970, **6,** 1–10.

Zanna, M. P., Kiesler, C. A., & Pilkonis, P. A. Positive and negative attitudinal affect established by classical conditioning. *Journal of Personality and Social Psychology,* 1970, **14,** 321–328.

13

An Attributional Analysis of Some Social Influence Phenomena

George R. Goethals

Williams College

This chapter explores several aspects of the social influence process. It considers how the opinions and judgments of one person affect those of another person. Although social influence processes, persuasion, and attitude change are central topics in social psychology, and although attribution has become a major theoretical focus, there has been little application of attribution to social influence and persuasion processes. This is somewhat ironic in light of the fact that the first substantive area treated in Kelley's (1967) germinal paper on attribution in social psychology was information dependence and influence. It is in this section, in fact, that Kelley outlines his famous analysis of variance analogy.

Kelley's perspective, as outlined briefly in his 1967 paper, is utilized extensively in the present chapter. Kelley suggested that whenever a person holds an opinion, he is concerned about whether it correctly reflects the entity about which the opinion is held or whether it is attributable to his own personal characteristics. In other words, what is the locus of causality of the opinion—the entity itself or the biasing personal characteristics of the person judging it? Kelley suggested that the consensus dimension is critical in making a person versus entity attribution. An attributor needs to know how others judge the entity, since a judgment of high consensus is generally attributable to the entity, while a judgment of low consensus is attributable to the person. To this point, Kelley's formulation is consistent with other social influence literature in suggesting that one's sense of being correct is enhanced when other people agree. In Festinger's (1950) theory of informal social communication, Asch's (1951) work on conformity, and Festinger's (1954) social comparison theory, it is also suggested that the opinions of other people affect one's confidence in his own opinion. However, Kelley pointed to additional complexities in the social influence process and commented that in addition to being concerned about the locus of causality of his own opinion, the person needs to be concerned about the locus of causality of any particular other person's judgment as well. That is, before increasing or decreasing his confidence on the basis of another's agreement or disagreement the person has to make attributions about the reasons for the other's judgment.

In this chapter, I will incorporate Kelley's perspective in the following way. First, it will be recognized that people make attributions about the opinions of other people. In the first part of the chapter, I deal with research that considers the attributions a person makes about the opinions of another (a) when the other agrees or disagrees with the person's opinion and (b) when the other is similar or dissimilar to the person in terms of values related to the opinion. Next, the ways in which a person's confidence in his judgments is affected by another's judgments will be considered. The predictions made about influence are based on the assumption that the opinions of others affect the attributions a person makes about his own opinions and thus his confidence in his own opinions. Finally, I will consider people's judgments or attributions about the influence process itself, their own feelings about how much they have been influenced, and the inferences they make about various influence agents.

It should be pointed out that while research on the attributions subjects make about others' opinions will be discussed before research on the influence those opinions have, the studies have actually been carried out in the reverse order. I have been primarily interested in testing predictions derived from attribution theory regarding the role of similarity in social influence processes. Although it has been theoretically implicit that special social influence effects are contingent on attributions made about the locus of causality of these opinions, until now, no research had been done to investigate these attributions. The rationale for presenting the newer attribution material first is the assumed priority of attributions over influence effects in the temporal sequence.

ATTRIBUTIONS REGARDING THE OPINIONS OF OTHER PEOPLE

It is the thesis in the present analysis that the individual will make attributions about the causes of another person's judgment. He will decide essentially whether that judgment represents an accurate appraisal of the entity being judged or whether it is attributable to what Kelley (1967) called "irrelevant" causes, such as the person's own biases or motives. In this portion of the chapter, I will discuss research conducted to investigate the attributions of causality that subjects make about other people's opinions.

First, an experiment will be described which was concerned simply with the attributions subjects make about opinions that agree or disagree with their own. In this experiment the subjects knew nothing about the characteristics of the person who agreed or disagreed. They simply received an opinion from an unknown person. A second experiment will then be described in which subjects were given some information about the person whose opinion they received; specifically, they knew whether his standing on values relevant to the opinion issue was similar or dissimilar to their own.

Attributions about Agreeing and Disagreeing Opinions of Unknown Others

Predictions about the attributions that people will generally make about agreeing and disagreeing opinions can be borrowed rather directly from Heider (1958). Heider suggested that "the person tends to attribute his own reactions to the object world, and those of another, when they differ from his own, to personal characteristics in *o* [p. 157]." Thus, it could be predicted that an agreeing opinion might be seen as being more attributable to the object, that is, facts and evidence, since it is the same as the person's own judgment, which Heider implies is attributed to the object. Conversely, following Heider directly, it could be predicted that a discrepant opinion should be attributed to the person, that is, to the disagreer's values.

These predictions were tested in an experiment by Goethals and Ryan (1975b) in which subjects were asked to make a judgment and then were asked to consider someone else's judgment. Specifically, the subjects watched a videotape of two high school seniors who were being asked questions about their plans for college and their various academic and extracurricular interests. The subjects simply watched the tape and indicated a preliminary judgment as to which of the two seniors they would admit to college if they were on an admissions board. After the subjects had made their own judgments, they were given the alleged judgment of another person and told they might want to consider it before making a final choice.

After the subjects had considered the judgment for several moments, they were asked to indicate to what extent the other person's judgment was influenced by the facts and to what extent it was based on the person's own values. They were also asked to complete a second form containing several questions regarding their perception of the other person and their reactions to the communication.

The results showed that the disagreer's judgment was seen as based more on values than was the agreer's judgment, but this difference was only marginally significant. There was no difference in the extent to which the agreer or disagreer was seen as having based his opinion on the evidence.

Heider's egocentric-attribution prediction, that divergent opinions will be attributed more to the person, was only weakly demonstrated in these data. But there were other findings which round out our view of subjects' reactions to agreement versus disagreement. First, one of the clearer findings was that subjects seem to have an expectation that others will agree with them. Subjects receiving the agreeing evaluation indicated that was about what they expected. Subjects receiving disagreeing evaluations indicated that those opinions were not expected. A second observation was that many subjects commented that they really could not assess the significance of the other person's opinion without knowing something about his personal characteristics. This comment seemed to occur more often in the disagree conditions where subjects were more perplexed. In making these comments, subjects supported our general attributional perspective that people make interpretations about the bases and meaning of other people's opinions and that knowledge of their personal characteristics is an important element in making these interpretations.

Somewhat at variance with the subjects' unwillingness to make definitive attributions about the bases of the other persons' judgments were data showing that they were quite willing to assign negative traits to the disagreer. The disagreers were seen as significantly less thoughtful, intelligent, and open-minded than agreers.

Attributions about Opinions of Similar and Dissimilar Others

The subjects whose data were described above argued that in order to make meaningful judgments about another person's opinions, whether that person agrees or disagrees, it is important that they know something about that person. As noted above, a second experiment was conducted to investigate the attributions subjects make about the opinions of others with known characteristics. Subjects in this experiment (Goethals & Ryan, 1975a) received agreeing or disagreeing opinions from others whom they knew were similar or dissimilar to themselves. The following predictions were proposed about the attributions that a person will make under these circumstances.

When a person finds that a similar person agrees, he should make attributions similar to the ones he makes about his own judgments. It was found in the experiment reported above that subjects felt, somewhat in contradiction to Heider's egocentric-attribution prediction, that both the facts and their personal values entered into their judgments about evenly, with a slight emphasis on values. So we should expect that subjects receiving agreement from a similar other to see that judgment as reflecting values and evidence about equally, but with a slight emphasis on values. That is, the opinion of a similar person that is similar to the subject's opinion will be attributed to the same factors.

One qualification of this hypothesis is suggested by Jones and Nisbett's (1971) theory of actor–observer differences in attribution. In general, they suggest that there is a tendency to attribute other people's responses to personal characteristics but one's own to the environment. However, this tendency is probably least operative when the other person and his response are similar to the attributor and his response. It should be noted at this point that actor–observer differences in attribution might be relevant in all four conditions of the present experiment. But as will be seen, only when the other disagrees does the Jones–Nisbett hypothesis seem persuasive.

What attributions would be made about a similar person's judgment which unexpectedly turns out to be at variance with the subject's judgment? We have seen already that there is a tendency for subjects to assume that disagreement is based on values. But the similar other has similar values, so simply dismissing his opinion as being a function of divergent values is somewhat illogical. On the other hand, an attribution to the facts is a little embarrassing to the subject. It implies that his own judgment incorrectly reflected those facts. Perhaps the only resolution would be to assume that the similar other did fall prey to his biases. The person can feel that he was able to judge the evidence objectively himself, in spite of his potentially biasing values, but the similar other was not.

Next, let us consider the attributions that a person might make when he finds that a dissimilar person agrees. Subjects probably feel that the facts are clear and expect, in general, that others will agree. Even the dissimilar person can be expected to reach the same conclusion, if the facts are clear enough and if he can suspend his biases. So if the dissimilar person does agree, the subject should make the attribution that the opinion is heavily influenced by the facts and not by values. He will give the person credit for suspending his values.

If a dissimilar person disagrees, a fairly strong attribution to values should result. Again, there is a general tendency to attribute disagreement to values. If subjects know further that the disagreer is dissimilar on values related to the judgment, the attribution to values should be facilitated

The experiment designed to test these predictions was very similar to the experiment described above. The first step in the procedure was for subjects to complete an "admission priorities scale" where they ranked the importance of 10 traits that might apply to college applicants. These characteristics included intellectual liveliness, athletic skill, and sense of humor. Then the subjects watched the videotape of two high school seniors used in the previous experiment and indicated their preliminary judgment as to which one they would admit to college.

At this point the experimenter told the subjects that they would be given information about another person's judgment, and he distributed to each subject two graphs on which the subject's and the other person's rankings were supposedly displayed. The priorities were not listed on the graphs, but the subjects were told that they could compare relative priorities by examining the rank-order correlation between their own scores and the other person's. The actual rank-order correlation between the fictitious scores was listed on the bottom of the subject's own profile sheet. The correlation was .78 in the similar condition and −.16 in the dissimilar condition. It was explained to the subjects that rank-order correlations of +.35 to +1.00 indicated similar admissions priorities, correlations from +.35 to −.35 indicated dissimilar (unrelated, different, but not opposed) priorities, while correlations of −.35 to −1.00 indicated opposite priorities. Thus, subjects were led to believe they were alike in the similar variation and different but not opposite in the dissimilar condition. Then they were given the agreeing or disagreeing opinion and asked to complete the questionnaires.

The pattern of results was most clearly confirming of the predictions on the "extent of evidence" variable. First, it was found that agreeing opinions were attributed more to evidence than were disagreeing opinions. Also, the ordering of means was as predicted. Agreement from the dissimilar other was attributed more to evidence, $\overline{X}$ = 4.91, (7-point scale) than was the agreement from a similar other, $\overline{X}$ = 4.73. However, this difference was not significant. In the disagree conditions, the subjects made the attribution that the dissimilar person paid significantly less attention to the evidence, $\overline{X}$ = 3.60, than the similar person, $\overline{X}$ = 4.50.

A nonsignificant trend toward a main effect for agreement was found for the scores on the value scale. Moreover, the means were not ordered exactly as predicted. It was predicted that the dissimilar disagreer would be seen as being most heavily influenced by values, $\overline{X}$ = 5.20, but our results show that the similar disagreer was seen as being slightly more influenced by values, $\overline{X}$'s = 5.25. However, the dissimilar disagreer was generally seen as more biased than the similar disagreer. For example, he was seen as being significantly less open-minded and thoughtful. The similar agreer was also seen as moderately influenced by values, $\overline{X}$ = 4.91. It was found, however, that the dissimilar agreer was perceived as having been influenced relatively little by values, $\overline{X}$ = 4.18.

The differences between attributions made to evidence and attributions made to values in the four conditions can also be considered. Agreement from similar others was attributed just about equally to values and evidence. Subjects in this condition seemed to feel that both the facts and the person's values were taken into account and affected the judgment.

In contrast, disagreement from similar others was attributed more, but not significantly, to values than to evidence. However, there was some interesting ancillary evidence as to how subjects tried to resolve the puzzle concerning the disagreement of the similar person.

First, the communicators in the similar–disagree condition received the highest rating for sincerity of any of the communicators in the four experimental conditions. Second, an interesting pattern of results occurred for an "intelligent–unintelligent" item. The agreers, similar and dissimilar, were rated as being equally intelligent and the dissimilar disagreer, not surprisingly, was seen as being significantly less intelligent than either agreer. But the similar disagreer was seen as being slightly more intelligent than the agreers. Subjects seemed to regard the similar disagreer as a superior intellect who worked to discern the truth. Or perhaps they felt that his active values (the highest attribution to values was made in this condition) led to a sincere but discrepant opinion. Thus, while the dissimilar disagreer was dismissed as biased, close-minded and stupid, the similar disagreer was seen as being generally less objective than his agreeing counterpart but, still, highly intelligent and sincere–maybe wrong, but still worthy of respect.

Our expectation was that unexpected agreement from dissimilar others would be seen as being attributable mainly to facts. Consistent with the expectation, only in this condition was the rating of influence of evidence higher than the rating for the influence of values. In fact, of the six experimental conditions in these two experiments, this was the only one where the judgment was attributed more to evidence than to values. (The overall tendency to attribute the opinions of others more to values than to the evidence is consistent with Jones and Nisbett's (1971) theory of actor versus observer differences in attribution. However, the tendency was very slight, and there was no evidence that subjects' attributions regarding their own opinions were any different.) Finally, as predicted, disagreement from dissimilar others was attributed significantly more to values than evidence.

ATTRIBUTION AND SOCIAL INFLUENCE

Now that we have considered some of the causal attributions that are made about the opinions of other people and the ways in which these attributions depend on (a) whether the other's opinion is in agreement or disagreement with the individual and (b) whether he is similar or dissimilar, we can proceed to a consideration of the actual effects of these opinions on the opinion of the person himself. Our basic perspective will be that the person's confidence that his opinion reflects the entity or situation being judged rather than irrelevant personal characteristics will be affected by other people's opinions and the attributions he makes about the causes of their opinions.

Cases of Agreement

I have conducted several studies in an attempt to support the prediction derived from attribution theory that a dissimilar agreer is more influential to the individual than is a similar agreer. These and several other recent studies (Goethals, 1972; Goethals & Nelson, 1973, Mills & Kimble, 1973; Wheeler & Levine, 1967), suggest that such a phenomenon occurs at least under some circumstances. Of course, these findings are somewhat at variance with results in other literature on social influence which emphasize that similar others are better liked and more influential. Byrne has done extensive work relating attitude similarity to interpersonal attraction (see Byrne, 1971). Festinger, in theoretical work on social influence (1950, 1954) and dissonance theory (1957), suggested that similar others will influence people more than dissimilar others. In his version of Heider's (1946, 1958) balance theory, Newcomb (1953) also emphasized that similar others are more influential. Several experimenters have supported this proposition (Back, 1951; Berscheid, 1966; Brock, 1965; Burnstein, Stotland, & Zander, 1961; see Collins, 1970, for a concise review of much of this work). Most of these investigators concentrate on situations where the communicator disagrees with the recipient and attempts to change his opinion. Only Berscheid (1966) has dealt with influence produced by opinion agreement. Nonetheless, there is nothing in this literature that supports the prediction that dissimilar agreers are more influential. Therefore, I will first consider how the prediction that a dissimilar agreer is more influential than a similar agreer derives from Kelley's attribution theory. Then the evidence will be considered.

Kelley (1967) emphasized that the individual who makes a judgment is aware of the fact that he may be mistaken (that his judgment may reflect biasing personal characteristics rather than the entity itself). His confidence that his response is attributable to the entity is affected by its distinctiveness and its consistency, but perhaps most importantly by considerations of consensus. That is, if other people agree with the individual, he can be more certain that he is correct. On the other hand, if others disagree the individual becomes less certain. While these general rules of consensus apply, Kelley made no differentiation between various kinds of consensus. Yet, certainly the personal characteristics of those composing the consensus need to be assessed and their opinions evaluated in light of those traits.

Uncertainty about the individual's own judgment reflects his concern that it might be person caused. While the agreement of others shows the response is consensual to some degree, if a particular consensus person is similar then it is possible that he has also fallen prey to the same biases and distortions implied in a person caused response. However, if a dissimilar other agrees, his judgment cannot be caused by a shared personal characteristic. By definition, the dissimilar other shares little with the individual. The causes for the dissimilar other's judgment could be either the entity itself or some personal characteristic different than the one which possibly generated the individual's judgment. Thus, it is only the entity which is sufficient to account for both the individual's and the dissimilar other's response, since it is only the entity which covaries with the judgment. Therefore, the likelihood of its being the cause of the response is subjectively increased (following Kelley's, 1971, covariation principle), and the individual's confidence can also be increased.

Two further points can be made here. First, the agreement of a dissimilar other is helpful because it demonstrates that the individual's personal characteristics are not necessary to cause the response. A person without those characteristics is known to respond in the same way. Second, the dissimilar other's agreement shows the response to be consistent across two different modalities (if perceivers can be regarded as modalities) which can be taken as evidence that the response is due to the entity rather than the characteristics of any particular observer.

In sum, the information provided by the agreement of a dissimilar other should increase the subjective probability that the entity is the cause of the response. The person who receives agreement from a similar other has no data pattern which similarly reduces the subjective probability that the response is person caused and increases the probability that it is entity caused.

Being more influenced by a dissimilar person, another who does not share one's values or have the same interests or goals, is referred to as the triangulation effect (Goethals & Nelson, 1973) because it seems analogous to the navigational principle where the position distance of an object can be accurately ascertained only when it is triangulated from two different perspectives. The current use of the term is nicely illustrated in the following quote from a book review by Larsen (1973) of Herber and Wooten's *Soldier*: "But whatever weaknesses Herbert's (obviously one-sided) case against his two immediate superior officers may contain, the evidence he presents against the Army's conduct as such is overwhelming, triangulated as it is by all the press reports, the My Lai trials and the PX scandals that have come before [p. 74]." In sum, a triangulated judgment is one which is supported by others with a different perspective relevant to the question at issue.

Evidence for the Triangulation Effect

A series of experiments has been conducted to demonstrate the triangulation effect and to delineate its applicability and limits. This research will be illustrated by a brief discussion of an experiment by Goethals and Nelson (1973). They demonstrated the triangulation effect and also showed the interaction of another independent variable, the nature of the issue, with the similarity variable.

Goethals and Nelson predicted that the logic of triangulation would apply when subjects were asked to make a judgment of belief about the actual characteristics of an entity or entities. Here the possibility of person caused error is present. Thus, according to the logic above, when a dissimilar other agrees, subjects should increase their judgmental confidence more than when a similar other does so.

On the other hand, it was predicted that when a subjective, evaluative judgment about the entity was called for, triangulation would not occur. Subjects would explicitly be making judgments reflecting their personal characteristics. Their confidence in their subjective evaluation should be increased more if a similar person, who values and enjoys the same things, agrees. The agreement of a dissimilar other should be of little value in increasing confidence about one's own subjective appraisal of the entity.

In sum, it was predicted that if subjects held a belief about an entity, their confidence in that belief would be increased more by a dissimilar than a similar agreer. Conversely, it was predicted that subjects' subjective evaluation of an entity would be increased more by a similar than a dissimilar agreer.

These predictions were tested in an experiment where subjects had to make judgments about two high school students who were allegedly being interviewed for college admission. In the belief conditions, the subjects were asked to decide which of the two students they thought had actually been more successful academically after 3 years at the college. Thus, they were making a judgment about empirically determined characteristics of the entities. In the value conditions, the subjects watched the same videotape but were merely asked to indicate which student they personally liked better.

After the subjects indicated their choice and their confidence in that choice, they were given an eveluation of the two students supposedly written by someone who had previously participated in the study. One-half of the subjects were led to believe that the other person was similar while half were led to believe that he was dissimilar. In this study, similarity was minipulated by giving subjects false feedback from a test which allegedly measured interpersonal judgment style. For this purpose they were given a profile of their own and the other person's scores. The scales on the profiles were not identified, but the subjects were able to determine how similar or different their own profile was from the other person's. After the subjects considered the profiles, they were asked to read the other person's evaluation. Later in the experiment, the subjects again indicated their choice and confidence. The confidence change scores constituted the main dependent measure in the study.

The results showed that subjects in the belief conditions raised their confidence significantly more when the agreeing other was dissimilar than when he was similar. However, as predicted, just the opposite result was obtained in the value conditions. Here, the subjects raised their confidence more when the agreeing other was similar.

It can be noted parenthetically that in the second attribution study described earlier in the chapter (Goethals & Ryan, 1975a), subjects were asked to indicate which student they would accept for admissions. Thus, they were required to make a choice which was based on both beliefs and values. Subjects were explicitly aware of this, especially when their values pushed them one way and their beliefs another. For example, some subjects felt that while they personally liked one student more, they judged the other to be more likely to perform better academically. Interestingly enough, confidence change data that were collected in this study resembled an averaging of the confidence change data in the belief and value conditions of the Goethals and Nelson (1973) experiment. The mean increase in confidence was identical in similar–agree and dissimilar–agree conditions. However, the variance in the dissimilar–agree condition was significantly larger. These is some suggestion that the larger increases in confidence in the dissimilar–agree condition occurred for subjects who felt the issue was one of facts while the smaller ones occurred for subjects who thought the judgment should be made more subjectively.

Other evidence for the existence of a triangulation effect in the kind of procedure used by Goethals and Nelson was obtained in a study by Goethals (1972). He also showed that when a person makes a factual judgment, his confidence in that judgment can be more increased when a dissimilar other agrees than when a similar other agrees. Like the belief-value study, however, it also demonstrated some of the limits of the triangulation effect. Specifically, it was found that when the subject and the other person had exactly the same information about the entity being judged, dissimilar agreers were more influential than similar agreers. However, it was predicted and found that when the other person was basing his judgment of the entity on information which the subject did not have, the other person would be more influential when he was similar.

This latter finding was predicted based on the reasoning that the subject can safely assume he would respond to the different information in the same way as a similar other, but he cannot assume that he would view it as did a dissimilar other. In fact, he might assume the opposite. Thus, the agreement of a similar other based on different information would provide data suggesting that the entity manifests itself consistently across modalities. The dissimilar other's opinion does not necessarily have this meaning since the individual cannot assume that he would respond as the dissimilar other did. Incidentally, the data suggest that confidence increases more when an agreeing similar other has different rather than the same information. This finding supports Kelley's statement that confidence in an entity attribution is increased when one's responses to the entity are assumed to be consistent across modalities.

A third experiment by Goethals, Darley, and Kriss (1974) relates more directly to the link between the attribution that a person makes about the locus of causality of another person's opinion and the impact of that opinion in terms of social influence. In this study, after reading some relevant material, subjects made a judgment about the best of three policies for controlling research and safeguarding individual rights. They then heard a tape-recording of another subject discussing his judgment about the same question. The subject knew that the other person was similar or dissimilar on values (such as individual freedom, truth and knowledge) which seemed related to the issue. The other person's opinion was always the same as the subject's. What differed was the reason the person gave for selecting his policy recommendation. He stated either that the policy made the most sense given his values but that those with different values might see things differently (person caused), or he stated that his opinion was based on the facts and evidence in the matter, which, he stated, should make it clear to most people that the policy he was choosing was the best (entity caused).

It was predicted that when the other person reported that his judgment was entity caused, a triangulation effect would obtain. A dissimilar other who feels that the facts clearly support the policy the subject chose more effectively suggests to the subject that his own reading of the facts is unbiased. On the other hand, if an agreeing dissimilar other reports his judgment to be based on his own values and interests and suggests that others with different values might see things differently, the subject might feel that he has come to a conclusion which is inconsistent with his interests. Thus, he might actually become less confident in his judgment. Berscheid (1966), in her study on communicator–communicatee similarity–dissimilarity, indicated that the individual would experience dissonance under these circumstances and might even reduce it by changing his opinion. A similar other's agreement based on values should raise confidence.

In sum, it was thought that a dissimilar other's communication of agreement could be structured so as to suggest that the individual has correctly read the evidence in an unbiased way or that he has adopted a position which may not be in his own best interests. In the first case, a triangulation effect should be produced. In the second case, dissonance should produce a decrease in confidence and perhaps even a change in position.

The data clearly supported the predicted interaction between the other person's similarity and the reported locus of causality of his opinion. The dissimilar other was more effective in raising confidence, $\bar{X} = 15.63$ (100-point confidence scale), than the similar other, $\bar{X} = -.73$, when he reported his opinions to be factually based. It can also be pointed out that the dissimilar other was more influential when his judgment was described as factually rather than value based.

The results do not show that the similar other was significantly more influential than the dissimilar other $\bar{X}$'s = 6.25 and 4.50, respectively, when the communications were person caused, although the difference was in the predicted direction. Several subjects commented that the dissimilar other's coming to the same conclusion on the basis of highly dissimilar values "didn't make any sense." Some subjects decided that in spite of the communicator's stated reasons for taking his position, he really must have done it because the facts were so clear. In other words, it appears that these subjects coped with the dissonance that seems to have been produced in the dissimilar/person caused condition by distorting the communication so that the dissimilar other was perceived to be agreeing on the basis of the evidence.

It is interesting to note in relation to the earlier predictions about the attributions that people will make about others' opinions that subjects in both communication variations who received agreement from a dissimilar other perceived that agreement to be more influenced by evidence (as opposed to values) than did subjects who received agreement from a similar other. This effect occurred on an item designed to check the communication manipulation and provides evidence that is supportive of the attribution predictions.

The studies cited above summarize the evidence that has been collected to date dealing with the triangulation effect. They suggest that the effect is obtained under specific circumstances. A triangulation effect can occur (a) when a person is making a factual judgment; (b) when he perceives the other person to be making a factual judgment rather than a judgment as to what is in his own best interests; and (c) when the individual has the same information about the entities as the other, so that he can be sure that he agrees with that other's assessment of the entities in the modality in which the other is transacting with them. Furthermore, there is evidence that subjects distort the locus of causality of judgments of dissimilar persons so as to perceive them to be more entity caused than they actually are described to be. Once this distortion of the locus of causality of the other's judgment occurs, the high increase in judgmental confidence suggested by the logic of triangulation seems to be manifested. While it may be that the agreement of a dissimilar other can produce dissonance, it seems that one way of reducing the dissonance is through distortion of the causal basis of the judgment toward the entity pole. This mode of reducing dissonance can ultimately lead to triangulation and an increase rather than a decrease in confidence, or even change in judgment, which would be, according to Berscheid, another mode of dissonance reduction.

Cases of Disagreement

While the effects of communicator similarity seem to be fairly complex when the communicator agrees, we find a much simpler conceptual and empirical picture in cases of disagreement. It would seem that no matter what kind of issue is being judged, the disagreement of a dissimilar person can be dismissed. It is, we have seen, attributed to a biased assessment of the issue which takes relatively little account of the facts. Thus, it does not have to be taken seriously, and our prediction is that it will have little impact on judgmental confidence. The divergent opinion of a similar other cannot be so easily dismissed. As noted earlier, it presents something of a puzzle to the person, given his expectation that others in general, and similar others in particular, will agree. But whatever resolution the individual comes to about the meaning of the similar other's divergent opinion, it can be predicted to have some negative effect, on the average, on judgmental confidence. In other words, the predictions from attribution theory about the most effective source characteristics in producing opinion change away from one's original position and towards a newer position advocated by a communicator are congruent with the predictions from social comparison theory. Similar others are seen as more influential in both theories.

The effects of disagreement have been investigated in two studies. There were disagree conditions in the Goethals (1972) study, and there were disagree conditions in the Goethals and Ryan (1975a) study reported above. In both of these studies, it was found that the dissimilar disagreer had very little measurable effect on confidence. In neither study was there a significant reduction in confidence when a dissimilar other disagreed. In contrast, it was found in both studies that a similar disagreer does reduce confidence.

The Impact of Aggressive Modeling

The social influence principles outlined above have been applied to the problem of aggression. In several recent studies, investigators have indicated that aggressive models can affect the magnitude of subsequent aggressive behavior of observers (for example, Bandura, Ross, & Ross, 1961). These researchers have considered the characteristics of the model to be an important determinant of the extent to which the observer follows the model's behavior (see Baron & Kepner, 1970). In a study of verbal aggression, Wheeler and Levine (1967) obtained a somewhat counterintuitive finding in this area. Opposite from what was predicted, they found that a dissimilar model was more effective in inducing the contagion of aggression than was a similar model. While Wheeler and Levine were uncertain as to the explanation for their finding, it seems to fit nicely with the research on the triangulation effect.

In the Wheeler and Levine experiment, naval recruits listened to two of their peers (actually confederates) expound on several topics of current political and social interest. The first confederate, the instigator, took extremely socially undesirable positions in a fairly obnoxious manner, in order to ensure that the subjects were highly instigated. Then when his turn came to speak, the second confederate, the model, verbally attacked the first. Finally, when the subject had a chance to speak, Wheeler and Levine measured the extent to which the subject followed the model's lead in being verbally aggressive toward the instigator.

If the subject is highly instigated to aggress, and then observes another person, similarly instigated, modeling aggressive behavior, the model's behavior can be considered as implying a judgment of the appropriate way of responding to the instigator. If the subject's judgment is the same, but his response is restrained by his uncertainty as to the appropriateness of aggression, the situation is perfectly structured for obtaining a triangulation effect. The aggression of the dissimilar model helps the subject rule out the possibility that his view of the situation, as one where some aggressiveness is called for, is simply a reflection of personal bias or idiosyncrasy. The modeling of the similar model is less helpful in this regard.

However, the model's aggression can only be regarded as an instance of agreement with the subject's view of the situation when instigation is high. If instigation is fairly low, aggressive behavior from the model may seem inappropriate. It may show disagreement with the subject's view of the situation and his judgment as to the appropriate response. As noted above, when another person disagrees, he is more influential when he is similar than when he is dissimilar.

This reasoning led to the following predictions. Under conditions of high instigation, an aggressive model may be implicitly agreeing with the subject as to the appropriate behavior toward the instigator. Here triangulation would come into play, and it is predicted that the Wheeler and Levine finding will be replicated. The dissimilar model will induce more aggression from the subject than the similar model. Under conditions of low instigation, where the model's behavior may constitute implicit disagreement, he should be more influential and induce more aggression if he is similar.

This predicted interaction was tested in an experiment by Goethals and Perlstein (1975) modeled closely after that of Wheeler and Levine. The subjects were high-school students who reported for an experiment described as a study of group discussion. They were taken to a cubicle equipped with a speaker and microphone and told that two other subjects were in similar adjoining cubicles. Subjects were told that in order to allow participants to know something about each other there would be an exchange of "special interests" forms describing their interests and activities. Each subject, always designated A, received forms from the others, designated B and C. B's form was always the same and typical for the subjects' population. The form for C was completed along prepared guidelines to give the impression that C was either very similar or very dissimilar to the subject in his interests and tastes.

A discussion procedure was explained in which the participants were to read the case history of a delinquent boy and present their views over an intercom system of the most effective treatment for him. After the subject spoke, he heard one of two prerecorded tapes representing B, the instigator. In the high-instigation condition, B took a highly socially undesirable position, aggressing strongly and personally against A. In the low-instigation condition, the speaker on the tape took a milder position, aggressing only slightly and apologizing frequently. Then the tape for C was heard. It contained a strong attack against B. The subjects were then asked to speak again. The degree of verbal aggression expressed against B was the major dependent variable. It was measured by two raters whose aggression ratings were closely correlated and were thus averaged.

The predicted interaction between similarity and instigation was shown on an analysis of variance performed on the ratings. Within the high-instigation conditions, there was significantly more aggression expressed toward the instigator when the model was dissimilar, $\overline{X} = 1.57$ (0 to 3-point scale), than when he was similar, $\overline{X} = 0.50$. Within low-instigation conditions, there was, as predicted, more aggression directed toward the instigator when the model was similar, $\overline{X} = 0.88$, than when he was dissimilar, $\overline{X} = 0.07$, but the difference was not significant.

Although the predicted interaction was obtained, the data were odd in one respect. The average level of aggression behavior was slightly higher in the low-similar than in the high-similar condition. Although this difference did not approach significance, any difference between the two conditions was expected to be in the opposite direction. Why should the level of aggression be so low in the high-instigation similar model condition?

During debriefing, high-instigation subjects commented that they were quite angry, and since the situation was rather highly charged, they were very cautious. They wanted to be even more careful about speaking harshly than they ordinarily were and only felt comfortable in expressing their anger when they had really good evidence that such expression was appropriate. Subjects in the high-dissimilar condition felt that they had such evidence. When the dissimilar model was aggressive, they felt certain that it was appropriate to be hostile toward the instigator. Subjects in the high-similar conditions indicated that they expected the model to be angry, as they were, and perhaps to be aggressive, but they could not really be sure it was appropriate. Only the rather unexpected support from dissimilar models seems to have been effective in releasing angry subjects from contraints on their verbal behavior which may have been more severe than usual.

These results suggest that the logic of triangulation is particularly salient to subjects who are making judgments about the appropriateness of behavior which is relatively low in social desirability. In these circumstances they are sensitive to the fact that they may act inappropriately because of their own personal biases.

RESPONSES TO SIMILAR AND DISSIMILAR OTHERS: INFLUENCE VERSUS EVALUATION

I have been discussing the attributions that subjects make about the opinions of other people and the extent to which they are influenced by those people. The findings suggest that dissimilar agreers are seen as fairly objective and highly influential. Yet there are indications that a rather mechanical bias is maintained against persons who are different from oneself. Even though these persons can be highly influential, they are given little credit for being so. This problem shall be discussed briefly.

The Nonrecognition of Influence

One of the ironical findings of the Goethals and Nelson (1973) study was that even though the dissimilar other was significantly more influential than the similar other in the belief conditions, when subjects were asked how much they were influenced by the other person's judgment, there was a nonsignificant effect in the opposite direction. The Goethals and Ryan (1975a) study contained similar results. In terms of actual influence, as measured by changes in confidence, there was no difference between similar–agree and dissimilar–agree conditions. The means were identical. However, in response to an item about the extent to which the other person's opinion influenced them, subjects in the similar–agree condition reported being much more influenced than subjects in the dissimilar–agree conditon. Again, the influence of the dissimilar agreer relative to the influence of the similar agreer was not perceived. That is, the triangulation effect does not seem to be recognized by the subjects for whom it occurs. This is odd since it seems that raising one's confidence on the basis of a triangulated judgment would be performed with full awareness of the logic of the triangulation argument.

The Stability of the Bias against Dissimilar Others

One of the major research traditions in social psychology which seems to be challenged by the research outlined in the present chapter is that of Byrne and his associates (see Byrne, 1971). Byrne has conducted a number of studies which show quite clearly that people are attracted to others who have similar attitudes. Although there is clearly a difference between liking someone and being influenced by him, it seems somewhat counterintuitive that subjects could be influenced so much by someone they like so little. Of course, part of the apparent paradox is resolved simply by noting that the influence was not recognized. At this point I would like to go further and indicate some other dimensions to which the usual bias against those with different values and priorities extends.

As mentioned above, in the Goethals and Ryan (1975a) study subjects were equally influenced by similar and dissimilar agreers but perceived the dissimilar agreer as being less influential. Furthermore, he was seen as being more biased, less thoughtful, less objective, open-minded, and careful than the similar agreer. This occurred in spite of the fact that subjects tended to attribute his agreement more to evidence while the general tendency was to perceive opinions as being based more on values. In other words, the subjects appeared to take readily and seriously the agreeing opinions of dissimilar others in triangulating their own. But they did not generally recognize this utilization of the other's judgment and continued to assign negative traits to them in ways that are consistent with the Byrne research.

In contrast, it was found in the study of aggressive modeling (Goethals & Perlstein, 1975) that subjects were very pleased to receive the unexpected social support of the dissimilar model in that rather highly charged situation. In that study they did not denigrate the dissimilar model, but there was a tendency to change their perception of the person's dissimilarity if they admitted the impact of his behavior and assigned positive traits to him.

Taken together these studies indicate that the dissimilar other who has triangulated a person's judgment and afforded him much increased judgmental confidence can still be viewed rather negatively. This does not always occur, but its nonoccurrence seems to depend on subjects changing their view of him so as to see him as more similar. Only if subjects can change their minds about his degree of similarity can they view him positively.

SUMMARY AND CONCLUSION

The attributions that a person makes about others' opinions and the influence that those opinions have on the person's own were discussed in this chapter. Some caution is necessary in accepting any generalizations about these phenomena since the results discussed here tend to be weak at the level of some of the individual studies. However, it seems that one study, to borrow a phrase, triangulates another, and that the data are mutually supportive and consistent. The results regarding attributions about others' opinions are weaker than the social influence results. In other words, we seem to have confirmation of attribution theorists' predictions about social influence in the absence of strong evidence that the subjects are really making hypothesized mediating attributions. Nevertheless, certain generalizations can be proposed. There is a tendency to attribute agreeing opinions to the entity and disagreeing opinions to the person. These tendencies are stronger when the opinion is received from someone who is dissimilar. In terms of influence, the agreement of dissimilar others provides especially persuasive data that the consensus is broad, that the facts are clear, and that one's opinion is not biased. Nevertheless, subjects continue to see the dissimilar agreer as a person who lacks objectivity and open-mindedness. He is used but not appreciated. In cases of disagreement, subjects are more consistently negative in their responses to the dissimilar person. They infer that his judgment ignores the evidence, and is person caused. He is seen as being unintelligent, biased, and thoughtless. The generalization that similar people are more influential should probably be fine-tuned in light of the data showing that dissimilar people are at times more influential. But the generalization that people respond more positively to similar others stands firm.

In conclusion, the material discussed in this chapter suggests that a process of judgment, attribution, and social influence occurs in something like the following sequence: (a) the individual makes judgments about reality but remains aware of the possibility of person caused error and, consequently, shows an interest in the opinions of others; (b) on receiving an opinion from another person, the individual considers the other's personal characteristics and his opinion, in relation to his own characteristics and opinion, and makes attributions about the causes of the other's opinion; (c) the individual's confidence is changed by the other's opinion in relation to the attributions he makes about the causes of the other's opinion and the implications these attributions have for the attributions he makes about his own opinions. Essentially, this process involves deciding to what extent another person's opinion provides evidence for one's own correctness or error.

ACKNOWLEDGMENTS

The work reported in this chapter was facilitated by National Institute of Mental Health Grant No. MH 23527.

REFERENCES

Asch, S. E. Effects of group pressure upon the modification and distortion of judgment. In H. Guetzkow (Ed.), *Groups, leadership, and men.* Pittsburgh: The Carnegie Press, 1951.

Back, K. W. Influence through social communication. *Journal of Abnormal and Social Psychology,* 1951, **46,** 9–23.

Bandura, A., Ross, D., & Ross, S. A. Transmission of aggression through imitation of aggressive models. *Journal of Abnormal and Social Psychology,* 1961, **63,** 575–582.

Baron, R. A., & Kepner, C. R. Model's behavior and attraction toward the model as determinants of adult aggressive behavior. *Journal of Personality and Social Psychology,* 1970, **21,** 84–92.

Berscheid, E. Opinion change and communicator–communicatee similarity dissimilarity. *Journal of Personality and Social Psychology,* 1966, **4,** 670–680.

Brock, T. C. Communicator–recipient similarity and decision change. *Journal of Personality and Social Psychology,* 1965, **1,** 650–654.

Burnstein, E., Stotland, E., & Zander, A. Similarity to a model and self-evaluation. *Journal of Abnormal and Social Psychology,* 1961, **62,** 257–264.

Byrne, D. *The attraction paradigm.* New York: Academic Press, 1971.

Collins, B. E. *Social psychology.* Reading, Massachusetts: Addison–Wesley, 1970.

Festinger, L. Informal social communication. *Psychological Review,* 1950, **57,** 271–282.

Festinger, L. A theory of social comparison processes. *Human Relations,* 1954, **7,** 117–140.

Festinger, L. *A theory of cognitive dissonance.* Evanston, Illinois: Row, Peterson, 1957.

Goethals, G. R. Consensus and modality in the attribution process: The role of similarity and information. *Journal of Personality and Social Psychology,* 1972, **21,** 84–92.

Goethals, G. R., Darley, J. M., & Kriss, M. The impact of opinion agreement from value-dissimilar others as a function of the grounds for agreement. Unpublished manuscript, Princeton University, 1974.

Goethals, G. R., & Nelson, R. A. Similarity in the influence process: The belief–value distinction. *Journal of Personality and Social Psychology,* 1973, **25,** 117–122.

Goethals, G. R., & Perlstein, A. L. Level of instigation and model similarity as determinants of aggressive behavior. Paper presented at the meeting of the Eastern Psychological Association, New York, April 1975.

Goethals, G. R., & Ryan, S. R. Attributions about the opinions of similar and dissimilar others. Unpublished manuscript, Williams College, 1975. (a)

Goethals, G. R., & Ryan, S. R. Attributions about the opinions of unknown others. Unpublished manuscript, Williams College, 1975. (b)

Heider, F. Attitudes and cognitive organization. *Journal of Psychology,* 1946, **21,** 107–112.

Heider, F. *The psychology of interpersonal relations.* New York: John Wiley & Sons, 1958.

Jones, E. E., & Nisbett, R. E. *The actor and the observer: Divergent perceptions of the causes of behavior.* Morristown, New Jersey: General Learning Press, 1971.

Kelley, H. H. Attribution theory in social psychology. In D. Levine (Ed.), *Nebraska Symposium on Motivation.* Lincoln: University of Nebraska Press, 1967.

Kelley, H. H. *Attribution in social interaction.* Morristown, New Jersey: General Learning Press, 1971.

Larsen, J. After the battle. *Time,* 1973, **101**(7), 74.

Mills, J., & Kimble, C. E. Opinion change as a function of perceived similarity of the communicator and subjectivity of the issue. *Bulletin of the Psychonomic Society,* 1973, **2,** 35–36.

Newcomb, T. M. An approach to the study of communicative acts. *Psychological Review,* 1953, **60,** 393–404.

Wheeler, L., & Levine, L. Observer–model similarity in the contagion of aggression. *Sociometry,* 1967, **30,** 41–49.

14

An Attributional Analysis of Helping Behavior

William John Ickes
Robert F. Kidd

University of Wisconsin

The question, "Buddy, can you spare a dime?" immediately brings to mind a situation we consider to be prototypic for the study of helping behavior—a situation in which someone who possesses a relatively large amount of some resource is asked for help either by or on behalf of someone whose supply of the resource is relatively small. In describing this situation as prototypic, we do not mean to imply that other kinds of helping situations do not exist or are not important. Obviously, there are instances in which two people who are both down-and-out will help each other, and there are other instances in which exploitive types will take advantage of the generosity and good will of people already less fortunate than themselves. Our statement is merely meant to suggest that when a person decides he wants help in obtaining some resource, he will usually seek help from someone whom he believes to possess more of the resource that he does—a strategy which is probably sound for a number of reasons.

While it would be interesting to discuss the various factors that might motivate one person to seek help from another (for reviews see Berkowitz, 1972; Krebs, 1970), our primary concern in this chapter will be to explain why the second person, once approached, would be willing to give help to the first. In other words, the major question we will address is: Under what conditions will someone who is relatively "successful" with respect to his ownership or control of some resource be responsive to a request for help from someone who is relatively "unsuccessful" in this regard? Our answer to this question will necessarily be a limited one, as it will largely ignore some of the variables traditionally associated with helping (for example, liking for the supplicant, the need to maintain equity), and focus instead on variables which have been integrated under the general heading of attribution theory. Our attempt will be to show how principles of Heider's (1958) "naive analysis of action," Weiner's (1972, 1974) theory of achievement motivation, and Rosenbaum's (1972) classification of outcome causes can be applied to account for helping behavior in situations structured along the lines of the prototype just described. More specifically, we will examine in some detail the manner in which the potential helper's explanations about the causes of his own outcome and the outcome of a dependent other will affect his subsequent responsiveness to the other's need for help.

Outcome attributions will occupy a central theoretical role in our analysis as they are held to be the primary variables mediating between the supplicant's request for help and the potential helper's response to that request. To provide a general conceptual framework in which the role of outcome attributions may be better understood and appreciated, we will first briefly trace the development of some relevant theorizing in this area and then proceed to examine the results of a number of recent help-giving studies in terms of this theoretical background.

THEORETICAL APPROACHES TO THE STUDY OF OUTCOME ATTRIBUTIONS

One of the earliest and best attempts to elaborate the perceived causes of personal outcomes is represented by Heider's (1958) naive analysis of action. Heider's analysis is not naive in the sense that it is shallow and unsophisticated; on the contrary, it is one of the most profound and insightful works in the history of contemporary social psychology. It is "naive" only in that it deals with what is typically called "common-sense psychology," the system of relatively unformulated concepts and principles which ostensibly guides the everyday behavior of the man on the street. Heider (1958) contends "that scientific psychology has a good deal to learn from common-sense psychology [p. 5]," and the intent of his work is to systematize this knowledge and translate its basic concepts into a language better suited to the scientific investigation of interpersonal behavior.

Heider points out that in common-sense psychology, as in scientific psychology, the outcome which a person experiences is seen as deriving from two sets of circumstances: factors within the person and factors within the environment. The within-person factors which affect a person's outcome are referred to collectively as his "effective personal force." The two major components of effective personal force are a power factor, usually represented by ability, and a motivational factor, represented by effort ("trying"). Ability is generally characterized as dispositional in nature; it refers to a relatively stable attribute which the person possesses and can use to influence his outcomes. Effort, however, is viewed as a variable or unstable attribute which fluctuates according to the person's momentary intentions and the amount of exertion which he manifests.

As conditions of action, both ability and effort are assumed to be necessary while neither alone is sufficient. This means that in any situation in which the person seeks to obtain an outcome at least partially dependent on his effective personal force, his ability and effort must both take some value greater than zero or else the value of his effective personal force will also be zero. In other words, the personal constituents of action—power and trying—are related as a multiplicative combination rather than an additive one. Thus, a person with sufficient ability but no motivation will exert no personal force toward obtaining a goal, and the same will be true of a person with high motivation but no ability.

However, even when the value of a person's effective personal force exceeds zero, he will not necessarily obtain his goal, for the final outcome of the situation may also depend upon the amount of effective environmental force which facilitates or hinders his goal attainment. According to Heider, the two major factors which combine to determine this effective environmental force are the difficulty of the task in which the person is engaged and the adventitious variation of more temporary influences within the situation. The first of these factors is referred to as task difficulty, while the second is called luck. Task difficulty is seen as a relatively stable, dispositional property of the environment, just as ability is a relatively stable property of the person. Similarly, luck, as a variable environmental factor, may be compared to effort, a personal factor which is also variable.

In fact, it was the awareness of these parallel relationships which led Weiner to propose that Heider's four major outcome determinants could be classified on the basis of a stability dimension (stable versus unstable) as well as a causal locus dimension (internal–personal versus external–environmental). In Weiner's 2 × 2 taxonomic scheme, ability is seen as a stable, internal cause; task difficulty is stable and external; effort is unstable and internal; while luck is unstable and external (Weiner, 1972; Weiner, Frieze, Kukla, Reed, Rest, & Rosenbaum, 1971). The explicit identification of the stability dimension proved to be an important theoretical step in the analysis of outcome attributions, as it soon led to the discovery that stability and causal locus effects had been confounded in some of the previous research on expectancy reported by Rotter and his associates.

Early research on the determinants of expectancy indicated that the expectancy of success generally rises following as experience of success and declines following a failure. Rotter (1954) suggested that these shifts in a person's expectancy may be due not only to the evaluation applied to his past outcome but also to the perception that the outcome derived from internal versus external causes. More specifically, Rotter (1966) proposed that shifts in expectancy due to evaluative feedback would be minimal following an outcome perceived as externally caused (due to chance or luck), but maximal following an outcome perceived as internally caused (due to ability or skill). It should be evident that in the above examples, the variation of the causal locus dimension is confounded with variation of the stability dimension, a problem which also characterized the early experimental work in this area (for example, Phares, 1957). However, in more recent experiments by Fontaine (1974) and Rosenbaum (1972), it was demonstrated that when stability and causal locus are manipulated as separate factors, expectancy shifts are found to be related less to the locus of the attributed cause than to its stability. This means that if an outcome such as success or failure is attributed to a stable cause (ability or task difficulty), expectancy clearly shifts in a manner indicating an increase in the person's belief that the same outcome will occur again. But if the outcome is attributed to a variable cause (effort or luck), subjective doubt is greater that the same outcome will occur again, and the magnitude of the expectancy shift is measurably attenuated.

In addition to clarifying the role of attribution in expectancy shifts, Weiner et al.'s (1971) separation of the stability and causal locus dimensions has permitted the more definitive testing of other attributional effects as well. Perhaps the most important of these findings is the effect of the perceived locus of causality on affective responses to an outcome. Weiner (1974) cites data from studies by Feather (1967) and Lanzetta and Hanna (1969) which indicate that ascribing an outcome to an internal cause produces a greater affective reaction to the outcome than ascribing it to an external cause. In other words, more "pride" is expressed if a successful outcome is attributed to one's own ability or effort than to good luck or the ease of the task, while more "shame" results if a failure is attributed to a lack of ability or effort than to bad luck or task difficulty. Moreover, effort rather than ability seems to be the internal factor which plays the greatest role in this effect, as affective reactions to success and failure are more divergent following ascription of these outcomes to effort than to ability (Eswara, 1972; Rest, Nierenberg, Weiner, & Heckhausen, 1973; Weiner & Kukla, 1970).

However, despite the obvious conceptual and empirical gains that were stimulated by the two-dimensional taxonomy, eventually Weiner (1974) and his associates began to recognize some important deficiencies within the formulation:

> For example, effort is classified as an unstable cause, although at times we perceive ourselves and others as "lazy" or "industrious," and the intent to succeed may be quite invariant. Perceived effort, therefore, also has stable properties. Further, within this scheme it is not possible to explain perceived improvement or learning, inasmuch as both determinants of "can" (ability and task difficulty) are conceptualized as fixed. Finally, and perhaps of greatest importance, individuals make causal distinctions that are not possible to make within the [2 × 2] table. For example, both fatigue and effort are internal, unstable factors, but the "naive" person surely distinguishes between these causes. In a similar manner, task difficulty and the bias of a teacher may be classified as external, stable factors although intuitively they are very different "kinds" of causes [p. 6, brackets ours].

In an attempt to overcome some of the problems outlined above, Rosenbaum (1972) proposed the addition of a third causal dimension, intentionality, to Weiner's classification scheme. The resulting three-dimensional taxonomy of perceived outcome determinants is shown in Table 1. Under Rosenbaum's scheme, the ability factor can be viewed either as a stable disposition or as a more variable capacity. Similarly, the effort factor can take the form of a dispositional "trait" or a fluctuating, unstable "state." More important, perhaps, the scheme allows the causes of outcomes to be classified as either intentional or unintentional, a development which brings a new dimension to the set of internal causes and permits the behavior of other people to be included in the set of external outcome determinants.

We consider Rosenbaum's addition of the intentionality dimension to be significant for at least three reasons. First, it is obvious that "common-sense psychology" uses this dimension as a basis for classifying attributed causes; thus, there is a strong reason to include it in a taxonomy which purports to resemble that of the "naive" person. Second, because the causes identified as "intentional" are typically those also identified as "internal" to a person, the dimension of intentionality is probably more than latent in Rotter's (1966) concept of locus of control. The tendency to fuse the intentionality and causal locus dimensions is understandable, but separating them may be an important prerequisite to understanding why people are typically held more responsible for the effects of some internal causes (such as laziness) than for the effects of others (such as illness). Third, while attributions in general may be instrumental in defining *beliefs,* attributions of intent in particular may be instrumental in defining *wants* (or, to be more exact, *beliefs about wants*). The specific processes by which people infer the presence or absence of such wants in themselves and others have

TABLE 1
A Three-Dimensional Taxonomy of the Perceived Causes of Personal Outcomes[a]

	Intentional		Unintentional	
	Stable	Unstable	Stable	Unstable
Internal	Effort traits of self, e.g., industriousness, laziness	Effort states of self, i.e., intention and exertion	Ability traits of self, e.g., intelligence, coordination, and skill	Ability states of self, e.g., fatigue, mood, and fluctuations of skill
External	Effort traits of others	Effort states of others	Ability traits of others, task difficulty	Ability states of others, luck

[a]This table was adapted from a table in Rosenbaum (1972, p. 21).

hitherto been neglected in attributional research (see Maselli & Altrocchi, 1969; Weick, 1968, pp. 425–427). The recognition of intentionality as an important attributional dimension may help to stimulate a greater interest in these processes.

It should be noted, however, that the theoretical value of Rosenbaum's or any related formulation depends ultimately on its ability to account for the data of human behavior in situations in which outcome attributions are themselves causally determining. We have already defined one such situation as our prototype for the study of helping behavior. We will now consider some specific experimental examples of this situation in light of the various theoretical developments just reviewed. In our discussion we will deal first with the effects on helping of the potential helper's attributions about the outcome of a dependent other.

ATTRIBUTIONS ABOUT THE DEPENDENT PERSON'S OUTCOME

Our theoretical overview has indicated that the causal locus dimension was the first to emerge as important in the classification of outcome attributions. For this reason, it is not surprising that the influence of this variable was the first to be tested in studies in which an attempt was made to relate outcome attributions to helping behavior. The major focus of these studies was the perceived locus of the dependent person's outcome, and, as will be seen, predictions about the effect of this variable were based on the implicit assumption that the locus of the other's dependency was an important determinant of the extent to which the potential helper would hold the other responsible for his state of need.

In the earliest of these studies (Schopler & Matthews, 1965), subjects were requested to help a subordinate other who was dependent on them for help. Half of the subjects were led to believe that the subordinate's request for help was demanded as part of the experimental procedure (external locus of dependency), while half were told that the subordinate had decided to ask for help of his own accord (internal locus of dependency). The data from this experiment revealed that significantly more help was given when the subordinate's need was perceived as resulting from external demands than from his own "internal" decision to request assistance. Presumably, subjects felt that the request for help of an externally dependent person was more legitimate than that of an internally dependent person, since the person was seen as less responsible for his outcome in the former case than in the latter.

A similar study was conducted by Berkowitz (1969). In this study, subjects were told that they were free to help a fellow subject by completing some of his work for him. Half of the subjects were informed that the other subject needed help because he has mismanaged his work and fallen behind; the rest were told that the other needed help because the experimenter had delayed the other's work by initially providing him with the wrong materials. In addition to this manipulation of the locus of the other's dependency, the degree of his dependency was also varied by informing subjects that their help would have a significant, moderate, or only negligible affect on the other's chance to win a prize for his work. The results of this experiment again indicated that the response to the other's need was greater when it was perceived as externally caused than when it was perceived as internally caused. However, this effect was clearly evident only when the subject perceived the other to be highly dependent on him for aid.

Further evidence for this relationship is provided by Bryan and Davenport (cited in Berkowitz, 1975), who analyzed the contributions made on behalf of 100 people whom the *New York Times* described as needing help during one Christmas season. Their analysis was consistent with the previous findings in suggesting that *Times* readers gave more generously to people who were externally dependent, such as child-abuse victims, than to people whose outcomes could be attributed to internal causes like their own moral or psychological deficiency.

In discussing the results of these studies, the various authors imply that the person in need may be held responsible for his dependency to the degree that internal factors are seen as determining his outcome. In other words, the attributed responsibility for a given outcome is assumed to vary as a function of the causal locus dimension, with internal causes leading to greater attributions of responsibility than external causes. This view appears to be a plausible one, but it does not reflect an important distinction which can be illustrated by the following simple example.

Suppose that you are downtown, standing on a corner across the street from a building which displays a large electric sign showing the time. A man standing next to you turns and asks you what time it is, and you tell him to look up at the sign across the street. If his dependence on you for this information is due solely to a lack of effort, he might answer, "Would you read it for me? I can see that far but I'm not willing to try." However, if his dependence is due solely to a lack of ability or capacity, he might answer, "Would you read it for me? I'm willing to try but I can't see that far."

It should be obvious that the other's dependency is "internal" in both of the cases described above, according to the classification schemes of Weiner and Rosenbaum. However, we do not believe that the tendency to help the other would be identical in the two cases but would instead be somewhat greater in the second case than in the first. For although effort and ability have a common perceived causal locus, Rosenbaum's model suggests that they differ in terms of their perceived intentionality. Specifically, the attribution of an outcome to the effort factor should lead to the perception that the outcome was relatively intentional, whereas the attribution of the same outcome to the ability factor should lead to the perception that it was relatively unintentional. The same distinction is also made by Heider (1958), who says that "people are held responsible for their intentions and exertions but not so strictly for their ability [p. 112]." Since intention and exertion are viewed by Heider as the basic constituents of "trying," his statement is clearly intended to distinguish effort from ability on the intentionality dimension.

The separation of effort and ability ascriptions on the basis of intentionality may be an important element in the assessment of responsibility for a given outcome. This separation is justified not only in terms of the intuitive theoretical precedents set by Heider and Rosenbaum, but in terms of an informal linguistic analysis as well. As students of Fritz Perls are well aware, "responsibility" can also be written as "response-ability," denoting the *ability to respond.* When a person's dependency is due solely to a lack of effort, the person is responsible, for he still possesses "response-ability" in the literal sense of the word. However, when his dependency is due solely to a lack of ability, he is not responsible, for he has no capacity to act—no ability to respond.

Because assessments of responsibility depend upon attributions of intent, the important distinction to be made is not whether the locus of the other's outcome is internal or external, but whether the outcome is intentional or unintentional. If the other's dependency can be ascribed to an internal lack of effort, it will probably be seen as intentional. But if it can be ascribed to an internal lack of ability, it will probably be seen as unintentional. Applying this reasoning to the helping experiments we have just considered, it seems quite likely that the intentionality variable was covaried with the causal locus variable in the manipulations described. The internal dependency of the needy other was explicitly intentional in the Schopler and Matthews (1965) study and implicitly intentional in the Berkowitz (1969) study (although effort and ability attributions undoubtedly also covaried in the latter case).

These observations lead us to conclude that the potential helper's attributions about the other's outcome may have the greatest impact on his helping response when they permit him to make inferences about the other's intent. If the other's outcome is seen as intentional, he is held responsible for it, and the potential helper has no need to assume responsibility on his behalf. However, if the other's outcome is seen as unintentional, he is not held responsible for it, and responsibility for altering the outcome is more likely to be assumed by the potential helper.

This reasoning also allows us to predict that more help will be given when dependency results from a lack of ability than when it results from a lack of effort. As yet there is no direct evidence to support this prediction, but some indirect evidence is available. For example, in a field experiment conducted by Piliavin, Rodin, and Piliavin (1969), it was found that the fallen victim of a staged accident on a New York subway train was helped more quickly by his fellow passengers when he was thought to be a cripple than when he was thought to be a drunk. In other words, the tendency to help him was greater when his dependency was attributed to an unintentional lack of ability to stand upright in a jolting train car than when it was attributed to an intentional lack of effort to maintain sobriety.

More indirect evidence is provided in the data cited by Weiner (1974) which show that a person who receives an unfavorable outcome is evaluated most negatively when he is perceived as having the ability to achieve a more favorable outcome but has made no effort to do so. Conversely, the same data (Eswara, 1972; Rest et al., 1973; Weiner & Kukla, 1970; Zander, Fuller & Armstrong, 1972) show that a person is given the most favorable evaluation for his performance when he is seen as having made a real effort despite an apparent lack of ability. These findings, which have been replicated cross-culturally, have led Weiner (1974) to assert that "effort ascriptions have a profound influence on affective (evaluative) reactions to another person's outcomes [p. 37]." In the present context, they also suggest that the reduction in helping which results from perceiving the other's dependency as intentional may be mediated in part by a negative affective response.

ATTRIBUTIONS ABOUT THE POTENTIAL HELPER'S OUTCOME

Having given some indication of the manner in which helping may be influenced by attributions about the outcome of the person desiring help, we are now in a position to extend our analysis in order to show how similar processes can be used to explain the influence of the potential helper's attributions about his own outcome. The theoretical development here, as before, begins with the causal locus variable, which was invoked by Midlarsky in her 1968 review to account for some helping effects of this type.

Midlarsky maintained that a person will engage in more subsequent helping when his outcomes are internally attributed than when they are attributed to some external agent or source. In support of this contention, she cited Rotter's (1954) theory and the results of some empirical studies (Gore & Rotter, 1963; Strickland, 1965) which indicate that persons who see their own outcomes as self-determined are more likely to help others than those who do not. Midlarsky interpreted this difference in terms of the contrast between a competent versus a fatalist orientation—the belief that one's efforts either will or will not have an influence on the course of events.

Elaborating this notion in the context of our prototypic helping situation, we suggest that the perception of competence that results from the potential helper's self-attribution of success may also facilitate helping by increasing his self-esteem, lowering the perceived cost of helping, and increasing the subjective probability that his attempts to help will be successful. We will qualify this assertion later, but for now it will stand as a general statement which can be supported by a variety of findings in the help-giving literature.

For example, Berkowitz and Connor (1966), Isen (1970), Isen, Horn, and Rosenhan (1973) and Levin and Isen (1975) have all proposed that the "warm glow of success" may increase subsequent helping by elevating mood and self-esteem. As Weiner (1974) has indicated, such effects should be especially apparent when success is self-determined. Self-attributed success should also lead to a smaller perceived cost of helping, for the knowledge that one's ability has been sufficient to bring about the favorable outcome implies that under similar circumstances in the future the outcome will again be attainable. However, when the outcome is strictly contingent on some outside agent or on the laws of chance, the person has no guarantee that he will be able to reliably reattain it. Thus, the potential cost of helping may be greater when the locus of one's success is external rather than internal, since in the former case the person may be giving up a resource which he is unable to replace. There is little doubt that the cost factor is an important variable in help-giving, as several investigators (Schaps, 1972; Schopler & Bateson, 1965; Wagner & Wheeler, 1969) have demonstrated its impact. Finally, the perception of competence that results from self-attributed success should increase the subjective probability that one's attempts to help will be successful and therefore rewarded (Kazdin & Bryan, 1971; Midlarsky, 1968).

While all of the sources just cited above are consistent with the prediction that helping will be greater following success which is attributed internally rather than externally, we could find no experimental data to support this view directly (the studies cited by Midlarsky are correlational in nature). In fact, the only experimental study relevant to this issue, a dissertation study by Jones (1974), was actually based on a prediction opposite to the one stated above, that is, that self-attributed success would lead to *less* subsequent helping than success mediated by some external causal agent. The issues raised by Jones are sufficiently important to our analysis that we will consider them in some detail.

Self-Attributed Success and Decreased Helping

In Jones' experiment, subjects in groups of three were led to believe that some of them would later be required to perform a series of unpleasant taste discriminations. To reinforce this expectancy, the experimenter gave all subjects a sample taste discrimination at the beginning of the session in which they taste-compared a highly saturated sugar solution with a very bitter sample of quinine water. This trial offered them convincing evidence that the upcoming series of test trials would indeed be quite unpleasant.

Following the sample trial, the experimenter explained that not every subject would be required to participate in the series of test trials. She indicated that because she wanted them to be motivated to do their best on a quantitative (math) test they would be given next, "Those who worked hard and did well on the math problems wouldn't have to take the taste test, and could do another, not at all unpleasant task, instead [p. 67]." However, in order to counteract the systematic biasing of the subject sample resulting from this procedure, she explained that a coin-flip would be used in some cases to determine the task decision: "Here's the way it will work: after I've collected your quantitative test paper, I'll flip a coin. If it comes up 'heads', I'll grade your paper and determine whether you'll take the taste test on the basis of your score. If it comes up 'tails', I'll flip the coin again and determine by chance alone whether you'll take the taste test [p. 67]." Thus, the subjects were told that they could "escape" the unpleasant taste test either by their own efforts on the math test or by chance alone. It was further emphasized "that each subject's outcome was independent of the others' and that all or none of the subjects could conceivably escape the taste test during any experimental session [pp. 67–68]."

After checking to make sure that the subjects understood these instructions, the experimenter isolated each subject in a separate cubicle to take the quantitative test. The test was composed of several simple but time-consuming problems in addition, subtraction, multiplication, and division. The experimenter told each subject that when the alloted time for work on the problems was up, she would collect his paper, determine his task status, and then give him an index card which would indicate whether or not he had to take the taste test and the reason for this decision. The message on the index card which the subject received constituted the manipulation of his outcome attribution. In the escape-by-effort condition, the message read: "You don't have to take the taste test–you did well enough on the math test." In the escape-by-chance condition, it read: "You don't have to take the taste test–by the flip of the coin." The use of prepared index cards was justified to the subject as a time-saving procedure, but the actual purpose was to permit the experimenter to remain blind with respect to the subject's experimental treatment.

Each subject was informed that he alone had managed to avoid the unpleasant taste task and was therefore given a short paper-and-pencil rating task instead. Regardless of the time it took him to complete this substitute task, the experimenter remarked that it had not taken as long as usual. She then suggested that because he had extra time, he might want to consider helping one or both of the other subjects, since they were both required to complete the taste test trials. The reasons for the others' dependency were left unexplained, but it was explained that each of them had 20 taste comparisons to perform; and since it was not required that the same person complete all 20 comparisons, the subject could help by doing as many as he wished on behalf of the others. The total number of taste trials which the subject volunteered to complete was then taken as the measure of helpfulness.

The data from the experiment provided marginally significant support for Jones' hypothesis that a person who has escaped misfortune by chance will give more help to someone who cannot escape the same misfortune than will a person who feels he has escaped by his own efforts, $p < .10$. On the average, subjects in the escape-by-chance conditions volunteered to do about 6.6 of the taste comparisons for their fellow subjects, while the mean for the escape-by-effort condition was only 2.3.

An attributional bias explanation. Jones' explanation of her result is closely tied to a sterotype which suggests that there are conditions in which the usual success-helping relationship would not be likely to hold. This is the enduring stereotype of the "self-made man." As popularly conceived, the self-made man has much in common with Archie Bunker. He is an individual who feels that he has achieved success in life totally on the merits of his own ability and hard work, without the aid of external forces such as good luck or the assistance of others. He is a strong believer in internal control, and just as he attributes his success to his own special talents, he tends likewise to attribute the failure of others to such dispositional factors as a lack of motivation or ability (see Sosis, 1974). In short, he has what amounts to an "attributional bias," which leads him to assert that because external factors played no part in his success, they should therefore exert no influence on others' outcomes as well. Consistent with his view of a "just world" in which everybody gets exactly what he deserves on the basis of his own failings or merits, he is likely to refuse to help others with the justification, "I made it on my own–why can't they?" The question, of course, is purely rhetorical, as the self-made man neither wants nor is willing to accept an answer to it.

In line with this conception, Jones (1974) also argues that "making it on your own" may lead to decreased helping. However, in the context of her argument, "making it" is defined as escaping from unpleasant circumstances or a state of need rather than achieving a reward:

> By contrast to the recipients of good fortune or help, people who have previously rescued themselves from a state of need may tend to be unaltruistic. A person who has experienced relief by helping himself may be intolerant of another person's need for help. Because he has managed to help himself, he may see himself as particularly resourceful or industrious. By comparison, someone who cannot help himself may appear to him to be lazy or inept. He should be very unlikely to see another person as helpless or unresponsible for his misfortune; to do so would not allow a self-enhancing comparison. Because a person who has experienced self-help may tend to see other people as capable of alleviating and preventing their own discomfort, he may be intolerant of their dependency and feel that they deserve no help from him [p. 20].

Jones' reasoning here implies that as long as the self-helped person can attribute the others' dependency to a lack of effort rather than to a lack of ability, he will be able to see them "as capable of alleviating or preventing their own discomfort. . . . [and] may be intolerant of their dependency and feel that they deserve no help from him [p. 20]." We are in full agreement with this point, which appears to be consistent with our earlier prediction that dependency resulting from a lack of motivation or effort will inspire less helping than dependency resulting from a lack of ability. However, this is clearly a function of the perceived intentionality of the dependent person's outcome and *not* the perceived locus of the potential helper's outcome.

What is unique about Jones' argument is her suggestion that once the self-helper makes an internal attribution to explain his own favorable outcome, he will then become motivated to bias his perceptions in order to see the less favorable outcomes of others as also due to internal (implicitly intentional) causes. The motivation for this bias is, of course, the self-helper's desire to enhance his self-esteem by favorably comparing his achieved outcome with that of the dependent other. As such a comparison is meaningful only to the extent that both outcomes are perceived as internally caused, the self-helper is assumed to be willing to risk the correctness of his attributions for the opportunity to gratify his ego. This attributional bias presumably also permits the self-helper to see the other as responsible for his outcome, thus justifying an unwillingness to help him.

Although this hypothetical biasing process offers an intriguing explanation for Jones' helping data, there is no direct support for this process in the form of relevant self-esteem and/or attributional measures. Moreover, the generality of the phenomenon is suspect on other, less empirical grounds as well. For while the self-aggrandizement resulting from such a process might be immediately reinforcing, the continued maintainance of an attributional bias of this type would probably be maladaptive and punishing in the long run. It would not only require the self-helper to engage in frequent and extreme distortions of reality in order to justify his often untenable "world view," but it would also work to his painful disadvantage when he fails and the other succeeds. Finally, it should be noted that, clinically speaking, this process of self-enhancement at the expense of reality is nothing less than a neurosis. Though it is often tempting to cynically cast much of human behavior in this mold, it is perhaps more appropriate and parsimonious from a theoretical standpoint to first attempt to find an adaptive, nonneurotic explanation of the behavior in question.

A "causal taxonomy" explanation. We suggest that a more "adaptive" account of Jones' results can be made in terms of our earlier distinction between intentional and unintentional causes. Rather than positing the operation of an ego-enhancing form of attributional bias, this approach is based on the assumption that the person attempts to make correct attributions about his own outcomes by using the same intuitive taxonomy of causes that he applies to the outcomes of others. If Rosenbaum's taxonomy is indeed representative of this intuitive one, the escape-by-effort subjects in Jones' experiment should have viewed their outcome as more intentional than did subjects in the escape-by-chance condition. This distinction is important because the perception of intentionality further implies the presence of a want or desire. Thus, the attribution of the escape-by-effort subjects may have supported the inference that they *wanted* to escape the taste task, while the attribution of the escape-by-chance subjects may not have supported such an inference. If we assume that these attributions affected the strength of the subjects' perceived desire to free themselves from the obligation of performing the taste trials, it is not surprising that the escape-by-effort subjects were less willing to help share in the task than their escape-by-chance counterparts. In fact, asking the former group to help in this way was probably like adding insult to injury, for after having extricated themselves from the unpleasant task by their own efforts, they were then asked to negate their original intention by voluntarily submitting to it anyway.

Extending this application of Rosenbaum's model a bit, there may be reason to suspect that not all "internal" attributions of success would lead to the decreased helpfulness which Jones found. For despite the fact that such attributions share a common causal locus, the model indicates that they may still differ on the basis of their perceived stability and intentionality. From this observation, we suggest the possibility that, when contrasted with the effects of attributing a successful outcome to an external factor such as luck, an internal attribution of the same outcome to ability could actually lead to *more* helping, while an internal attribution to effort could lead to *less.* For although causal locus data suggest that both of these internal attributions would elevate the person's mood and self-esteem relative to an external (luck) ascription, certain negative influences on helping could also result from an effort ascription that might not result from an ascription to ability.

Specifically, because effort is usually seen as a relatively unstable and intentional "state," while ability is seen as a relatively stable and unintentional "trait," the person may perceive his costs in helping to be greater when success is attributed to the former than to the latter cause. The instability of the effort factor may not only weaken the person's belief that he can easily reattain his favorable outcome, but may impair his confidence in his capacity to help as well. Perhaps more important, the perception of intentionality in the case of an effort attribution should lead the person to infer that he wants and values the outcome he has obtained. This inference, which should attenuate in strength as the outcome is attributed more strictly to ability, may increase his reluctance to give up the benefits of his success in order to help the other. It may also cause the increased positive affect resulting from an "internal" success to actually work *against* helping insofar as the maintenance of this affect depends upon the person's retaining possession of these benefits.

In comparison to the effects on helping due to ability and effort ascriptions, the negative influences on helping due to a luck ascription should be those associated with its external causal locus (for example, a less positive affective response) and with its instability (impaired confidence that the favorable outcome can be reattained and that the attempt to help will be successful). These negative influences on helping may often be fewer than in the case of an attribution to effort, especially to the degree that the inhibitory influences of intentionality in effort also tend to negate the positive effects of an internal causal locus. However, these same negative influences may commonly be greater than in the case of an ability ascription, for in this case the inhibitions and other negative influences due to instability and intentionality should be minimal. In other words, depending on the relative situational emphasis given to the causal locus, stability, and intentionality dimensions, the negative influences on helping may be greatest in the case of an effort attribution, intermediate in the case of a luck attribution, and least in the case of an ability attribution.

From this reasoning, it can be predicted that Jones' finding would be reversed if an ascription of ability, rather than effort, were contrasted with an ascription of luck in accounting for the favorable outcome of a potential helper. Experimental support for this prediction would, of course, be evidence that under some conditions self-attributed success will lead to increased helping. It would also provide a means to reconcile Jones' data with those cited by Midlarsky by taking into account the effects of the stability and intentionality dimensions as well as those of the causal locus dimension.

Self-Attributed Success and Increased Helping

To test the prediction that success which is attributed to ability will lead to more subsequent helping than success which is attributed to luck or chance, a pilot experiment was conducted by Ickes, Kidd, and Berkowitz (1976). In this study, male undergraduates were led to believe that they were participating in a test validation task with another male student, who was actually an experimental confederate. The validation task required the subjects to try to match a list of first names (Robert, Margaret, etc.) with a series of photographs of college students. In order to roughly equate the subjects' motivation to succeed at this task, the experimenter explained at the outset that he would pay them 10 cents for each of the name–face matches they performed correctly. He then left the two participants to work independently on the matching task, returning later to "score" their response sheets and announce the results.

In all cases, the subject received false feedback which indicated that he had done very well on the task, earning $1.10 in change. However, the confederate's feedback was such that he always "failed" the task, earning only $.30. Half of the subjects were then informed that their successful performance was due almost entirely to their "relational ability," a capacity which the task supposedly measured (ability condition). The remaining subjects were told that their success on the task was determined by chance or luck and that personal factors were unrelated to the task outcome (chance condition). In both of these conditions, parallel attributions were made to explain the confederate's failure.

Following this manipulation of outcome attribution, the experimenter excused himself to obtain copies of a postexperimental questionnaire that he wanted the subjects to complete. In his absence, the confederate began idly thumbing through a library book he had brought with him, and suddenly "discovered" that an article he had planned to copy for a class was considerably longer than he had anticipated. Claiming time pressure (the book was checked out on 3-hour reserve) and insufficient funds, the confederate then asked the subject to help him by giving him as much change as he could spare to xerox the article. (As all subjects had just received \$1.10 in quarters, dimes, and nickels, it was ensured that their refusal to help could not be interpreted simply in terms of a lack of funds.) After the subject had responded in some manner to this requres for aid, the experimenter returned to administer a short postexperiment questionnaire, conduct a debriefing session, and restore to the subject any money he had contributed.

The results of the study revealed that the ability/chance manipulation was very effective in altering the subjects' outcome attributions. On a postexperimental item in which subjects were asked to estimate in percentages the degree to which ability, effort (hard work), and chance determined their outcome on the task, subjects in the ability condition indicated that ability was a significantly greater factor in their success than did subjects in the chance condition, $p < .001$. More importantly, however, subjects in the ability condition gave a significantly greater percentage of monetary help to the confederate than subjects in the chance conditions, $p < .05$. On the average, subjects gave 24% of their total funds to the other when their task outcomes were ostensibly determined by ability, but only 10% when the same outcomes were attributed to chance.

This latter result provided direct support for our prediction that the attribution of success to ability would lead to greater helping than an attribution of success to chance. Our reasoning further suggested that the attribution of success to ability should also produce more helping than a similar attribution to effort, but this hypothesis could not be tested directly since perceived effort was not manipulated in our design. It was possible, however, to obtain some indirect evidence for this notion by examining the correlations between subjects' ratings of the degree to which ability and effort determined their success and the percentage of monetary help which they gave to the other. The test of the difference between these correlations was significant, $p < .01$, and revealed–as predicted–that attributions of ability were positively correlated with helping, $r = .39$, while attributions of effort were not, $r = .03$.

Taken together with the results of Jones' study, these pilot data suggest that the kind of internal ascription the potential helper uses to explain his favorable outcome may be an important determinant of his subsequent helpfulness. However, a major drawback of the pilot study was that the potential helper's attribution about his own outcome was not manipulated independently of his attribution about the other's outcome. Since both the subject and the confederate were working together in a situation in which task outcome was defined either by ability or by chance, this situational definition was made to apply to both participants. Thus, because of the methodology employed, when the subject succeeded by chance, the other failed by chance; and when the subject succeeded by his ability, the other failed by a lack of it. For this reason, we cannot conclude definitely from these data that it was the subject's attribution about his own success that influenced helping rather than his attribution about the other's failure.

In order to separate the effects of the subject's attribution about his own outcome from those resulting from his perception of the other's outcome, we conducted a second experiment in which these attributions could be manipulated independently (Ickes, Kidd, & Berkowitz, in press). In this study, the ability/chance manipulation of the subject's success was crosscut by an additional manipulation which is best described as a variation of the intentionality versus unintentionality of the dependent other's outcome. As this latter variable had already been demonstrated to affect helping behavior in a predictably reliable manner, we expected that this manipulation would yield more clearly interpretable results than the ability/chance manipulation of the other's outcome used in the pilot experiment.

The subjects in the second experiment were male undergraduates who were asked to complete the same name–face matching task used in the pilot study. However, in this study the subjects were run individually and worked alone on the task, without a confederate present. When the experimenter returned and finished "scoring" their responses, all subjects were informed that they had done very well and were given $1.10 in change for their performance. Subjects in the ability condition were then told that their success on the matching task was due primarily to their relational ability, while subjects in the chance condition were told that their success was due to chance. Following this manipulation, the experimenter asked each of the subjects to report to an upstairs office where a research secretary would have him complete a posttest questionnaire and give him experimental credit for his participation in the study.

When the subject arrived at the office, the research secretary (actually an experimental confederate) asked him to fill out a short postexperimental questionnaire. When he had completed this form, the secretary gave him a copy of an interoffice memo and explained that she had been asked to have all subjects read it who had participated in experiments for which payment had been provided. This memo was ostensibly written by a psychology professor who, for purposes of the study, was given the title, Head of the Research Committee. The contents of the memo constituted the manipulation of the intentionality of the other's dependency.

In the intentional dependency condition, the memo indicated that because of a scarcity of funds for the payment of experimental subjects, money was not available to pay subjects who reported to an experiment for which they had signed up, but who then refused to complete the experiment. Since payment had been promised simply for reporting to the experiment and not for completing it, a recommendation was made that a fund be established to provide payment for these subjects. It was further recommended that contributions to this fund be solicited from subjects who had already received payment for their experimental participation. Thus, the subjects in this condition were led to believe that their contributions would be used to pay someone whose dependency had resulted from an intentional refusal to participate.

In the unintentional dependency condition, the memo subjects received was identical except that the potential recipients of the donations were described as subjects who had reported to the experiment and then failed to complete it through no fault of their own, but rather through some cause such as experimenter error, equipment breakdown, or scheduling problems. It was thus made clear that the contributions of subjects in this condition would be used to pay someone whose dependency on them for help was completely unintentional.

In all cases, the research secretary waited until the subject had finished reading the memo and then called his attention to a large glass jar which contained $2.00 in change and was clearly labeled Subject Payment Fund. She then made an explicit request for help by asking the subject if he would like to contribute to the fund. When the subject had responsed to this request, either by making a contribution or declining to do so, he was given a full debriefing, and any money he had contributed was restored to him.

As in the pilot experiment, the measure of helping was the amount of money the subject gave to the confederate relative to the total amount he had in his possession at the time of the request. Results for this measure indicated that the percentage of money given was highest in the condition where the subject's success was attributed to his ability, and the other's dependency was perceived as unintentional. The mean of the ability/unintentional dependency condition (24% of total funds given) was significantly different, p's $< .05$, from the means of each of the other three conditions (which were very similar and varied between 5 and 7%). In addition, subjects' ratings of the degree to which ability determined their success were again correlated with the percentage of monetory help which they gave. The Pearson r of .40 was strikingly similar to the value of .39 obtained in the first study.

An internal analysis of these data further revealed that the difference between the correlation of perceived ability with helping and the correlation of perceived effort with helping was also significant, $p < .01$. This effect replicated our earlier finding, suggesting that the attribution of a favorable outcome to ability is associated with greater helping, $r = .40$, while the attribution of the same outcome to effort may be associated with less, $r = -.12$. In fact, when subjects were divided by median split on the basis of their effort attributions, it was found that the subjects who saw their effort as having played a relatively large role in their success gave significantly less money to the other, $\overline{X} = \$.16$, than those who saw their effort as relatively unimportant in determining task outcome, $\bar{X}$ = \$.35, $p < .05$. The data for the percentage-help measure showed a similar trend for this effect but fell somewhat short of significance.

THE ROLE OF ATTRIBUTIONAL STRUCTURES IN HELPING BEHAVIOR

The data from our second experiment indicate that the greatest responsiveness to a request for help may occur when a person ascribes his own favorable outcome to his ability but views the dependent other's outcome as unintentional and beyond his control. It is tempting to try to simplify this statement further by concluding that helping is greatest when the potential helper makes an internal (dispositional) attribution to account for his outcome, but an external (situational) attribution to account for the other's. This conclusion is not justified, however, for Jones' data and the results of our own correlational analyses indicate that the facilitative effects on helping of a potential helper's internal attribution of success may hold only for ascriptions of ability and not for ascriptions of effort. Moreover, Rosenbaum's model suggests that the facilitative influence of the other's unintentional dependency may extend beyond external causes to include his internal lack of ability as well.

We can perhaps more readily define the optimal attributional structure for helping by first noting that, in the context of our prototypic situation, the attribution of an outcome to ability versus effort may lead to similar effects on helping regardless of whether these attributions are applied to the outcome of the dependent other or that of the potential helper. In other words, the potential helper's belief that the other's dependency is due to a lack of effort should lead to decreased helping, and the attribution of his own success to an expenditure of effort should produce the same result. Conversely, his belief that the other's relatively unfavorable outcome is due to a lack of ability should increase his tendency to help, and the same should occur when an attribution of ability is made to explain his own success. Assuming that the effects of these attributions are additive, helping should be minimized when the potential helper believes that he has expended effort to obtain his positive outcome but that the dependent other has not. On the other hand, helping should be maximized when the potential helper believes that he has the ability to achieve a positive outcome but that the dependent other does not.

Intuitively, helping another person makes the most "sense" when the potential helper is asked to do something for the other which the potential helper is capable of doing but which the other cannot do for himself. Helping makes much less sense when the potential helper is asked to expend effort on behalf of a person who is perceived as capable of achieving a positive outcome but unwilling to expend the effort to obtain it. Common-sense notions about when it is or is not appropriate to help are predicated upon outcome attributions such as these. We suggest that the reason such notions make "common-sense" is that the ascription of each person's outcome to a particular cause further implies its classification on the underlying attributional dimensions of causal locus, stability, and intentionality. The implicit classification of an outcome on these dimensions is assumed to directly mediate such influences on helping as the affective response to self and other, the perceived costs and rewards of helping, and the degree to which self and other are seen as responsible for altering the other's unfavorable outcome.

An attributional analysis of this type may easily be applied in the two specific instances contrasted above. In the first instance–the one in which the potential helper attributes his favorable outcome to his own expenditure of effort but sees the other's unfavorable outcome as due to a lack of effort–the causal locus data indicate that he should evaluate his own performance the most positively but the other's performance the most negatively. The relative instability of his own outcome should increase his perceived costs in sacrificing its benefits to help the other, while the instability of the other's outcome may make the perceived need for help less certain. Additionally, the perception that both outcomes were intentional should lead the potential helper to infer that he wants and values the benefits of his own outcome, but to question whether the other really wants the help he says he needs. Because the other apparently lacks only effort and not ability, he should be seen as responsible for his outcome in the sense that he is "able to respond" in order to change it. The inference that he intentionally does not do so may cast doubt on the legitimacy of his request and on his professed desire to improve his outcome.

In the second instance–the onc in which the potential helper ascribes his favorable outcome to his ability but ascribes the other's unfavorable outcome to a lack of ability–the causal locus findings suggest that he will evaluate his own performance somewhat less positively but the other's performance somewhat more positively. The relative stability of both outcomes should tend to decrease his perceived cost in helping the other but increase his certainty that the other's outcome will not improve unless he is given the help he seeks. Finally, the perception of both outcomes as relatively unintentional may minimize the potential helper's reluctance to share the benefits of his outcome with the other and cause him to define the other's desire for help as sincere. Because the other is seen as unable to respond to change his own outcome, the potential helper should consider it appropriate to exercise his "response-ability" on the other's behalf.

In both instances it can be seen that the effects of causal locus, stability, and intentionality act in concert to influence the resulting helping response. These influences work heavily against helping in the first instance, which embodies the attributional structure we have described as best characterizing that of the self-made man. Helping is unlikely for a number of reasons here, not the least of which may be extreme disparity in affect which the potential helper feels for himself in comparison with the other. Most of these negative influences on helping tend to coincide with those invoked as support for Jones' attributional bias argument, but, as we have indicated, the attributional structure they derive from may not be the inevitable by-product of an internal attribution of success.

The second instance probably represents the optimal attributional structure for help-giving for all of the reasons we have already given. Moreover, it closely resembles the attributional structure which led to the greatest helping in the second of our two reported experiments–that of the ability/unintentional dependency condition. The differences between the two should be minimal since they both represent cases of ability-attributed success and unintentional dependency. Although the locus of dependency is external in one case and internal in the other, the dependent other's capacity to respond on his own behalf is clearly limited in both.

The attributional structures formed by the potential helper's perception of the causes of his own and the other's outcome should not only affect his helping response in the specific instances just described but in other cases as well. The attributional approach we have elaborated here should easily be applied to predict behavior across a wide range of helping situations. It is intended to provide a broad theoretical framework in which such previously unrelated determinants of helping as affective state, perceived competence, and perceived cost of helping can be conceptually integrated. It may further serve to highlight the importance of additional factors such as perceived intent which may have been neglected by researchers in the past.

SUMMARY AND CONCLUSIONS

In this chapter we have attempted to show how the theoretical notions of Heider, Weiner, and Rosenbaum can be used to account for the influence of outcome attributions on help-giving. By applying their concepts to reinterpret past findings and to predict future ones, we have also had occasion to discuss some apparent conflicts or misconceptions in the existing literature on helping and to indicate how an attributional approach may help to resolve them. In particular, we have argued that the causal locus dimension is by itself not sufficient to account for all the effects on helping resulting from attributions about the outcome of a dependent other or a potential helper. We have therefore proposed that the additional dimensions of stability and intentionality must also be considered, and have presented data and arguments to support their inclusion in our theorizing.

What we have *not* done is all too obvious–at least to us. We have not yet fully tested the implications of our arguments across a range of subject populations and helping contexts. As initial steps toward this goal, we will first need to test for the differential effects on helping of ability versus effort attributions and then attempt to systematically manipulate the three major attributional dimensions in a completely crossed design. From a theoretical standpoint, there exists an additional need to extend our theorizing to other aspects of the helping relationship–help-seeking, for example. This should not be difficult, as the same attributional processes that would cause a person to give help should also influence his decision to seek it (see Tessler & Schwartz, 1972). Beyond the analysis of helping behaviors, it should be possible to apply this attributional approach to the broader study of moral decision-making (see Deinstbier, Hillman, Lehnhoff, Hillman, & Valkenaar, 1975) and to other interpersonal behaviors as well.

ACKNOWLEDGMENTS

The preparation of this chapter was supported by a National Institute of Mental Health Grant (MH 26646) to William Ickes and Robert Kidd. The authors would like to thank Bernard Weiner, Robert Cialdini, and John Harvey for their comments on an earlier draft of this manuscript.

REFERENCES

Berkowitz, L. Resistance to improper dependency relationships. *Journal of Experimental Social Psychology,* 1969, **5**, 283–294.

Berkowitz, L. Social norms, feelings, and other factors affecting helping and altruism. In L. Berkowitz (Ed.), *Advances in experimental social psychology* (Vol. 6). New York: Academic Press, 1972.

Berkowitz, L. *A survey of social psychology.* Hinsdale, Illinois: Dryden Press, 1975.

Berkowitz, L., & Connor W. H. Success, failure, and social responsibility. *Journal of Personality and Social Psychology,* 1966, **4**, 664–669.

Deinstbier, R. A., Hillman, D., Lehnhoff, J., Hillman, J., & Valkenaar, M. C. An emotion-attribution approach to moral behavior: Interfacing cognitive and avoidance theories of moral development. *Psychological Review,* 1975, **82**, 299–315.

Eswara, H. S. Administration of reward and punishment in relation to ability, effort, and performance. *Journal of Social Psychology,* 1972, **87**, 129–140.

Feather, N. T. Valence of outcome and expectation of success in relation to task difficulty and perceived locus of control. *Journal of Personality and Social Psychology,* 1967, **7**, 372–386.

Fontaine, G. Social comparison and some determinants of expected personal control and expected performance in a novel task situation. *Journal of Personality and Social Psychology,* 1974, **29**, 487–496.

Gore, P. M., & Rotter, J. B. A personality correlate of social action. *Journal of Personality,* 1963, **31**, 58–64.

Heider, F. *The psychology of interpersonal relations.* New York: John Wiley & Sons, 1958.

Ickes, W. J., Kidd, R. F., & Berkowitz, L. Attributional determinants of monetary help-giving. *Journal of Personality,* 1976, **44**, 163–178.

Isen, A. M. Success, failure, attention, and reaction to others: The warm glow of success. *Journal of Personality and Social Psychology,* 1970, **15,** 294–301.

Isen, A. M., Horn, N., & Rosenhan, D. L. Effects of success and failure on children's generosity. *Journal of Personality and Social Psychology,* 1973, **27,** 239–247.

Jones, C. *The effects of prior experience on empathy and helping behavior.* Unpublished doctoral dissertation, University of Texas–Austin, 1974.

Kazdin, A. E., & Bryan, J. H. Competence and volunteering. *Journal of Experimental Social Psychology,* 1971, **7,** 87–97.

Krebs, D. L. Altruism–An examination of a concept and a review of the literature. *Psychological Bulletin,* 1970, **73,** 258–302.

Lanzetta, J. T., & Hanna, T. E. Reinforcing behavior of "naive" trainers. *Journal of Personality and Social Psychology,* 1969, **11,** 245–252.

Levin, P. F., & Isen, A. M. Further studies on the effect of feeling good on helping. *Sociometry,* 1975, **38,** 141–147.

Maselli, M. D., & Altrocchi, J. Attribution of intent. *Psychological Bulletin,* 1969, **71,** 445–454.

Midlarsky, E. Aiding responses: An analysis and review. *Merrill–Palmer Quarterly,* 1968, **14,** 229–260.

Phares, E. J. Expectancy changes in skill and chance situations. *Journal of Abnormal and Social Psychology,* 1957, **54,** 339–342.

Piliavin, I. M., Rodin, J., & Piliavin, J. A. Good Samaritanism: An underground phenomenon? *Journal of Personality and Social Psychology,* 1969, **13,** 289–299.

Rest, S., Nierenberg, R., Weiner, B., & Heckhausen, H. Further evidence concerning the effects of perceptions of effort and ability on achievement evaluation. *Journal of Personality and Social Psychology,* 1973, **28,** 187–191.

Rosenbaum, R. M. *A dimensional analysis of the perceived causes of success and failure.* Unpublished doctoral dissertation, University of California, Los Angeles, 1972.

Rotter, J. B. *Social learning and clinical psychology.* New York: Prentice–Hall, 1954.

Rotter, J. B. Internal versus external control of reinforcement. *Psychological Monographs,* 1966, **80**(609).

Schaps, E. Cost, dependency, and helping. *Journal of Personality and Social Psychology,* 1972, **21,** 74–78.

Schopler, J., & Bateson, N. The power of dependence. *Journal of Personality and Social Psychology,* 1965, **2,** 247–254.

Schopler, J., & Matthews, M. The influence of perceived causal locus of partner's dependence on the use of interpersonal power. *Journal of Personality and Social Psychology,* 1965, **2,** 609–612.

Sosis, R. H. Internal–external control and the perception of responsibility of another for an accident. *Journal of Personality and Social Psychology,* 1974, **30,** 393–399.

Strickland, B. The prediction of social action from a dimension of internal–external control. *Journal of Social Psychology,* 1965, **66,** 353–358.

Tessler, R. C., & Schwartz, S. H. Help seeking, self-esteem, and achievement motivation: An attributional analysis. *Journal of Personality and Social Psychology,* 1972, **21,** 318–326.

Wagner, C., & Wheeler, L. Model, need and cost effects in helping behavior. *Journal of Personality and Social Psychology,* 1969, **12,** 111–116.

Weick, K. E. Systematic observational methods. In G. Lindzey & E. Aronson (Eds.), *The handbook of social psychology* (Vol. 2, 2nd ed.). Reading, Massachusetts: Addison–Wesley, 1968.

Weiner, B. *Theories of motivation: From mechanism to cognition.* Chicago: Rand–McNally, 1972.

Weiner, B. Achievement motivation as conceptualized by an attribution theorist. In B.

Weiner (Ed.), *Achievement motivation and attribution theory.* Morristown, New Jersey: General Learning Press, 1974.

Weiner, B., Freize, I., Kukla, A., Reed, L., Rest, S., & Rosenbaum, R. M. Perceiving the causes of success and failure. Morristown, New Jersey: General Learning Press, 1971.

Weiner, B., & Kukla, A. An attributional analysis of achievement motivation. *Journal of Personality and Social Psychology,* 1970, **15**, 1–20.

Zander, A., Fuller, R., & Armstrong, W. Attributed pride or shame in group and self. *Journal of Personality and Social Psychology,* 1972, **23**, 346–352.

15

Sex: A Perspective on the Attribution Process

Kay Deaux

Purdue University

The recent storm of research on attribution processes has covered considerable ground, as testified by the range of topics covered in the present volume. My aim in the present chapter is to consider one particular parameter of these judgment processes: sex. In and of itself, the sex variable does not define a separate area of study nor constitute a theoretical model. Yet researchers have shown it to be an important contributing variable, and thus it can serve to clarify and enlarge the domain of existing attribution models. In the present chapter, I will consider three areas in which sex has proved critical in the attribution process: (a) sex as a characteristic of the object being evaluated, (b) sex as a subject variable which moderates causal judgments, and (c) sex as a defining characteristic of a task.

SEX AS AN OBJECT CHARACTERISTIC

The role of sex as an object characteristic in the attribution process is a special case of a more general judgment process. While fuller accounts of these general processes have been described in Chapter 17 and elsewhere (cf. Azjen & Fishbein, 1975; Jones & Davis, 1965; Kelley, 1967), the concentration here will be on those aspects relevant to differential judgments of male and female actors.

For this discussion, the following assumptions will be made. When an observer is called upon (or voluntarily chooses) to interpret or causally attribute the behavior of an actor, two general types of information are used in arriving at the final explanation. The first type of information is the observed behavior itself—what the actor is doing in a given situation. This information is then evaluated together with a second set of information: the expectancies which the observer had for the behavior of that individual. Causal attributions are a function of the match or mismatch between these two sets of information.

The expectancies which the observer holds for the actor in question may derive from a number of sources, for example, prior observations of that actor, information provided by an experimenter, or categorical assumptions which the observer makes about the actor as representative of a particular group. In Chapter 17 of the present volume, Jones and McGillis have made a distinction between target-based expectancies and category-based expectancies, and it is this latter category which is most pertinent in understanding the function of sex as a variable in the attribution process. We are assuming that, in general, observers have expectancies for the behavior of an individual male or female which derive from the stereotyped assumptions made of men and women as groups. Consequently, the behavior of the female or male is judged in conjunction with this set of stereotyped expectancies, and the resultant attributions differ to the extent that the stereotyped expectancies differ.

Brief Review of Sex-Role Stereotypes

Because the stereotypes of men and women are critical to the present discussion, it is necessary to summarize briefly the relevant literature. The study of ethnic stereotypy has an extensive if rather undisciplined history within the social sciences (Allport, 1954; Brigham, 1971; Harding, Proshansky, Kutner, & Chein, 1969). More recently, with the surge of the Feminist movement, stereotypes of men and women have become a frequent target of study. The evidence from these studies points to clearly delineated conceptions of what women and men are typically like. Broverman and her colleagues (Broverman, Vogel, Broverman, Clarkson, & Rosenkrantz, 1972; Rosenkrantz, Vogel, Bee, Broverman, & Broverman, 1968) have isolated two distinct clusters of traits which are seen as distinguishing women from men. The first cluster contains traits reflecting competence, such as independent, competitive, objective, dominant, active, logical, ambitious, and self-confident. These traits are perceived as being characteristic of men, while the opposite pole of each trait (dependent, noncompetitive, etc.) is associated with women. A second cluster of traits which distinguish men and women centers on the characteristics of warmth and expressiveness. Women are typically seen as tactful, gentle, aware of the feelings of others, and able to express tender feelings, while men are viewed as blunt, rough, unaware of others, and unable to express their own feelings. In addition, Boverman and others have found that more of the traits which are valued in U.S. society are contained in the competency cluster than in the expressiveness cluster (McKee & Sherriffs, 1957; MacBrayer, 1960). While some exceptions to this general set of stereotypes have been reported, it is clear that in general these stereotyped conceptions are being sustained in the present decade.

To the extent that these stereotypes are held, they constitute a set of expectancies for the individual male or female performer. If, as is generally the case, the observer is provided little information about the actor's past performance, then these stereotypes should weigh heavily in the subsequent judgment. In contrast, if the observer has had considerable opportunity to observe the behavior of the actor in question, then the stereotypes should be less influential than specific target-based expectancies.

Causal Attributions of Performance

The range of alternative explanations which an observer can offer for the performance of an actor was originally discussed insightfully by Heider (1958). In the simplest terms, Heider suggested that we can look for the cause of an action in either the person or the situation and that both of these possible causes can vary in stability. More recently, Weiner and his colleagues (Weiner, 1974; Weiner, Frieze, Kukla, Reed, Rest, & Rosenbaum, 1971) have specifically proposed that causal attributions may be categorized along at least two dimensions: the first, an internal–external dimension, and the second, a temporary–stable distinction. Exemplars of this 2 × 2 matrix are ability (stable and internal), effort (temporary and internal), luck (temporary and external), and task difficulty (stable and external).

While the neatness of this proposed categorization is appealing, there are some problems with the inclusion of task difficulty as a stable and external characteristic. First of all, the stability of task difficulty can be inferred only when we have consensus and consistency information (cf. McArthur, 1972). Consequently, unless this information is provided by the experimenter (e.g., Frieze & Weiner, 1971), the observer must infer task characteristics from other evidence. Returning to the fertile ground of Heider, we find that inferences about the difficulty of a task are directly related to ability and effort attributions. Specifically, Heider has proposed that ability and effort are inversely related to the simplicity of a task in the judgment of the observer. Furthermore, the "stability" of task difficulty is limited to the specific situation and does not allow any certainty in predictions to other tasks, For these reasons, task difficulty will be considered a temporary characteristic in the present analysis. (In support of this categorization, we have obtained scaled ratings by subjects on a variety of possible causal factors on the two dimensions in question. These ratings indicate that subjects do indeed perceive task difficulty to be a temporary rather than stable characteristic).

Given this framework, let us outline the basic process by which causal attributions of performance are made. The observer is presented with evidence about the outcome of a performance and is asked to supply reasons for that performance. In most experimental cases, the outcome is clearly defined as either a success or failure. Additional information about the actor (either specific to the actor or generalized from stereotypical beliefs) is matched against the observed behavior in an essentially logical information-processing approach.

In this judgment process, the following principles will be assumed:

1. Performance by an actor which is consistent with the expectations for that actor will be attributed to a stable rather than a temporary cause. Specifically, in most performance situations, ability or skill will be the choice. Expected success will be attributed to ability, and expected failure will be attributed to lack of ability.

2. Performance by an actor which is inconsistent with the expectations for that actor will be attributed to a temporary cause, which may range from internal to external in nature. The choice of a particular temporary explanation may be influenced by a number of factors, including the nature of the task. This latter point merits some amplification.

Most performance situations contain their own limits for attribution. For example, luck is one possible cause for performance on a multiple-choice examination; in contrast, the career of a successful neurosurgeon is implausibly connected to luck. Thus, the internality or externality of the temporary cause will be limited by the task itself. (These defining characteristics of a particular situation will be considered in more detail later.)

Attributions for Male and Female Performance

To understand the operation of sex-role stereotypes as a source of expectations and their effect on the attributions made for male and female performance, let us examine several recent experiments in which both success and failure outcomes were considered.

The case of success. In a study by Deaux and Emswiller (1974), subjects were told that they would be listening to another student (either male or female) perform a task in which he or she was required to recognize a common object and select the correct response from two alternatives. While there were two different classes of objects in this experiment, let us consider for the moment only the masculine objects, which included lug wrenches and Philips screwdrivers. The subject, listening on headphones, heard the student answer 16 out of 25 items correctly, a performance which was superior to the normative performance which had been previously placed at 12.3. Following this performance, the subject was asked to make a series of evaluations, including one judgment on a scale ranging from "performance mainly due to skill" to "performance mainly due to luck."

At the time this study was conducted, we were operating on a more simplified model of attribution which used the single luck–skill dimension as previously used by Feather (for example Feather & Simon, 1971a). In terms of Weiner's (1971) analysis, this scale confounds the two dimensions of stability and internality, in that ability or skill is both internal and stable while luck is external and temporary. However, the problem is not as serious as it might appear. First of all, we are assuming that the expected performance will always be attributed to ability to a greater extent than will an unexpected performance, when consensus information is absent or when the information about success or failure defines the performance as other than the average. Consequently, when ability is paired with any other attribute, including luck, a greater attribution of ability should be made to the performance which is consistent with expectancies.

Beyond this rationale, however, we have conducted experiments where each attribute is measured on a single scale (for example, high to low ability, high to low luck). In these instances (Deaux & Farris, in press), ability ratings show the same patterns of greater utilization when the behavior is expected. At the same time, at least one of the temporary attributes (dependent on the task) is invoked more strongly when the performance outcome is unexpected.

Given the stereotyped belief that males are more competent, particularly on a masculine task, we would predict that subjects would be more likely to attribute the male's performance to ability (use of a stable concept consistent with expectations for the performance). To the extent that the observed performance was inconsistent with the expectations for females (based on stereotypes of female incompetence), an unstable explanation–in this instance luck–should be used. These predictions were confirmed. Significantly more ability was attributed to males than females on the masculine task. Male and female subjects did not differ in their attributions, suggesting that similar stereotypes are held by both sexes (an assumption borne out in much of the stereotype research).

It was initially predicted that performance on a similar task with objects that were feminine in nature (for example, whisks, colanders) would show the reverse results. Observers should expect women to succeed on a feminine task and attribute their success to ability, while the male's success, being somewhat more unexpected, should be attributed to more temporary causes. While the differences were in the appropriate direction relative to the masculine task (males were credited with less ability, women with more ability), the differences between judgments of male and female actors were not significant. One possible reason for this nonparallelism is the character of tasks, and it will be dealt with in greater length in later discussion. A second explanation relies on the assumption that expectations are always higher for males than for females, at least in task-oriented situations. Loading the deck by making the task "feminine" will reduce this discrepancy, but the headstart that observers accord to male performance cannot be overcome.

A similar demonstration of differences in the performance attributions for male and female actors is reported by Feldman–Summers and Kiesler (1974). In this case, the performance was a successful medical career, and the four attributions were measured separately. Unlike the Deaux and Emswiller study, in this second investigation sex of subject differences were found. Male subjects showed a pattern of greater ability attribution for the male physician, while they attributed the female physician's task to a combination of greater effort, an easier task, and (nonsignificantly) more luck. Women, in contrast, did not attribute greater ability to the male, though they also saw greater effort as a cause of the female doctor's success. Both male and female subjects, however, attributed relatively more effort than ability to the female doctor in comparison to the male. Overall, task and luck were used as explanations far less often than were ability and motivation.

In a third study by Etaugh and Brown (1975), a similar pattern of findings emerges. Subjects asked to explain the successful performance of male and female actors on a mechanics task attributed the expected male success to ability. In contrast, an unexpected success by a woman on a mechanics task was attributed to the more temporary cause of effort.

To account for the results of these experiments, let us refer back to the postulated attribution process. I have assumed that our observer has an initial set of assumptions about the capability of males and females in general, which is applied to the particular male or female in question. Given the assumption that males are more competent than females, a successful performance by a male should be consistent with the expectation, and result in a stable and internal attribution to the actor. Ability is the likely choice. All subjects in the Deaux and Emswiller and in the Etaugh and Brown studies, and male subjects in the Feldman–Summers and Kiesler study, showed this attributional pattern. If the observer assumes that women are less competent, then a successful performance by a woman is inconsistent with expectations and requires a search for some temporary reason for the success. The choice of temporary unstable reasons allows a number of possibilities and is hypothesized to range along a dimension of external to internal. This range is, however, limited by characteristics of the particular task. Although luck may be a perfectly credible reason for success when an actor is choosing between two possible answers on an ambiguous test, it is less believable as an explanation for a successful medical career. Thus, while the explanation for the woman doctor is equally temporary, luck is not used as it was in the Deaux and Emswiller study.

It is important to note that the choice of attributions is directly related to the prior assumptions and expectations which are held. To the extent that an individual does not accept the stereotype that men are competent and women incompetent, ability attributions to males should be less predominant. It is quite possible that the women in the Feldman–Summers and Keisler study were more positive in their stereotype of women. Although such a suggestion can be only speculative, it would be possible to establish a priori the rigidity of an individual's stereotypes, and on the basis of this measure to predict stable or unstable attributions.

The case of failure. If the proposed sequence of events is correct, it should apply to failure experiences as well. Thus, if the same stereotyped assumptions that males are competent and females incompetent are accepted, then failure should be consistent with the expectations for females and inconsistent for males. Consequently, a stable internal reason should be given for the female's failure (specifically, a lack of ability) while more temporary reasons would be sought for the male's failure. Feather and Simon (1975) provide evidence that such is the case. In their study, subjects were asked to attribute the cause of success or failure of a male or female character in a series of different occupational fields. In line with the previous discussion, success by a male was significantly more often attributed to ability than was the success of a female. In contrast, failure by a female was more often attributed to lack of ability than it was for males, supporting the contention that stable internal attributes will be used to explain expected outcomes.

Other possible attributions offered to subjects by Feather and Simon also support the argument that temporary reasons will be used for a woman's success and a man's failure. Two possible attributions in this study were cheating in the case of success and an accusation by the examiner of cheating as a possible cause of failure. In accord with our assumption, cheating was used to explain the success of a woman more often than the success of a man, while alleged cheating was used more often to explain the failure of a man.

Etaugh and Brown (1975) also reported that an expected failure by a woman (in this instance, on a mechanical task) will be attributed to lack of ability moreso than will a man's. In contrast, the difficulty of the task (considered a temporary explanation in this discussion) was more likely to be used for a man's failure than for a similar failing performance by a woman.

Generality of the Model

While in the preceding discussion the focus has been on the influence of stereotypes about men and women, it is reasonable to assume that the same principles would hold for any stereotyped group. If, for example, members of a particular ethnic group are considered lower in competence than members of another ethnic group, then the expectations for the two groups should operate in a manner similar to sex. Success by members of the negatively valued group will be explained by temporary reasons, while success by the favored group will be credited to inherent ability. Thus, sex is merely one example of a more pervasive characteristic of the attribution process, whereby stereotypes and expectations influence the attribution which is selected.

Furthermore, despite the often faulty basis of the stereotypes themselves, the subsequent process of comparing expectancies to performance is essentially a rational processing of information. If, for example, we know that person X has always succeeded, then an attribution to the stable internal quality of ability is logical. Explaining failure by some transitory event is equally logical. Consistency cues, as demonstrated by McArthur (1972), are critical in making attributions to the person. With reference to stereotypes, however, the error in the system lies in the initial expectancies which are based on somewhat distorted information. The expectancy is not necessarily a valid one, and consequently the resultant attributions reflect what appears to be a bias in favor of men. As pointed out earlier, this bias in attributions will exist only in so far as different stereotypes of men and women persist. Changes in the beliefs of large numbers of people would eliminate most of the findings reported above.

Yet stereotypes have a record of longevity, and stereotypes of some group or another will probably persist or develop even if sex biases diminish. Furthermore, despite a potential decline in sex stereotypes, the more important contribution of research in this area is the development of the concept of expectancy as a mediator in causal attribution.

SEX AS A SUBJECT CHARACTERISTIC

In the preceding section, I have discussed the attributions made by observers as they focus on the behavior of two different types of actors–specifically, males and females. In this section I will discuss the attributions make by actors, again emphasizing sex as a critical variable. Paralleling the preceding section, research in this area indicates that (a) attributions made by male and female actors for their own performance generally show distinct differences, (b) these attributional differences can be related to differences in expectancies or self-stereotypes, and (c) the use of sex as an investigatory parameter can affirm and extend more general assumptions of an attribution model.

One of the most pervasive findings in the literature on sex differences is the lower expectations which females hold for their performance as compared to males. This consistent finding has been reported for a variety of tasks and age groups (Crandall, 1969; Deaux & Farris, in press; Montanelli & Hill, 1969). The parallelism of these self-evaluation tendencies to the stereotyped conceptions of males and females would lead us to conclude that the males and females themselves have accepted the stereotyped views of the society.

These lower expectations in fact serve as a self-stereotype, constituting an anchor point against which subsequent experience is judged. In line with the previous discussion, it would be expected that men, who have a higher expectancy for their performance, should find success consistent with their expectations and subsequently attribute that success to their higher ability. Failure, in contrast, would be an unexpected event for the male who has a high expectancy, and the explanation of that failure would be sought among a variety of plausible temporary causes. For women, in contrast, a typically lower set of expectations would result in a success being discrepant with the set. As a consequence, stable internal explanations would not seem appropriate and explanations would rely on one or more temporary reasons. Failure, if more consistent with the self-stereotype, should in turn be attributed to a stable internal attribute, most typically a lack of ability.

Attributions by Male and Female Actors

In general, these assumptions have been borne out in research. In a typical experiment of this type, Deaux and Farris (in press) asked subjects to work on an anagram task, and at the completion of the task, the male and female subjects were asked to attribute the causes of their performance. Two independent variables were manipulated in this experiment: the outcome of the subject's performance and the sex-linkage of the task. Outcome was controlled by giving subjects either very easy or very difficult anagrams to solve, thus ensuring either a successful or unsuccessful performance, respectively. Sex-linkage of the task was established by descriptions provided to subjects prior to task performance, wherein subjects were told either that women typically perform better on the task or that men are typically better. (In fact, anagram performance shows no sex differences.)

As in the Deaux and Emswiller (1974) study, differences were more pronounced when the task was defined as a masculine one. Differences in expectancies and attribution patterns were minimized but not reversed on the task when it was labeled feminine, again suggesting the pervasive character of male expectancies in this type of task-performance situation. For the moment, discussion will be confined to the results for the masculine task, though there will be occasion to return to the problem of task characteristics.

When the anagram task was defined as masculine, males expected to be significantly more successful than did females. In both success and failure conditions, the distinctions between female and male attributions were clearest on the ability and luck dimensions (measured independently), while effort and task difficulty showed less variation between the sexes. Consistent with our earlier predictions, men were more likely to claim ability as a cause of their success than were women. Conversely, the poor performance by a woman was more readily attributed to a lack of ability. The woman who succeeded was more likely than the man to attribute her success to luck, again supporting the assumption that temporary causes will be sought for unexpected outcomes. The explanation used most frequently by men for their failure was the difficulty of the task. These results are quite consistent with our predictions, based on differential expectancies of the two sexes and a use of stable versus temporary reasons to explain expected and unexpected outcomes, respectively.

The only real divergence in the data which would not be predicted from this expectancy model was the women's additional use of luck to explain failure as well as success, a finding also reported by Bar-Tal and Frieze (1974). The reason for the use of luck is not clear, but one suggestion is that women may generally see less connection between their behavior and outcomes than do males, at least in tasks such as the frequently used anagrams. Additional research is needed in this regard (see Deaux, White, & Farris, 1975, for another view of this question).

Results similar to the Deaux and Farris findings have been reported by Nicholls (1975). In this study, fourth-grade boys and girls were asked to account for their success or failure on an angle-matching task. A dependent measure in which the four attributions are made in percentages which must add up to unity was used. Following failure on the first trial, girls felt that a lack of ability was far more accountable for their performance than did boys (42.5 versus 29.1%). In contrast, boys claimed ability as a cause for their success to a greater extent than did girls. Nicholls did not obtain expectancy measures until after this first trial, and at that point they were significantly higher for boys than for girls. On the basis of previous research (Crandall, 1969; Montanelli & Hill, 1969), one could speculate that the boys' expectations would have exceeded those of girls prior to this performance as well. The patterns for luck attributions showed the opposite pattern. Boys were far more likely to claim bad luck as the cause of their failure than were girls (33.6 versus 15.4%, respectively). Sex differences were not found on effort or task difficulty attributions.

Attributions to effort. While it could be hypothesized that women would be more likely than men to use the temporary cause of effort in the case of success, in experiments such as those by Nicholls and by Deaux and Farris, little evidence has been found for sex differences in the attribution of effort. Two explanations can be offered for this circumstance. The first relies on demand characteristics. For the most part, subjects in an experiment feel an obligation to try their best and acknowledging little effort or a lack of effort would be in contradiction to an attempt by the subject to look good. Consequently, effort attributions are generally very high by both sexes under a variety of experimental conditions. A second reason, suggested by Nicholls (1975), is that the actor who is performing a task is aware of how much effort he or she has exerted. Unlike ability, task difficulty, and luck, all of which must be inferred from the outcome, the actor has direct knowledge of how much effort has been exerted. Effort thus becomes a unique kind of attribute for the actor in that he or she has a control which is lacking in the other causes. (Such a distinction is less important from the observer's point of view, as the observer must infer rather than directly experience effort.) If it is assumed that most experimental subjects will try to solve the task at hand, then effort can be used to explain success with some confidence. The subject knows that he or she exerted effort and knows that the outcome was successful. In contrast, failure is less explainable by an effort attribution. If the actor was trying but failure still resulted, then another inference must be sought. And indeed, lack of effort is rarely used to explain an actor's failure (Deaux & Farris, in press; Feather & Simon, 1971b). Furthermore, because of the extensive use of effort to explain success by both male and female actors, a ceiling effect may prevent the detection of sex differences in the use of this attribute.

However, despite the circumstances mitigating against differential effort attributions, there are at least two studies which suggest that women do use effort more than men do, particularly in comparison with the ability attribution. In both cases the sex differences were found not among a random sample of women, but for selected subgroups of the female population.

In the first of these studies, Bar-Tal and Frieze (1974) selected male and female college students who were either high or low on the Mehrabian measure of achievement motivation. The task was again one of solving anagrams, and causal attributions were made following success or failure. Achievement level clearly had a mediating effect on the attribution patterns. The pronounced use of ability to explain success was found only among high- as opposed to low-achievement males (but not among high-achievement females, thus continuing to support the sex difference in the use of this stable internal attribute). While avoiding an ability ascription, high-achievement females did not show a strong tendency to invoke luck (though their ascription to this factor remained higher than comparable males), but instead relied most heavily on effort to explain their success. However, these same women were more prone than high-achievement men to explain their failure by a lack of ability.

In a second study, Deaux (1974) asked men and women who held first-level management positions within a series of organizations to identify the causes for either their success or failure in specific job related situations. Women in this sample (presumably higher than average on achievement motivation) made heavy use of effort in explaining their successful performance. And again, men used ability more than effort relative to the women in the sample.

The findings from these two studies provide an interesting parallel to the studies of observer attributions discussed previously. Whereas observers could not reasonably attribute a surgeon's career to luck, and instead invoked effort as a causal explanation for the women's successful performance, the woman actor who is high in achievement strivings and/or has a successful career will also find luck an unrealistic cause of her success. However, these women still show a tendency to reject the stable attribute of ability and instead seek out an internal cause which lies on the temporary dimension.

Generality of the Model

Reliance upon the notion of expectancies as a critical determinant in the attribution processes of women and men is a highly parsimonious explanation, for in this way the literature on sex differences can be related to the more general research on expectancies and attributions. For example, Feather and Simon (1971b) manipulated subjects' expectancies of success by pretrial performance (and in additional groups, simply categorized subjects on the basis of pretest expectancies) and then looked at the attributions made following an expected or unexpected success or failure. Subjects who expected to succeed saw ability as causal, while subjects who succeeded contrary to their expectations were more inclined to invoke luck as an explanation. Similarly, expected failures were attributed to a lack of ability, while unexpected failures resulted in relatively stronger attributions to bad luck. Similar results were found by Gilmor and Minton (1974). Subjects who approached the task with a high degree of confidence attributed failure externally and success internally, while these trends were reversed for subjects initially low in confidence.

The argument that sex differences in attribution patterns will appear only when expectancies differ was nicely demonstrated by McMahan (1973). In his study (with sixth graders, tenth graders, and college students as subjects), no differences in initial expectancies for performance on an anagram test were found. As would be expected, no differences between males and females on attribution measures were found either. McMahan did, however, provide added support for the proposition that expected success and failure are attributed to high and low ability, respectively, while unexpected outcomes are attributed to one of a variety of temporary factors (including effort and luck).

Thus, as in the case of observer attributions, it can be seen that sex is not a "deviant" variable but instead is conceptually linked to a major body of research in attribution theory. Furthermore, as argued by Miller and Ross (1975), the strategies of self-attribution can be interpreted as a logical information-processing strategy rather than an ego-defensive mechanism. Given the expectancies that men and women have, their subsequent attributions are logical products of the match between expectancy and performance.

Yet the expectancies which men and women hold may themselves be questioned, and consideration of these differences suggests an ego-defensive strategy may be operating in this initial baseline. Clearly, a considerable portion of one's expectancy is based on prior experiences of success or failure on a given task. However, in addition to this rational basis of expectancy, there may be an ego-defensive portion as well. For example, we may see the unrealistic student who, despite a history of failure, continues to expect to do well. When this individual fails, external reasons are sought. (In contrast, an observer may form expectations of that same individual on the basis of performance alone, and be far more likely to assume lack of ability as a cause.)

This ego-enhancing form of expectation is also apparent in an examination of the differences between the sexes. As previously noted, males have a consistently higher level of expectation than do females (though exceptions such as the McMahan study have been found, suggesting that these sex differences may be an endangered if not extinct species). Yet in many of these same situations, actual performance differences between women and men have not been found. The available evidence suggests that men are more guilty of overestimation than women are of underestimation, (though a margin of error exists in both instances). Consequently, success is rarely at odds with the expectancies of males, and females are more likely to find failure consistent with their expectations. In and of themselves, the explanations of ability or lack of ability may not reflect an accurate processing of information, but given the expectancies which have been established, the subsequent attribution process can be explained in logical terms.

A major question which remains unanswered in this area of research is the extent to which the expectancies can be altered. My own research suggests that one experience with success may not be sufficient to revise the expectancies of women substantially upward, and a single failure experience is even weaker as a catalyst for decreasing the man's expectation. Nor should one expect such a rapid change, given a presumably lengthy history by both men and women for their typical expectancy patterns. A further difficulty is raised by the cyclic nature of attributions and expectancies. Attribution of success to luck, by definition a temporary explanation, provides no basis for raising one's expectancy in the future. Thus, both women and men preserve their self-evaluation patterns in a somewhat vicious cycle. Presumably, however, a consistent set of success or failure experiences over some duration of time should allow the reformulation of expectancies to reflect past experience more accurately.

SEX AS A TASK CHARACTERISTIC

Some consideration should be given to a third aspect of sex which can influence the attribution process and which in turn may provide us with information about more general attribution patterns. This third characteristic is the specific sex-linkage of a task, which has been varied in only a limited number of studies.

The specific characteristics of a task have in general been ignored in studies of attribution. However, as suggested earlier, the nature of the task defines the range of attributions which will be reasonably used by most people in explaining task performance. Thus, for certain kinds of tasks, luck is a noncredible explanation, despite a wide divergence between a person's expectations and actual performance. The surgeon in the Feldman–Summers and Kiesler (1974) study is a case in point.

The experimenter may control this perception of the task requirements by specific instructions and thus set boundary conditions on the attributions which will be used (to the extent that the instructions are credible in terms of the specific task). For example, defining a game as one of ability will make ability attributions dominant; defining a game as one of chance will increase attributions to luck (cf. Deaux et al., 1975). Far more attention needs to be given to these task characteristics vis-à-vis the causal performance attributions supplied by subjects. (My own suspicion is that varied measures used in attribution research frequently confound attributions made by the actor to him/herself with definitions of the task. Questions phrased in the form "To what degree was ability a cause of your performance" may reflect perceptions of task requirements, while "How much ability do you have" may tap the actor's ascription of qualities to self. As an argument for the nonidentity of these two forms of questions, I have replicated data which show only moderate correlations between these two measures within subjects on the same task.)

In a similar fashion, defining the sex-linkage of a task appears to put a set of boundaries on the attributions which are used by subjects. Specifically, tasks labeled feminine or defined as ones in which women excel are perceived by subjects to be easier tasks. Evidence of this bias is available in a number of studies (see Deaux & Farris, in press; Taynor & Deaux, 1975). Heider (1958) has posited that ability (or power), effort (or exertion), and the difficulty of a task are directly related to each other. More exactly, Heider proposed that: ability $\times$ effort = task difficulty [transposed from Heider, 1958, p. 111]. Consequently, if one task is perceived as less difficult than another, then the judge should also assume that less ability and or effort are necessary for successful performance on that task. In terms of the present analysis, I would predict that feminine tasks, being viewed as easier, will be given lower ratings of ability and/or effort than will masculine tasks.

The evidence for such lower estimates is a weak but consistent. First, in the Deaux and Emswiller (1974) study, it was found that ratings of a woman's ability on a feminine task were lower than ratings of a man's ability on a masculine task. While the initial prediction was that these two attributions would be approximately equal, assuming that judges would have similar expectations for actors performing appropriate-sexed tasks, the failure to find these results can be explained post hoc. Subjects reported that performance on a masculine task was better than the equivalent performance on a feminine task. Although no direct measure of task difficulty was included in this study, the performance evaluation data are consistent with the proposed difference. If this difference is true, then it follows that ability estimates would be lower on the feminine task, in accordance with Heider's formulation.

More directly, Deaux and Farris (in press) found that subjects perceived a task that was labeled feminine to be easier than an identical task which was labeled masculine. The greater perceived difficulty of the masculine task was paralleled by greater ability and effort ratings for performance on the masculine task. Additional indirect evidence is provided by Feather and Simon (1975) in their study where occupations which varied in their masculine association (medicine, teaching, and nursing) were accorded differential ability and effort ratings. Both attributions were greatest in medicine and lowest in nursing, though the differences were significant only for effort.

If indeed masculine and feminine tasks are perceived to have different degrees of difficulty, then attributions for success and failure on these tasks should have different degrees of information for both actors and observers. Success on a masculine task should provide the most information about an individual's ability whereas failure on a feminine task should provide the strongest indication of the individual's lack of ability. While the available evidence for this prediction is not available, an indirect test of the general prediction is provided by Nicholls (1975). Varying the attainment value or importance of the test, Nicholls found that attributions by the actor to ability were greater following success on a high-valued task than on a low-valued task. Similarly, attributions to a lack of ability were greater following failure on a low- than on a high-value task. If the perceived difficulty of a task is conceptually equivalent to the attainment value of that task, then equivalent predictions can be made for sex-linked tasks as well. These are the kinds of studies which must be done if sex is to be incorporated within existent social psychological theory.

CONCLUSIONS AND IMPLICATIONS

In the preceding discussion, I have attempted to serve two purposes: (a) to describe the role of sex in the attribution process; and (b) to point to ways in which sex, as a marker variable, can be related to more general conceptual frameworks within attribution theory.

Consideration of sex as a variable in social behavior has accelerated in recent years, following an unduly long period of neglect, and it is evident that the sex differences are available to the seeker within the area of attribution theory. While from the existent findings, a reasonably high degree of consistency is shown, several practical and theoretical problems remain. At a theoretical level, we need to know considerably more about the origin of sex differences in attribution. Differences have been observed in children as young as age 10 which suggests an early divergence between sexes in the modes of explaining own performance. Patterns of reinforcement may well differ for young girls and boys, causing a biased use of stable versus unstable explanations. At present, however, our knowledge of these patterns is sparse, and research on the genesis of attribution patterns is desperately needed.

It is also important to understand the ways in which expectancies can be altered. If, as has been suggested, an unexpected event is attributed to some temporary factor, then future expectancies may remain unchanged, producing a self-perpetuating cycle. The consequences of this cycle in terms of behavior is a question of both theoretical and practical import. Although the connections between attributions and behavior have received somewhat limited attention, these potential connections may prove to be the most significant aspect of attribution theory.

With specific reference to sex, I would predict that an attribution of a woman's performance to luck would lead an observer to hold less promise for the future performance of that woman. Practically, this pattern could be responsible for lower salaries, less frequent promotions, and other quite concrete manifestations within a work setting. Indeed, Terborg and Ilgen (1975) have reported some suggestive evidence that women whose performance is explained by luck are in turn assigned more routine tasks within a simulated organizational setting, although the causal chain for such a process has not been firmly established.

Consideration of the role of defined task characteristics suggests one route by which expectations may be altered, both on the part of the actor and the observer. To the extent that the requisite attributes for a task are defined, the attributor may have little choice but to invoke the defined attribute. Thus, if only ability is believed to contribute to success, a successful performer may not be able to look to luck as a convincing explanation. In a recent study (Deaux et al., 1975), we have found supportive evidence for this possibility, but considerably more work remains in order to test the limits of applicability as well as the duration of the attribution pattern.

Although this kind of attention to sex is critically important, both practically and theoretically, it is perhaps even more important for researchers to move beyond the variable of sex to more encompassing theoretical systems. The payoff is in both directions. Subsuming sex within existent structures makes those theories increasingly sound, and, at the same time, findings for sex-related variables may serve to illustrate gaps in the existent models, encouraging redevelopment and extension. The expectancy variable appears to be a viable choice for making this kind of bridge and has applicability far beyond the specific area of women and men.

ACKNOWLEDGMENTS

I am grateful to Elizabeth Farris, Brenda Major, and Arie Nadler, as well as to Editors Harvey, Ickes, and Kidd, for their thoughtful suggestions on an earlier version of this chapter.

REFERENCES

Allport, G. W. *The nature of prejudice.* Reading, Massachusetts: Addison–Wesley, 1954.

Ajzen, I. & Fishbein, M. A Bayesian analysis of attribution processes. *Psychological Bulletin,* 1975, **32,** 261–277.

Bar–Tal, D., & Frieze, I. Achievement motivation and gender as determinants of attributions for success and failure. Unpublished manuscript, University of Pittsburgh, 1974.

Brigham, J. C. Ethnic stereotypes. *Psychological Bulletin,* 1971, **76,** 15–38.

Broverman, I. K., Vogel, S. R., Broverman, D. M., Clarkson, F. E., & Rosenkrantz, P. S. Sex–role stereotypes: A current appraisal. *Journal of Social Issues,* 1972, **28,** 59–78.

Crandall, V. C. Sex differences in expectency of intellectual and academic reinforcement. In C. P. Smith (Ed.), *Achievement-related motives in children.* New York: Russell Sage, 1969.

Deaux, K. Women in management: Causal explanations of performance. In V. O'Leary (Chairperson), *The Professional Woman.* Symposium presented at the meeting of the American Psychological Association, New Orleans, September 1974.

Deaux, K., & Emswiller, T. Explanations of successful performance on sex-linked tasks: What is skill for the male is luck for the female. *Journal of Personality and Social Psychology,* 1974, **29,** 80–85.

Deaux, K., & Farris, E. Attributing causes for one's own performance: The effects of sex, norms, and outcome. *Journal of Research in Personality,* in press.

Deaux, K., White, L., & Farris, E. Skill vs. luck: Field and laboratory studies of male and female preference. *Journal of Personality and Social Psychology,* 1975, **32,** 629–636.

Etaugh, C., & Brown, B. Perceiving the causes of success and failure of male and female performers. *Developmental Psychology,* 1975, **11,** 103.

Feather, N. T., & Simon, J. G. Attribution of responsibility and valence of outcome in relation to initial confidence and success and failure of self and other. *Journal of Personality and Social Psychology,* 1971, **18,** 173–188. (a)

Feather, N. T., & Simon, J. G. Causal attributions for success and failure in relation to expectations of success based upon selective or manipulative control. *Journal of Personality,* 1971, **39,** 527–541. (b)

Feather, N. T., & Simon, J. G. Reactions to male and female success and failure in sex-linked occupations: Impressions of personality, causal attributions, and perceived likelihood of different consequences. *Journal of Personality and Social Psychology,* 1975, **31,** 20–31.

Feldman–Summers, S., & Kiesler, S. B. Those who are number two try harder: The effect of sex on attributions of causality. *Journal of Personality and Social Psychology* 1974, **30,** 846–855.

Frieze, I., & Weiner, B. Cue utilization and attributional judgments for success and failure. *Journal of Personality,* 1971, **39,** 591–605.

Gilmor, T. M., & Minton, H. L. Internal versus external attribution of task performance as a function of locus of control, initial confidence, and success–failure outcome. *Journal of Personality,* 1974, **42,** 159–174.

Harding, J., Proshansky, H., Kutner, B., & Chein, I. Prejudice and ethnic relations. In G. Lindzey & E. Aronson (Eds.), *Handbook of social psychology* (Vol. 5). Cambridge, Massachusetts: Addison–Wesley, 1969.

Heider, F. *The psychology of interpersonal relations.* New York: John Wiley & Sons, 1958.

Jones, E. E., & Davis, K. E. From acts to dispositions: The attribution process in person perception. In L. Berkowitz (Ed.), *Advances in experimental social psychology,* (Vol. 2). New York: Academic Press, 1965.

Kelley, H. H. Attribution theory in social psychology. In D. Levine (Ed.), *Nebraska Symposium on Motivation.* Lincoln: University of Nebraska Press, 1967.

MacBrayer, C. T. Differences in perception of the opposite sex by males and females. *Journal of Social Psychology,* 1960, **52,** 309–314.

McArthur, L. A. The how and what of why: Some determinants and consequences of causal attribution. *Journal of Personality and Social Psychology,* 1972, **22,** 171–193.

McKee, J. P., & Sherriffs, A. C. The differential evaluation of males and females. *Journal of Personality,* 1957, **25,** 356–371.

McMahan, I. D. Relationships between causal attributions and expectancy of success. *Journal of Personality and Social Psychology,* 1973, **28,** 108–114.

Miller, D. T., & Ross, M. Self-serving biases in the attribution of causality: Fact or fiction? *Psychological Bulletin,* 1975, **82,** 213–225.

Montanelli, D. S., & Hill, K. T. Children's achievement expectations and performance as a function of two consecutive reinforcement experiences, sex of subject, and sex of experimenter. *Journal of Personality and Social Psychology,* 1969, **13,** 115–128.

Nicholls, J. G. Causal attributions and other achievement-related cognitions: Effects of task outcome, attainment value and sex. *Journal of Personality and Social Psychology,* 1975, **31,** 379–389.

Rosenkrantz, P. S., Vogel, S. R., Bee, H., Broverman, I. K., & Broverman, D. M. Sex-role stereotypes and self-concepts in college students. *Journal of Consulting and Clinical Psychology,* 1968, **32,** 287–295.

Taynor, J., & Deaux, K. Equity and perceived sex differences: Role behavior as defined by the task, the mode, and the actor. *Journal of Personality and Social Psychology,* 1975, **32,** 381–390.

Terborg, J. R., & Ilgen, D. R. A theoretical approach to sex discrimination in traditionally masculine occupations. *Organizational behavior and human performance,* 1975, **13,** 352–376.

Weiner, B. *Achievement motivation and attribution theory.* Morristown, New Jersey: General Learning Press, 1974.

Weiner, G., Frieze, I., Kukla, A., Reed, L., Rest, S. & Rosenbaum, R. M. *Perceiving the causes of success and failure.* Morristown, New Jersey: General Learning Press, 1971.

16

Attributional Conflict in Young Couples

Bruce R. Orvis
Harold H. Kelley
Deborah Butler

University of California, Los Angeles

The attributional perspective in social psychology assumes that people are interested in the causes for various events and particularly in the causes for behavior. Although the validity of this assumption has not been fully explored, it hardly seems debatable for the behavior engaged in by persons with whom the individual is closely interdependent. In such circumstances it is particularly important for one to know what the behavior of one's associates means–why they act as they do, under what circumstances a particular behavior can be expected to recur, and what changes in causal factors will be necessary for different behavior to occur in the future. It may be argued, further, that close relationships not only increase the importance of determining the causes behind other persons' behavior but also strengthen the need to present clearly the reasons for one's own actions.

If it is assumed, pursuing the line of reasoning above, that close relationships are characterized by strong concern about both accurate presentation and veridical attribution of causes, it might seem to follow that closely related persons will generally agree about the causes for their respective actions. However, there are good reasons to expect that different interpretations will be placed on behavior, even under these apparently ideal conditions. The true causes for behavior are usually quite complex. They encompass numerous factors present at the time of the behavior, each further traceable along a tangled skein of causality to prior factors. Given this complexity, any single causal explanation is bound to be selective, and it is easy to imagine the differential selectivity possible between two persons who, even though closely interdependent, have different information and needs.

In the present paper, we report an investigation of differences in causal attributions occurring between the members of young, heterosexual couples. They were asked to describe instances in which their explanations for the behavior of one member of the couple differed. It was not our purpose in this procedure to determine the incidence of attributional disagreement relative to attributional consensus. Although this is an important question for future research, we chose to focus exclusively upon instances of disagreement. Our purpose was to explore the domain of attributional conflict, and, most particularly, to identify the kinds of causes most commonly placed in opposition. Incidentally, we expected our investigation to provide valuable information about the types of causes the layman uses to explain behavior, a kind of datum that is rather lacking in current attribution research. Of course, our evidence on this last point may be somewhat limited by both the nature of our subject sample (couples in a university community) and our sample of causes (based on cases of disagreement). Yet, the procedure of eliciting examples of *different* explanations has some justification as a means of obtaining varied and extreme types of causes and, thereby, of revealing the entire domain of the layman's causal repertoire.

Finally, we hoped our results would stimulate thinking about the role of attributions in interpersonal processes, although in our study we did not directly deal with this matter. Our respondents' reports of occasions when the two of them "have" different explanations probably refer to instances in which there has been communication between the young couple about their attributions. This moves us outside the mainstream of current attribution research, in which attributions are "made" simply to be transmitted to the investigator. In relationships such as those of our young couples, communicated attributions undoubtedly serve many purposes, for example, to attack or influence one's partner or to defend or justify one's own behavior. It is possible that the very reason for which an attribution is to be communicated itself affects the attribution that is "made." It was this sort of phenomenon, related to the social context and its possible effects upon the attribution process, to which we hoped to call attention.

COLLECTION OF EXAMPLES

Couples were recruited through a classified advertisement in the UCLA daily newspaper. When prospective respondents called to make an appointment, they were told that the experimenters were studying behavior of young couples. It was indicated that they would be given a questionnaire requiring approximately an hour and a half and that they would be paid $7.00 for their participation. The conditions for participation–both members of the couple over a minimum age of 18 and either dating for at least a year or married or living together for at least 6 months–were also stipulated.

Forty-one couples participated in this study. Of those 41, 11 were living together, 5 were married, and 21 couples were dating. Approximately half the couples had been together 2 or more years. Most respondents (69) were students, and a large majority were between the ages of 18 and 26.

When a couple arrived, the interviewer[1] described the nature and purpose of the questionnaire. Respondents were told that we were interested in studying behavior in young couples–specifically, behavior for which each person had a different explanation. Emphasis was placed on our desire for examples of *behaviors* rather than opinions. Two examples, deliberately selected from unlikely areas of intracouple conflict, were provided for clarification:

1. Suppose that a football team wins a game. The coach says they won because the other team played a bad game, but the players say they won because they played a good game. Notice that there is no disagreement about whether the event or behavior occurred (the team won the game). However, there is disagreement about *why* the behavior or event occurred (why the team won the game).
2. Suppose one of you voted for a particular candidate in the last election, and the reason you say you voted for this cadidate is different than the reason your partner says you voted for him. Again, notice that you agree that the behavior occurred, but disagree about why it occurred.

Respondents were then told that the behaviors they listed could be something that occurred only once or something that was done with some frequency. At this point we also explained our desire for examples that had some importance for the relationship. However, our concern about promoting conflict prompted us to warn the respondents not to write about highly sensitive material. They were advised that the mere structure of the questionnaire tended to elicit negative behaviors. Even though they would not see or discuss each other's answers, we pointed out that we could not guarantee that they would not want to discuss the questionnaire when the experiment was over. We suggested that, as an index of sensitivity, they ought to consider whether or not they would feel comfortable discussing the particular example with their partner. We urged them to think twice before writing an example down if they would not be willing to discuss it. We then reiterated our desire for examples that were not trivial ("like why one of you does not change shirts before going to the market"), but were not too sensitive for discussion.

The couples were then told that after they had listed as many examples as they could of both own behavior and partner's behavior, we would select one

[1] The interviewers were the first and third authors.

example of each for them to write about in greater detail. The examples they would elaborate would be ones they had both listed independently.

Finally, confidentiality was assured, each member of the couple was given a copy of the questionnaire, and they were conducted into separate rooms where they independently answered the questions.

The questionnaire began with the following written instructions:

> Usually, men and women in close relationships agree about the cause of each other's behavior. Occasionally, however, one partner will do something for which each of them has a different explanation. When has this occurred in your relationship? On page 1, list instances of disagreement about the cause of *your* behavior. On page 1A, list instances of disagreement about the cause of *your partner's* behavior. In each case, briefly describe the behavior, your explanation for it, and your partner's explanation for it.

On Page 1 of the questionnaire space was provided in outline form for five examples of (a) own behavior, (b) own explanation, and (c) partner's explanation. Page 1A differed only in that it asked for examples of partner's behavior. Each person filled out both pages, but the order in which the two pages were completed was randomized by couple. The last three pages of the questionnaire (which were not analyzed) asked the respondents to discuss particular instances of their reported disagreements (ones they had both reported).

Number of Examples

We were interested in how easy or difficult our respondents would find their assignment of providing examples of attributional disagreement. While our data do not directly assess the incidence of such disagreements, we thought they might provide indirect evidence about the matter. The 41 couples responding to the questionnaire provided a total of 691 examples, each consisting of a "triad" of (a) behavior, (b) own explanation, and (c) partner's explanation. In other words, on the average each person provided 8.4 examples. It is our impression that the generation of this many examples suggests that the respondents found it relatively easy to think of instances of attributional disagreement. They apparently have many such instances in their experiences from which to draw examples.

The preceding speculation may also find support in our experience in obtaining "matched" examples, used for the last pages of the questionnaire. Our desire was to identify instances where both members of a couple had independently given the same triad of behavior and causes. Such instances were far less common than we had anticipated. One possible explanation for this lack of overlap between the two persons' sets of examples is that they are independently sampling from large populations of instances. Thus, by chance, they will not be likely to select the same examples. Only if the number of behaviors generating causal disagreement were small would one expect high intracouple agreement in examples recalled.

Analyses of the number of triads listed by sex of respondent and by whether the triad concerns own or partner's behavior revealed an unexpected sex difference. The females gave 4.1 examples of their own behavior and 3.9 examples of their partner's behavior, whereas the males gave 4.2 examples of their own behavior and 4.7 examples of their partner's behavior. Statistical analysis by couple (chi-square with Yates' correction) shows that males gave more examples than females, $p < .01$. This difference was significant, however, only for examples of partner's behavior, $p < .01$. Several interpretations might be suggested for this difference. Compared with females, males involved in close heterosexual relationships may be more analytical or evaluative of their partner's actions and, hence, more aware of disagreement about the causes of those actions. A second interpretation is that females are less willing to reveal their partner's negative behaviors to the investigator, for example, they might be more protective. A third possibility is that women more often explain their own behavior to their partners, thus providing the partners with more instances of attributional conflict.

ANALYSIS OF THE CAUSES

In coding the causes, we attempted to construct a set of categories that faithfully reflected the data and was minimally influenced by our preconceptions of types of causes. The latter were represented only be a general scheme distinguishing causes identified with the actor, partner, relationship, and external environment. Specific categories evolved from an examination of the similarities and dissimilarities between causes within each general cluster. Several specific categories suggested the additional broad heading of "activity."

Various problems soon developed, requiring the creation of special coding rules. One problem was that a given explanation might refer to several causes. For example, an explanation for drinking alcohol was: "He really likes drinking. It relaxes him and he enjoys the taste." This obviously refers to two different types of causes, a first associated with the actor and a second associated with the activity. With such items, our rule was to code the most prominent attribution. This particular example was placed in the "activity is enjoyable" category. We also encountered explanations containing causal chains. For example, the cause of behavior might be given as a characteristic that had been instilled by family upbringing. In these instances, our rule was to code the cause in the chain that was closest to the behavior. Finally, we came across a number of items that were not categorizable because they were not causal interpretations. Generally, they were reactions to the behavior, rather than explanations for it. These items were placed in the category "uncodeable."

In performing the coding, the three parts of the behavior–own explanation–partner's explanation triad were separated. Reference was made to the original triads in cases where a brief statement of a cause proved unintelligible out of context. In most cases, own and partner's explanations were coded independently.

The initial coding, conducted jointly by the first and third authors, resulted in 35 causal categories (including "uncodeable"). Each category was labeled, defined, and illustrated with an example. Given a list of these definitions and examples, a third person independently recoded the causes. A comparison was then made between the original coding and the recoding. Over the 35 categories, the percentage of instances coded identically was 44%. Although far from the degree of intercoder agreement one would like to see, this was judged to be adequate in view of the large number of categories (the chance rate of agreement was 6%). (In later analyses, the specific causal categories were grouped into 13 classes. Therefore, we also calculated the intercoder agreement for these larger causal classes. This was 55%, in contrast to the 13% agreement expected by chance.) However, the comparison also revealed that several undesirable discrepancies existed with respect to the categorization of small groups of related causes. Therefore, each explanation was recoded, using 29 of the original categories and defining 3 new ones. These 32 more precisely defined *categories* were then grouped, on the basis of related content, into 13 broad *classes.*

The classes and categories with their definitions and examples are given in Appendix A. The labels of the classes and categories are listed in Table 1. The percentage figure given in the second column of this table is the percentage of the 1382 examples falling in the particular class or category.

Relation of Categories to Existing Causal Classifications

It is ironic that in an area of psychological research as concerned with thought and phenomenology as is attribution theory, there has been so little attention given to the causal distinctions our subjects make. (One exception is Frieze's, 1973, investigation of unprompted explanations given for success and failure.) Certainly one important line of work for the future is the assessment of subjects' untutored causal categories and the determination of the dimensions or topography of these categories. Our analysis provides a useful place to begin this research, at least for explanations of interpersonal behavior.

Meanwhile, it is useful to ask how our empirically derived categories map onto the current a priori classifications. It is clear that the terms in the classic internal–external distinction are well-represented in our classes. It is interesting, in fact, to dichotomize them in these terms, placing actor's characteristics, preferences/beliefs, concerns, intention to influence partner, negative attitude toward partner, and state (though the last may lie between internal and external) on the one side, and circumstance/environment, people/objects, partner is responsible, and properties of the activity (Classes 10, 11, and 12) on the other side. If we compare the percentages of these two clusters, the often-noted bias toward internal or person-explanations is apparent (internal = 58.4% and external = 39.0%). This is true even though we have an equal number of actor and observer explanations.

The cross-cutting dimension of stable versus unstable causes (Weiner, Frieze, Kukla, Reed, Rest, & Rosenbaum, 1972) is also readily discerned among our categories. On the internal side, the characteristics class clearly represents stable person properties; and the state class, unstable properties. However, one wonders how the respondents view a number of the actor categories. We might expect intentions and concerns to be on the unstable side, but what about attitudes, preferences, and beliefs? On the external side, the stable and unstable factors correspond roughly to a "stimulus" versus "circumstance" distinction (McArthur, 1972, after Kelley, 1967), these being represented in our data by the people/objects and circumstance categories, respectively. However, our categories and classes suggest that the external causes can be distinguished not only in terms of stability, but also in terms of what might be called focal versus contextual causes. The four subtypes are as follows: (1) stable, focal: people/objects; (2) stable, contextual: state of environment; (3) unstable, focal: partner is responsible; and (4) unstable, contextual: circumstance. As the examples show, focal causes are specific elements which serve as the elicitors or targets of the behavior. The contextual causes are the general background conditions or "field" which shape and delimit the behavior.

Our respondents often use positive properties of the activity as explanations for behavior. The behavior ensues because the activity itself is enjoyable or because it has good direct or indirect consequences. (The distinction here corresponds to one Kruglanski, in press, has recently emphasized, between an act that is endogenously attributed–when judged to constitute an end in itself–and an act that is exogenously attributed–when judged to serve as a means to some further end.) Activity explanations have an interesting status in current attribution research. For example, in the McArthur paradigm, the activity is sometimes the effect that is being explained and sometimes part of the stimulus. The nature of the behavior constrains where the cause can be located. Liking and other expressions of attitude cannot be explained in terms of the attractiveness of the activity (liking). Nonproductive expressive behavior (dancing, piano playing) can readily be explained in terms of the activity itself. Productive and consummatory behavior (carpentry, eating) can be explained in terms of the objects produced or consumed. To a considerable degree, then, the nature of the behavior determines whether an explanation must be sought outside it or whether the behavior itself (the activity) can serve to explain itself. Kanouse (1972) has made a very similar point in his discussion of the implications of different verbs.

One of the discontinuities in attribution theory has been between the focus on *intentions* resulting from consequence-weighting decision making processes (Jones & Davis, 1965), on the one hand, and the focus on *conditions* present at

TABLE 1
Frequency of Use of Causes by Actors and Partners

Causal class or category	Total (%)	Actors (%)	Partners (%)	*p* value
1. Circumstance/ environment	5.4	9.0	1.7	.001
a. Circumstance	2.1	3.5	.07	–
b. State of environment	3.2	5.5	1.0	.001
2. People/objects	6.4	8.1	4.8	.02
a. People	4.8	5.1	4.6	–
b. Objects	1.6	3.0	0.1	.001
3. Actor's state	5.6	8.1	3.2	.001
a. Psychological state	3.8	5.1	2.5	.02
b. Physical state	1.8	3.0	0.6	.01
4. Actor's characteristics	22.6	11.4	33.9	.001
a. Inability	8.0	3.0	13.0	.001
b. Ability	0.3	0.6	0.0	–
c. Characteristics	13.1	6.5	19.7	.001
d. Habit	1.2	1.3	1.2	–
5. Actor's preference/ belief	13.0	17.7	8.2	.001
a. Preference	7.8	9.4	6.2	.05
b. Belief	5.1	8.2	2.0	.001
6. Actor's concern	4.9	7.4	2.5	.001
a. Concern for partner	3.0	4.5	1.6	.01
b. Mutual benefit	1.2	2.2	0.3	.001
c. Concern for other people	0.6	0.7	0.6	–
7. Actor's intention to influence partner	4.2	4.2	4.2	–
a. To change partner's behavior	2.8	2.7	2.9	–
b. To define the relationship	1.4	1.4	1.3	–
8. Actor's negative attitude toward partner	8.1	3.3	12.9	.001
a. Lack of concern for partner	3.7	0.4	6.9	.001
b. Negative feelings toward partner	2.8	1.6	4.1	.01
c. Insecurity about the relationship	1.6	1.3	1.9	–

(continued)

TABLE 1 *(continued)*

Causal class or category	Total (%)	Actors (%)	Partners (%)	*p* value
9. Partner is responsible	6.1	6.7	5.5	–
a. Partner's fault	3.0	3.8	2.3	–
b. Partner's influence	0.9	1.0	0.7	–
c. Partner's characteristics	1.7	1.9	1.4	–
d. Partner's positive behavior	0.5	0.0	1.0	.05
10. Activity is desirable	11.4	15.8	6.9	.001
a. Activity is enjoyable	5.6	8.1	3.0	.001
b. Activity has good direct consequences	4.2	5.4	3.0	.05
c. Activity cuts costs	1.6	2.3	0.9	.05
11. Activity is undesirable	2.7	3.5	1.9	–
a. Not enjoyable	1.1	1.4	0.7	–
b. Bad direct consequences	1.6	2.0	1.2	–
12. Activity has desirable indirect consequences	7.0	3.3	10.7	.001
a. Good indirect consequences	5.3	2.2	8.4	.001
b. Response from other people	1.7	1.2	2.3	–
13. Uncodeable	2.7	1.6	3.8	.02

the time of the behavior (Kelley, 1967), on the other.[2] It is clear that our respondents recognize the importance of both factors. References to intentions, preferences and beliefs, and properties of the activity all imply cause–effect sequences mediated by discerning choice. References to characteristics, circumstances, states, and people/objects give more emphasis to pressures arising from situational or internal requirements. This suggests the necessity in the interper-

[2] These different emphases may have their causes in historical differences in scientific explanations for "personality." Brewster Smith (1971) reports Gordon Allport's Observation that for him the important question is what are the individual's intentions and values, whereas for Kurt Lewin the important question is, what is the situation? Jones comes from the Allport tradition; and Kelley, from the Lewinian, The development of an attributional model that emphasizes the layman's use of historical explanations awaits the entry into our field of the intellectual descendent of Henry Murray, for whom, according to Allport, personality is the life history.

sonal domain of the distinction that Rosenbaum (1972) has suggested for the success–failure domain, namely, that between causal factors subject to control by the intentions the person adopts and those outside such control.

Are any major types of causes pertaining to the layman's accounts of interpersonal behavior missing from our list? This seems probable because of the particular way in which the list was generated—from instances of attributional conflict. For example, one type we had expected, having to do with the relationship itself or the *pair* of persons, was not clearly in evidence. This may be due to our asking for instances in which the behavior of *one person* was the subject of explanatory conflict. If we had asked for joint events or experiences of the couple (for example, joint success or failure, miserable times they had gone through), it seems reasonable that we might have found more explanations of this sort, for example, "It was due to a bad thing going on between us."

THE ACTOR VERSUS THE PARTNER

The design of our questionnaire permits us to analyze differences in the use of each causal category according to (a) whether the respondent was reporting his own explanation or the explanation he believed to be his partner's, (b) whether the respondent was reporting an explanation for his own behavior or for his partner's behavior, and (c) whether the explanation was used by the enactor of the behavior or was used by the partner of the individual engaging in the behavior. The first and second factors considered separately made little difference. However, the third factor (which statistically is the interaction between the first two) had massive effects upon the distribution of causal explanations. We will refer to these effects as actor–partner differences. It must be kept in mind that we are not comparing the two persons' actual explanations for any given behavior. Rather, for each behavior, we are comparing the explanations that one (or the other) *says* the two of them gave.

The results of chi-square analysis of the actor–partner differences are given in Table 1 for both the general classes of causes and the specific categories.[3] It can

[3] In Table 1, the p values shown for actor–partner differences are derived from chi-squares determined by the method suggested by McNemar for change scores (Siegel, 1956, pp. 63–67). This invloves making a cross tabulation over the 691 explanations for each causal category to determine the frequency with which it was given by both actors and partners, by only one or the other, or by neither. Chi-square is calculated to test the null hypothesis that the frequency with which the cause was given only by the actor is equal to the frequency with which it was given only by the partner. It may be objected that calculating chi-square over the 691 examples violates the assumption of independence among events. In defense of this procedure, it may be noted that cross tabulations of the behaviors given as examples (a) for own and partner's behaviors and (b) by the male and the female show there is very little interdependence among them. In other words, there is no evidence that a given respondent tended to give multiple examples of the same type, nor that the two members of

be seen that the actor more often explains his behavior in terms of classes (1) circumstance or environment, (2) people or objects, (3) actor's state, (5) actor's preference or belief, (6) actor's concern, and (10) activity is desirable. Examined at the level of the more specific categories, it appears that the actor emphasizes external causes (state of environment and objects), temporary internal states, judgments of what is preferable or necessary (preference and belief), his concern for partner's welfare, and both the intrinsic properties of the activity and its direct consequences. In contrast, the partner more often gives explanations in terms of (4) actor's characteristics, (8) actor's negative attitude toward partner, and (12) activity's indirect consequences. The tone of these explanations is largely negative, as indicated by the ascription of inability, selfishness at the partner's expense, and the possible implications of ulterior purpose in reference to the activity's indirect consequences. It may be noted that many of the characteristics (Category 4c) ascribed by the partner were also negative (for example, lazy, forgetful, irresponsible, violent).

The above differences may reflect, in part, actor–partner disagreement concerning the stability or frequency of the behavior being explained. The partner tends to see the particular action as part of a more general pattern of behavior, whereas the actor may view it as a single incident. Thus, the causal classes used more often by the partner contain explanations implying relative stability, whereas those used more often by the actor include classes containing unstable explanations. This stability contrast is most apparent for Class 4 versus Classes 1 and 3.

Not all of the categories that one might think to be consistent with the general picture above (where the actor seems either to excuse or justify his/her action, see below) yield significant results. Explanations referring to other people outside the relationship (2a and 6c) and placing blame on the partner (9a, b, and c), while slightly more frequent for the actor, were not significantly so. The difference for Category 9d occurs in a set of instances where the partner says that partner's prior positive behavior provides the explanation for the actor doing something nice for the partner.

Actor–Observer versus Actor–Partner

At one level, the pattern of actor–partner differences confirm the Jones and Nisbett (1972) hypothesis concerning actor–observer discrepancies, that "there is a pervasive tendency for actors to attribute their actions to situational

the couple tended to give related examples. The proof of this independence is found in the fact that tests of the differences in Table 1 made on a couple–by–couple basis yield virtually identical results. Why, then, use the former method? In later analyses it would have been extremely complicated to use a couple–by–couple procedure. Therefore, it seemed preferable to use the example-based statistic throughout.

requirements, whereas observers tend to attribute the same actions to stable personal dispositions [p. 80, italics omitted]." The "situational requirements" attributions by actors here obviously include state of environment, objects, and properties of the activity–its intrinsic desirability and its good direct outcomes. It also appears that some of the "internal" explanations refer to causes within the person that mediate between situational requirements and behavior. Thus, psychological and physical states are often part of a causal chain linking external conditions with behavior. Similarly, actions caused by preferences and beliefs reflect the actor's assessment of properties or "requirements" of the environment. One indication of the marginal position of these types of causes is found in our experience in coding them. We repeatedly found the boundaries between the state of environment and the actor's state, and between the activity's desirability/consequences and the actor's preferences/beliefs to be blurred.

It seems clear from the different factors represented in these explanations that the actor explanations are not always those of a "pawn," who is the helpless victim of external circumstances and forces (de Charms, 1968). This might be a proper characterization of the environment and objects categories, but is not at all appropriate for the two most frequent actor classes, namely preference/belief and activity is desirable. The latter type of explanation was wisely anticipated by Jones and Nisbett (1972): "When the actor compares himself to others, we might expect him to believe that he differs chiefly in the priorities that he assigns to his goals and in the particular means he has devised to achieve them [p. 91]." In the perspective of the origin–pawn distinction, it appears to us that the actor more often *justifies* himself as a discerning origin than *excuses* himself as a helpless, pushed-around pawn. The action is "reasonable and appropriate" more often than it is "necessary and compulsory." Perhaps the term situational requirements in the Jones and Nisbett hypothesis should be replaced by the term situational contingencies.

Although our actor–partner differences conform in large part to the Jones and Nisbett hypothesis, we doubt that they are based on the process that Jones and Nisbett emphasize as underlying the actor–observer discrepancy. This process involves information differences that lead actors and observers to infer different causes for behavior. Our actors and partners are clearly much more than simply "actors" and "observers," and differ in more drastic ways than merely in the information they possess. Our actor is a person who, in a close relationship with another person has (in most cases reported, see below) behaved in an unfavorable or at least questionable manner. The relationship requires that he/she be concerned about justification and exoneration. Similarly, the partner is no mere observer. Having been affected negatively by the behavior, the partner is concerned about its meaning, about redress or retribution, and about preventing its recurrence. It seems, then, that the dynamics of this situation require intracouple communication of discrepant interpretations of behavior, such as those reported by our actors and partners.

The preceding comments might seem to imply that the present data are simply irrelevant to the Jones and Nisbett hypotheses. Thus, it might be argued that they deal with perception of causality (or "attribution" as it is used in social psychological research), whereas we deal with causal communication. Yet, a rereading of the Jones and Nisbett essay with this distinction in mind shows that they often draw their examples from the domain of public communication of explanations. Furthermore, they explicitly discuss the possible influences of the "actor's need to justify blameworthy action [p. 80]" and "the motive to maintain or enhance one's self-esteem [p. 92]." However, the analysis is neither clear nor thorough because a comparison of causal perception versus causal communication is never directly posed. Protection of self-esteem may "affect information processing," but it may also be merely a matter of excusing "reprehensible actions by blaming them on circumstances" (Jones & Nisbett, 1972, p. 92). This apposition, the closest they come to the one we emphasize here, is blurred over by treating the motivational effects as simply giving a "set back" or "boost" to the postulated overall actor–observer difference. As will be discussed below, we believe it will be necessary to deal with perception and communication as inseparable processes and, particularly, to view them in the context of relationship-maintaining processes.

SEX DIFFERENCES

We were also able to analyze the causal explanations in relation to sex differences. Comparisons were made for each causal category according to (a) whether the explanation appeared on the questionnaire of a female or a male respondent, (b) whether the behavior being explained had been enacted by a female or a male actor, and (c) whether the explanation was ascribed to a female or a male "explainer."

It was found that sex of respondent and sex of the person giving (or reported to give) the explanation made little difference in the causal explanations. However, there were a number of differences in explanations for female actors as compared with male actors. The behavior of the female is more often attributed to environment, $p < .05$; people, $p < .001$; inability, $p < .01$; and insecurity about the relationship, $p < .001$. The attributions are clearly consistent with common stereotypes of the woman, that is, that she is the more dependent and weaker person. The weakness appears to be translated into her susceptibility to external control in the form of influence by the environment and other people. Consistent with most studies of sex stereotypes (for example, Broverman, Vogel, Broverman, Clarkson, & Rosenkrantz, 1972), these differences are reflected in the reports of both the female and male respondents.

Only two types of causes are given significantly more frequently for male actors, namely Category 10b (activity has good direct consequences, $p < .05$)

and Class 12 (activity has good indirect consequences, $p < .02$). The implication of these two types of explanations seems consistent with the notion that the male more actively controls his life, exercising choice among actions in terms of their consequences.

THE COMMON ATTRIBUTIONAL CONFLICTS

One of our main purposes in this study was to determine whether certain types of causes are placed in opposition to each other more often than other types. Table 2 presents the frequencies with which each type of cause (as defined by the 13 classes) was paired with each other type of cause in the examples provided by our respondents. The causal classes have been grouped on the basis of the results presented in the preceding section, that is, according to whether they were used more often by actors ("actor-type" causes), more often by partners ("partner-type" causes), or about equally often by the two ("other" causes). The data in Table 2 were analyzed by determining the frequency expected in each cell by chance, and comparing this with the obtained frequency. A calculation of chi-square (with Yates' correction) for the entire table–though not entirely justified because a number of the theoretical frequencies were low–suggests that the null hypotheses (that causes are contraposed at random) can be rejected, $\chi^2(66) = 163, p < .001$. It appears that certain types of attributional conflict are more common than others.

The positive and negative signs in Table 2 indicate the instances in which the obtained frequencies deviate most sharply above or below the expected ones. (In each of these cells, the contribution to the overall chi-square is greater than 3.85, that is, $p < .05$ for the individual term.) It can be seen that these deviations reflect in large part the tendency for actor-type explanations to be contraposed to partner-type explanations, for example, circumstance/environment versus actor's characteristics, actor's state versus negative attitude toward partner, actor's preference/belief versus characteristics, and activity is desirable versus activity has good indirect consequences. At the same time, the cells in which the obtained frequencies are unexpectedly low all represent cases where two actor-type causes or two partner-type causes are contraposed. Both of these results are further reflections of the massive differences in our data between actors' and partners' uses of the various causes. The reader will recall that every triad provides us with both an actor's and a partner's explanation. Therefore, to the degree that different types of explanations are used more frequently by one or the other party, the contraposition of actor- and partner-type causes (for example, preference/belief versus characteristics) will be common, and the contraposition of two actor-type or two partner-type causes (for example, negative attitude toward partner versus activity's indirect consequences) will be rare.

TABLE 2
Frequency of Pairing of Causes

	Actor-type						Partner-type			Other			
Causal class	1	2	3	5	6	10	4	8	12	7	9	11	13
Actor-type													
1. Circumstance/ Environment	2												
2. People/objects	4	8+											
3. Actor's state	4	2	5										
5. Actor's preference/belief	4	12	8	10									
6. Actor's concern	2	2	0	10	3								
10. Activity desirable	7	4	5	8–	3	22+							
Partner-type													
4. Actor's characteristics	26+	27	22	62+	9	35	40						
8. Actor's negative attitude to partner	9	4	13+	18	8	7	11–	8					
12. Activity's indirect consequences	7	9	2	11	9	25+	5–	1–	6				
Other													
7. Actor's intention to influence partner	2	1	3	3	7+	5	7	9	2	6+			
9. Partner is responsible	3	4	6	8	9+	7	16	9	5	6	5		
11. Activity undesirable	1	3	1	4	2	3	8	3	4	1	0	3+	
13. Uncodeable	1	1	1	11+	1	4	5	4	5	0	1	1	1

Note. + indicates significantly larger than expected frequency. – indicates significantly smaller than expected frequency.

The contribution of the actor–partner differences to the causal opposition pattern is even more dramatic if we determine, for each opposition in the actor–partner section of Table 2, how often the actor-type cause was given by the actor and the partner-type cause by the partner. In every one of the 15 actor–partner cells, such instances account for a large majority of the contrapositions. Of the 303 oppositions in that section of the table, only 38 are instances in which the actor used a partner-type cause and vice versa.

It will be noted that four of the cells with larger than expected frequencies lie on the diagonal of Table 2. These cells represent pairings of causes within a given class. This is disturbing because it raises questions about whether our causal categories capture the distinctions that our respondents make among types of causes. Inasmuch as they were, presumably, trying to provide instances in which *different* explanations were given for the same behavior, the contraposed causes should fall in different categories. Of course, the data in Table 2 reflect *classes* of causes which include discriminably different *categories* of causes. For example, Class 10 includes activity is enjoyable, activity has good direct consequences, and activity cuts costs. It is quite possible for a respondent to view these as distinct causes. In fact, 14 of the contrapositions in this class are pairings of two different causal categories within Class 10. However, an examination of contrapositions in terms of causal categories also reveals a surplus of cases in which the two causes (that the respondent gives as "different") fall into the same category. Therefore, it is also apparent that we did not preserve all the distinctions made by the respondents at this finer level.

ANALYSIS OF BEHAVIORS INVOLVED IN ATTRIBUTIONAL CONFLICT

The behaviors included in the respondents' examples were coded in order to permit us to examine the relationships between type of behavior and type of cause or type of attributional conflict. The coding procedure for the 691 behaviors was similar to that followed in the categorization of the causes. However, the behavior data seemed more amenable to classification in small categories with highly specific content. Because the necessity of interpretation was considerably less than in the coding of the causes, the behavior categorization was reviewed but not independently verified.

Fifty-two behavior categories were formed. These content-specific categories were then grouped on the basis of general content similarity into 19 larger categories. The categories and subcategories they encompass follow. The percentage figure after each category indicates its rate of occurrence in the 691 examples.

1. *Actor performs behavior benefiting partner* (3.9%): behaves unselfishly toward, does things with partner.

2. *Actor fails to behave warmly toward partner* (2.4%): does not behave affectionately, expresses uncertainty about relationship.
3. *Actor is insensitive or unyielding to partner* (3.8%): insensitive or belligerent, will not yield to partner's preferences.
4. *Actor criticizes or places demands on partner* (10.1%): criticizes, teases, places demands on partner, attempts to dominate relationship.
5. *Actor is too involved in outside relationships and activities* (10.8%): engages in activity without partner, close to own family or friends, behaves unfavorably toward partner in social situations, flirtatious or friendly, unfaithful to partner.
6. *Actor fails to conform to partner's sex role expectations* (2.3%): sex role behavior is excessive or insufficient.
7. *Actor's behavior is inept or situationally inappropriate* (3.8%): socially inappropriate, unable to perform some activity adequately.
8. *Actor fails to fulfill specific responsibilities* (2.9%): fails to fulfill responsibilities, perform household chores.
9. *Actor's behavior inconveniences others* (5.5%): late for appointments, inconsiderate, does not pay enough attention to others.
10. *Actor has distorted view of self or others* (4.9%): displays attitude of self-superiority, poor attitude toward world.
11. *Actor maintains distance from or dislikes other people* (4.0%): unsociable, distant from own family or partner's family or friends.
12. *Actor behaves emotionally or aggressively* (7.1%): displays bad temper, behaves roughly, emotionally.
13. *Actor is quiet or passive* (1.9%): quiet, passive.
14. *Actor is unmotivated or untidy* (7.4%): messy, neglects own hygiene or appearance, lacks motivation, sleeps too much or at wrong times, procrastinates
15. *Actor has undesirable practices* (4.9%): smokes tobacco, marijuana, drinks liquor, has annoying eating preferences, TV watching.
16. *Actor has annoying habits* (2.2%).
17. *Actor engages in or wants to engage in activity* (9.0%): engages in miscellaneous activity (including working late, guitar lessons, reading, talks about particular topic, performs household activities), sports activities, likes particular activity.
18. *Actor avoids particular activity* (8.5%): avoids particular activity, cannot or will not engage in serious talks, reluctant to try new things, does not perform self-help behaviors.
19. *Actor performs or wants to perform activity in particular way* (4.5%): performs activity in particular way (including carefully, with deep involvement, with attention to appearance of self or belongings), has undesirable ways of spending money.

Behavioral Differences

It is quite clear that the overwhelming majority of behaviors provided by our respondents are negative in character. Only the few instances in Category 1 (especially la) can clearly be considered positive. In general, the list of behaviors seems to be a nearly complete summary of "ways to aggravate, frustrate, threaten, anger, or embarrass your partner."

The design of our questionnaire made it possible to analyze the behaviors according to the sex of the actor, the sex of the respondent, and the identity of the actor (respondent versus partner). The second analysis showed that the sex of the respondent made little difference. The analysis by sex of actor shows that male actors were more frequently described as engaging in (3) insensitive or unyielding behavior toward their partner, $p < .05$, and (17b) sports activity, $p < .01$. On the other hand, female actors were more frequently described as (18) avoiding an activity, $p < .05$, especially (18d) self-help behaviors, $p < .01$; as engaging in (12c) emotional behavior, $p < .01$; and as (1b) doing things with their partners, $p < .05$. The consistency of these differences with the stereotype that men are active and aggressive whereas women are passive and emotional is apparent.

Two respondent versus partner differences are especially interesting because negative behavior is listed more often for own behavior than for partner's behavior. Respondents more frequently reported themselves as (2) not behaving warmly toward their partners, $p < .05$, and as (4a) criticizing their partners, $p < .05$. This tendency for the respondent to admit to more negative behavior than he/she gives as examples for the partner seems intimately related to the social context of the explanation for the behavior, and is reflected in the differential use of Causal Classes 1 and 3 and Categories 6a, 8a, and 8b, reported earlier, The partner views the above behaviors in terms of (8a) actor's lack of concern for partner and (8b) actor's undesirable attitude toward partner. In reporting examples of disagreement, the respondent's emphasis on his/her role as the enactor of these behaviors may stem from the desire to counter these negative interpretations. Our data show that the actor "writes off" failure to behave warmly to situational factors: (1) circumstance/environment or (3) physical/psychological state. In a similar vein, actor places on "criticizes partner" the positive interpretation of (6a) actor's concern for partner.

EXPLANATIONS FOR DIFFERENT BEHAVIORS

The preceding example illustrates that a given type of behavior may be especially associated with certain classes of explanations. It also suggests that examining these relationships separately for actors and for partners can reveal the nature of the causal disagreements generated by each type of behavior. In analyzing these relationships, we thought it desirable to form larger behavior classes which would increase the expected frequencies to more reasonable levels. To accomplish this, we first computed separately for actors and partners the rates of use of the 13 causal classes within each of the 19 behavior categories. This procedure yielded a causal profile of 26 percentages for each behavior category. The profiles were then intercorrelated for each pair of behavior categories. Categories with high intercorrelations were grouped together, resulting in nine *classes* of behavior. Three categories were retained individually because they had relatively unique causal profiles.

The reader should not assume that the member categories of a given behavior class are characterized by similar content. Rather, the behavior categories are grouped together because they are explained similarly. Caution in inferring content similarity is further dictated by the unfortunate fact that the inductive procedure capitalizes on chance similarities among causal profiles generated by random error in the earlier content codings of the behaviors and explanations.

Table 3 summarizes the results of an analysis of the association between these behavior classes and the causal classes. The base rate of a given explanation over all behaviors and the total frequency of the particular behavior were used to determine the expected frequency of each causal class within each of the nine behavior classes. A chi-square test was then used to determine whether the observed frequency for a cell deviated significantly from its expected value. All deviations with $p < .05$ are given in Table 3.

The most striking feature of the profiles used to determine the nine behavior classes represented in Table 3 is their extensive similarity to the general pattern of actor versus partner causal explanation discussed previously. As we have already seen, Causal Classes 1, 2, 3, 5, 6, and 10 tend to be given by actors, whereas Classes 4, 8, and 12 tend to be given by partners. In view of the magnitude of these differences, it is hardly surprising that two-thirds of the actor–partner contrasts in Table 3 can be interpreted as specific instances of them. However, Table 3 also shows the ways in which the actor–partner contrasts are accentuated, weakened, or even partially reversed within the different classes of behavior.

Among those behavior classes showing the actor–partner contrasts, the particular explanations accentuated differ in several important ways. In Classes II, poor affectionate or sex-role behavior, and IV, irresponsible or annoying behavior, the actor is especially likely to excuse his/her behavior, explaining it in terms of (1) circumstance/environment or (3) physical/psychological state. The partner is especially likely to view these behaviors in serious terms, interpreting them in terms of (8) negative attitude toward the partner (Class II) or (4) actor's characteristics (Class IV). It may be noted that Behavior Class V contains the most extreme accentuation of excusing versus serious explanations, that is, (3) state and (9) partner is responsible versus (4) actor's characteristics. Here, the actor is especially likely to blame emotional or aggressive behavior on his

TABLE 3
Causes Given with Greater or Lesser Frequency Than Expected for Each Class of Behavior

Behavior class	Original behavior categories	Causes given by: Actors	Partners
I. Quiet/passive	13	More frequent than expected: (4) Characteristics	(8) Negative attitude to partner
II. Poor affectionate or sex role behavior	2, 6	More frequent than expected: (1) Circumstance/environment (3) State	(8) Negative attitude to partner
III. Insensitive or unyielding to partner	3	More frequent than expected: (11) Activity undesirable	(11) Activity undesirable
IV. Irresponsible or annoying behavior	8, 9, 16	More frequent that expected: (1) Circumstance/environment (3) State	(4) Characteristics
		Less frequent than expected: (10) Activity desirable	
V. Emotional/aggressive	12	More frequent than expected: (3) State, (9) Partner responsible	(4) Characteristics
		Less frequent than expected: (4) Characteristics	(10) Activity desirable
VI. Avoids or is particular about activities	14, 18, 19	More frequent than expected: (1) Circumstance/environment	(4) Characteristics

		(3) State (5) Preference/belief, (11) Activity undesirable	
		Less frequent than expected: (4) Characteristics (8) Negative attitude to partner (12) Indirect consequences	(8) Negative attitude to partner
VII. Situationally inept or socially rejective behavior	7, 10, 11	More frequent than expected: (2) People/objects (5) Preference/belief	(2) People/objects (4) Characteristics, (12) Indirect consequences
		Less frequent than expected:	(10) Activity desirable
VIII. Involved in activity disliked by partner	5, 15, 17	More frequent than expected: (10) Activity desirable	(8) Negative attitude to partner, (10) Activity desirable, (12) Indirect consequences
		Less frequent than expected: (4) Characteristics, (9) Partner responsible	(5) Preference/belief
IX. Benefits or places demands on partner	1, 4	More frequent than expected: (6) Concern, (7) Influence partner (9) Partner responsible	(7) Influence partner (8) Negative attitude to partner (9) Partner responsible
		Less frequent than expected: (4) Characteristics	(10) Activity desirable

partner. Behavior Class VI, avoids or is particular about activities, is relatively similar to Classes II and IV, accentuating (1) circumstance/environment and (3) state, for the actor, and (8) negative attitude toward the partner, for the partner. However, in this case a new component (and a strong one) appears on the actor's side. This is an explanation that implies he takes responsibility for the behavior, namely (5) actor's preference/belief.

Behavior Classes VII and VIII differ from the behaviors discussed above in several important aspects. Class VII, situationally inept or socially rejective behavior, is unique among the behaviors in accentuating (2) other people or objects, for both actors and their partners. In the actor's case, this seems to be a different "excusing" explanation. He is also likely to give the "responsibility-taking" explanation of (5) preference/belief. On the other side, the partner is not only likely to explain the behavior in terms of (4) actor's characteristics, but also in terms of (12) indirect consequences of the activity, an interpretation less critical of the actor. In addition, the accentuation of Causal Class 2 represents a reversal of the general pattern of actor versus partner explanation. Class VIII, involved in activity disliked by partner, resembles Class VII in that both more and less damning partner-type explanations are accentuated for the partner, namely (8) actor's negative attitude toward partner and (12) indirect consequences of the activity. Further, Class VIII is unique in the accentuation of (10) activity is desirable, for both actors and partners. This appears to be a different responsibility-taking explanation for the actor. The accentuation of Causal Class 10 for the partner again represents a reversal of the general actor–partner pattern. The clearest reversal of the general actor–partner explanation pattern occurs for Behavior Class I, quiet or passive. The only explanation accentuated for actors is (4) actor's characteristics. Further, this is the only instance of accentuation of a partner-type explanation for actors.

Behavior Classes III and IX stand out in their tendency to accentuate the same causal classes for both actors and partners. Yet, this consensus is more apparent than real. A closer examination of Class III, insentitive or unyielding behavior, reveals that partners are considerably more likely to explain this behavior in terms of (5) actor's preference/belief than by (11) activity is undesirable, although this deviation does not reach significance. For Behavior Class IX, benefits or places demands on partner, the heterogeneous causal class (9), partner's influence, is used quite differently by the actor and partner. The actor invokes negative influence, blaming critical or demanding behavior on the partner (in a similar manner to its use for Class V, emotional or aggressive behavior). In contrast, the partner invokes positive influence, taking credit for beneficial behavior.

The instances in which a type of causal explanation is used less frequently than expected seem to include several rather different phenomena. First, some causes are simply not logical or plausible as explanations for certain kinds of behavior. This seems to be the case for (1) partner is responsible in relation to VIII, activity is disliked by partner; and for (10) activity is desirable as the partner's explanation for VII, situationally inept or socially rejective behavior; or for the part of IX consisting of criticizing and placing demands on the partner. Second, some explanations are probably not given by the actor because they associate him in a derogatory way with a negative behavior. Examples are (10) activity is desirable as an explanation for IV, inconsiderate or annoying behavior; and (4) actor's characteristics as an explanation for V, emotional or aggressive behavior (or for VI, avoids activities). Third, some of the infrequent usage of causes probably reflects a separation of certain domains of activity from the rest of the couple's relationship. Thus, the actor's particular orientation toward or avoidance of various activities (Behavior Class VI) is infrequently attributed to (8) negative attitude toward partner.

The various causes are accentuated for considerably different numbers of behaviors. This difference may, again, reflect both the plausibility and desirability of giving particular types of explanations for specific kinds of behavior. For example, (1) circumstance/environment and (3) psychological/physical state are accentuated for a great variety of behaviors by actors. Apparently, actors find these explanations, used almost interchangeably, to be both relevant and acceptable for many different kinds of actions. Perhaps for the same reasons, (4) actor's characteristics and (8) actor's attitude toward partner are frequently accentuated by partners. On the other hand, (6) actor's concern for partner is accentuated by actors only for behavior benefiting or placing demands on their partners, Class IX. In this context, it seems to associate them in a favorable way with behavior that might otherwise be undesirable. In a similar vein, (9) partner's influence is accentuated by partners only for behavior which benefits them. This allows partners to take credit for the actor's positive behavior, as mentioned earlier.

VARIATIONS IN CONTRAPOSED CAUSES

In the preceding analysis, we highlighted the causes that are accentuated or deemphasized for different behaviors, by both actors and partners. The results imply how the contraposed causes may vary for different behaviors, but we will now analyze this phenomenon directly.

We first made an analysis for each behavior class (see Table 3) of the degree to which actors give actor-type causes and partners, partner-type causes. For six of the nine classes, the number of such instances was significantly larger than the number of "reversals," that is, instances in which the actor gives a partner-type explanation and the partner, an actor-type explanation. Classes I, II, and III are the classes that depart from the general pattern of actor–partner contrapositions. Examination reveals that they do so for different reasons. Class I, consisting only of Category 13 (actor is quiet or passive), tends to be given a partner-type explanation by both parties, namely (4) actor's characteristics versus (8) actor's negative attitude toward partner. The results for Class II mainly reflect the fact that category 6 (failure to comform to partner's sex-role expectations) tends to be given various actor-type explanations by both parties. As previously noted, Class III, insensitive or unyielding to partner, tends to be explained by the actor in terms of (11) activity is undesirable, while partners are more prone to invoke (5) actor's preference/belief. In general, this analysis shows that there is considerable variation among the classes of behavior in the degree to which they elicit the actor–partner contrapositions portrayed by the average trends in our data.

A second approach to identifying variations in contraposed causes is to determine the behavior for which specific contrapositions occurred most often. The summary that follows is limited to common contrapositions that occurred significantly frequently for a given behavior class. (The latter was determined by comparing obtained frequencies with the expected frequencies derived from the base rates of the two causes and the number of triads in the behavior class.) What follows, then, is a summary of the types of unique combinations of behavior, own explanation, and partner's explanation that appeared most frequently in our sample. (This should not be taken to imply that any of these triads appeared with high frequency. The reader will appreciate that our sample of 691 triads is small relative to the number of triads possible (9 behavior classes $\times 13^2$ causal contrapositions = 1521). Consequently, the expected frequency of any given triad was rather small.)

1. *Circumstance/environment versus actor's characteristics.* This contraposition occurred 11 times with Behavior Class VI, avoids or is particular about activities. For example, actor explains having quit competition in sports with the comment: "Too time consuming, I am too busy." The partner's explanation: "Too lazy, no competitive spirit any more."

2. *Actor's preferences/beliefs versus actor's characteristics.* This was the *most frequent* contraposition of causes in our sample, and, as shown in Table 2, occurred more often than would be expected by chance ($p < .001$ for the one-cell comparison). There were 26 instances in which this contraposition was paired with Class VI, avoids or is particular about activities. In an example similar to one above, the actor explains not competing in intramural sports by: "I believe sports programs are foolish." The partner's explanation: "He is uncompetitive."

This causal contrast also occurred 15 times for Class VII, situationally inept or socially rejective behavior. For example, the actor says, in reference to his extreme pessimism in looking at the world: "I'm being realistic and merely conveying facts." Partner says: "He is close-minded in that he only sees what is wrong."

3. *Activity is desirable versus actor's characteristics.* This was an especially frequent contraposition (17 instances) for Class VIII, involved in activity disliked by the partner. An example is provided by the different explanations for smoking marijuana and getting drunk. The actor says: "It's a release and I enjoy

the times and activities while under the influence." Partner: "He's addicted and has no self-control."

4. *Activity is desirable versus activity has good indirect consequences.* This was also a common contraposition for Class VIII, involved in activity disliked by partner (11 instances). One typical example contraposes desirability with appearance as explanations for eating health foods. The actor eats them "because they are good for you." The partner says: "She wants to project an 'organic' image."

In 19 instances, both parties invoked "activity is desirable" explanations for Class VIII. However, as the following example shows, the actor and partner give different types of explanations within the causal class. Flirting with girls is explained by the actor as, "I enjoy the tease and the challenge, but it's all in the verbal communication." The partner's interpretation: "He's trying to pick up the girl."

5. *Partner is responsible versus actor's characteristics.* Seven of the 13 instances of this opposition occur in relation to Class V, emotional or aggressive behavior. These cases of disagreement about where blame is to be placed within the pair are illustrated by an example in which the actor yells and swears in the course of arguments. Actor's explanation: "I'm so mad because he can't see how wrong he is and that he can't understand me." Partner's explanation: "She can't stand to be quiet and, above all, lose an argument."

6. *The contraposition of different actor characteristics.* Causal Class 4 includes a wide range of properties of the actor and was often used by both the actor and partner for the same behavior (see Table 2). This pairing occurs for 13 instances of Class IV, irresponsible or annoying behavior. The actor often explains failure to do something, for example, by his/her forgetfulness, while the partner invokes inconsiderateness.

7. *Contrapositions involving the actor's attitudes towards the partner.* Causal Categories 6, actor's concern for partner, and 8, actor's negative attitude toward partner, are used with moderate frequency by the actor and partner, respectively. However, as seen in Table 2, they are not contraposed to each other with great frequency nor are they frequently contraposed to any particular other causes. The first fact reflects their uses for different classes of behavior. As mentioned earlier, actor's concern is used by the actor particularly for Behavior Class IX, benefits or places demands on partner. In contrast, actor's negative attitude is particularly used by the partner to explain Class II, poor affectionate or sex-role behavior. In each case, the other person's explanation takes a variety of forms.

THE SOCIAL CONTEXT OF ATTRIBUTION

We have considered the implications of each of our main results as we have presented them. Here, we wish to consider the more general implications of this type of data and of our experience in gathering it.

We believe that we have provided a view of what closely related persons say to each other about why they behave in a particular way. We did not, of course, directly investigate their conversations about these matters (a study of that sort would be very valuable), nor did we ask them to report what they say to each other. Yet, we believe that the reported attributional conflicts contain explanations they have communicated to one another, and not simply imagined or projected disagreements. One indication of this is the close similarity between the distribution of causes provided as "own explanations" and the distribution of the causes provided as "partner's explanations." If there were little communication about these matters, the partners would have to speculate about the nature of each other's explanations. Thus, we might expect systematic differences between the explanations the respondent gives and those he/she believes the partner gives. Further support for this belief comes from the fact that the overwhelming majority of the 691 behaviors were undoubtedly negatively valued by one or both of the parties. Intuitively, problem behaviors would seem likely to prompt causal analysis and discussion.

The indications that our data reflect the consequences of a communication process, together with the apparent readiness with which examples of attributional conflict were provided, discussed earlier, lead us to believe that communication about divergent causal interpretations of behavior is a common and important part of the interaction within these couples. Several reasons can be given for the importance of attributional communication within close relationships. For the individual, such communication is obviously useful in justifying one's own actions and questioning those of the partner. From the perspective of the relationship itself, such communication serves an important maintenance function. Overt explanations serve to define, emphasize, and acknowledge the basic understandings within the relationship about the condition under which various behaviors may and may not occur.

It is not our purpose, here, to analyze *why* attributional communication is important, but, rather, to consider *the implications of assuming that it is important*. In much of the thinking about the attribution process, it has been assumed that this process serves the purpose of providing a causal understanding of the world.[4] This view asserts that the person makes causal inferences in order to understand his environment and to predict what future events are likely to occur (for example, Shaver, 1975). This assumption accords to the man in the street the status of a quasiscientist and lends plausibility to the use of paradigms of scientific attribution as models of his interpretive processes (for example, Kelley's use of ANOVA, 1967; Trope's use of Bayes' theorem, 1974). Now,

[4] Heider (1958): ". . . we try to make sense out of the manifold of proximal stimuli . . . [p. 296]." Jones and Davis (1968): "The perceiver seeks to find *sufficient reason* why the person acted and why the act took on a particular form [p. 220]." Kelley (1967): "The theory describes processes that operate *as if* the individual were motivated to attain cognitive mastery of the causal structure of his environment [p. 193]."

what if this view is essentially wrong? What if the person learns and is motivated to make attributions not for some abstract understanding of the world, but, rather, to explain his own actions and to attempt to control the actions of his close associates? In other words, what are the implications of assuming that the attribution process is originally learned and subsequently maintained primarily in the social context of justification of self and criticism of others? A brief discussion of several apparent implications follows.

Implication 1: The attributional process will be evoked primarily in situations of conflict of interest. The individual will muster good reasons to justify those things he wants to do that conflict with others' preferences. Clear lapses in his behavior will occasion his identification of exonerating excuses, in order to avoid negative sanctions. In short, attributions will be derived by actors in order to escape the constraints of social control, that is, to deal with interpersonal conflict. On their side, partners will place negative causal interpretations on behavior they find disagreeable or contrary to their own interests. Affected detrimentally by inadequate or malevolent behavior, they will emphasize personal responsibility and thereby exert pressure on the actor to exercise voluntary control and avoid such behavior in the future.

From the above we seem to be implying that attributions will be made more often for what is commonly regarded as "bad" behavior than for "good" behavior. However, they will also be made for instances of good behavior for which issues of credit or reciprocity might arise. In referring to these as the occasions on which the attributional process will be "evoked," we mean that they are the times when the person thinks about causation, considers information bearing on it, and draws causal conclusions. The question of "why" is not raised unless the behavior entails a conflict of interest.

The essential point here is that social interactions corresponding to our triads of behavior and contraposed examples may form the basic context in which the attribution process is learned. Our categories of behavior provide a fairly extensive list of instances of conflict of interest, and the different explanations given by actors and partners reflect their attempts at justification and excuse, on the one hand, and responsibility-placing criticism, on the other.

Implication 2: Attributions are an integral part of the interpersonal evaluative process. The *reasons* for behavior are learned as part and parcel of the *evaluation* of behavior. Attributions are not simply handy auxiliary interpretive tags that permit the person to make adjustments in his evaluation of behavior, for example, shifting the judgment of theft toward the positive side if it occurred out of dire need. Rather, they enter into the very description of the behavior itself. For the layman, the cause will be required in any adequate description of behavior in conflict of interest settings.

This implication might seem to be belied by our ability to separately code the behaviors and causes provided by our respondents. However, the point refers to the layman's sense of what is a complete description of behavior, and not ours. We would predict that if actors and partners are asked to describe their behaviors in detail, their descriptions will be in terms corresponding to the general pattern of actor–partner differences.

Implication 3: The attribution process will be constrained by conventions as to plausibility and rationality. If the explanations involved in justification, excuse-making, placing responsibility, etc., are to serve their various purposes, they must be plausible. While egocentric in purpose, the communicated reasons may not be autistic in quality. They must meet commonly accepted standards of evidence, logical use of rules of inference, and consistency with general experience. Some of the aspects of appropriately linking cause to behavior were covered in the discussion of Table 3.

Even if not wholly convincing to the listener, he must at least find it credible that the explanation is truly believed by its pronouncer. Otherwise, the self-serving nature of the attribution will be out in the open, its usefulness nil, and the explainer subject to irrefutable attack. Further, if the listener is to believe that the explainer believes his own explanation, it is probably desirable that the explainer does believe it. Thus, it can be argued that communicated attributions must generally meet social and personal standards of quality, including those presented in current attribution theories.

Implication 4: The attribution process will be selective, both with respect to causal hypotheses and to information. How can attributions serve conflicting personal interests and still meet the criteria of plausibility? The answer is that ordinary events, including interpersonal behavior, readily lend themselves to different interpretations. The stream of causality upon which any given behavior surfaces is both long and broad. Everyday life presents a sharp contrast to our laboratory settings in which the attributor has only three or four kinds of information and in which another person can readily identify what aspects of it he overlooks or accentuates. Furthermore, we suspect that most attributional conflict concerns not which causal factors were *present,* but which one was *crucial.* "Sure, he thinks that sports are a waste of time, and perhaps they really are. But the real reason he doesn't participate is that he's simply lazy." In the identification of the "real" reason (that is, the key one, the one that "makes the difference"), the everyday attributor has great freedom to select from many different kinds of information and causes.

Implication 5: Attributional conflicts will generally be irresolvable. The reasons for this assertion are contained in the preceding discussion. The complexity of the evidence relevant to any given action, the fact that much of it is "history" and essentially irrecoverable, the multiple interpretations to which such complex and ambiguous evidence is subject, and the continuing conflict of interest regarding the action itself–all these make it improbable that the actor and partner can come to see eye-to-eye in the matter.

If so, why are the conflicting explanations raised? The answer is that they serve as elements in a continuous give-and-take about the issues involved in the conflict of interest (loyalty, responsibility, control of emotion, dominance, fairness, etc.). The explanation is not presented so much to change the partner's mind about the particular incident as to change what he will do or how he will view the behavior next time. At a more general level, as noted earlier, explanations acknowledge and reinstate the basic normative structure of the relationship regarding behavior. Thus, the discussion of causes, itself, acts as a causal influence on later behavior.

Implication 6: Meta-attributions will figure importantly in the evaluation of persons. When a given explanation is not credible to its hearer, he is likely to wonder why it was given. This raises questions about the speaker's candor, perceptiveness, and rationality. To the extent that the inadequacies of explanations can be attributed to properties of the explainer, serious doubt may be cast on the future of the given relationship. For how can one count on stable and productive interaction with a person who locates his behavior dishonestly or absurdly in the causal structure of the world? Hence, meta-attributions–explanations for explanations–will be a basic and important part of the evaluation of individuals. Further, the issue of why a given attribution is made may often, itself, become a matter of open discussion and conflict between close associates.

Implication 7: In general, the actor will genuinely believe both that he has an accurate understanding of the causes of his behavior and that they justify it. This follows from several of the preceding ideas–that the evidentiary basis of most action is sufficiently complex and ambiguous to support various inferences, that it is desirable for the explainer to believe his explanation if it is to receive a serious hearing, and that causal interpretation and evaluation of behavior are inextricably interwined. With regard to the last factor, there is undoubtedly a strong cognitive balance effect in one's assessment of his own behavior and its causes. Believing that I am a highly worthy person, my good actions are linked in a unit relation to myself ("caused by" me), whereas my bad actions are dissociated from my "self."

Space does not permit drawing further implications, nor, unfortunately, can we examine the research and methodological consequences of those given above. The attributional contraposition we state here, between "causal understanding" and "social context" explanations for the attributional process, is not original with us. Within social psychology, a roughly parallel distinction was made by Jones and Thibaut in their 1958 paper on inferential sets in interpersonal perception. Their first assumption was that "interpersonal perception can most fruitfully be treated as both instrumental to social interaction and conditioned by it [p. 152]." One of their classifications of interactions involved a distinction between those in which (a) the perceiver is in a *value-maintenance* set, attempting to facilitate the attainment of his personal goals, and (b) the perceiver is in a *causal-genetic* set, attempting to make a causal analysis of the partner's personality. The first situation corresponds to that of the partner in our "social context" model, inasmuch as it "clearly involves the attribution of substantial cause to the other actor, who is thereby seen as in some personal sense responsible for abetting or thwarting the perceiver" (Jones & Thibaut, 1958, p. 159). The causal–genetic set exists for the perceiver who is not evaluating the consequences of the actor's behavior for himself, but, rather, is seeking to identify (in a quasiscientific way) the ultimate causes for the behavior. Under these conditions, (according to Jones and Thibaut, 1958), the perceiver will inhibit the "seductive practice of inferring intent from consequences [p. 165]," and his conclusions "are apt to be more systematic, consistent, and logical than those of the observer in a value-maintenance set [p. 166]." The latter set is characterized as "the most natural, primitive egocentric orientation [p. 171]," a view not far from the one emphasized here.

The social context of attribution has also been explored in sociology by Lyman and Scott (1970), although primarily from the actor's viewpoint. They assert that social actors must present an "account" "to explain unanticipated or untoward behavior [p. 112]." Accounts "prevent conflicts from arising by verbally bridging the gap between action and expectation [p. 112]." According to Scott and Lyman, these explanations are of two basic types, functioning either as excuses or as justifications for the questioned behavior. While there is usually a settlement regarding the style in which an account will be presented, typically one account leads to questions which lead to a further account, and so on. Accounts which are inappropriate with respect to the "gravity" of the behavior, to acceptability of motivation, or to norms dealing with reasons for engaging in particular types of conduct are likely, in themselves, to require subsequent accounting for.

Our line of thinking is admittedly highly speculative, there being little in our results that compels us to present this model of the attribution process. Furthermore, our thinking is highly tentative. Although we have tried to state the "social context" view in a strong and unequivocal style, we would not be ready to abandon the prevailing "causal understanding" model in its favor. A more reasonable judgment would be that the truth lies somewhere between the two models. However, for the moment it may be more constructive to pose clearly the alternative models, rather than to move immediately to some ambiguous mixture of them. The reader will certainly appreciate the unique research directions that are suggested by the social context assumption; directions toward an understanding of the intercommunication of explanations, the situations eliciting such communication, and the significance of the intercommunication for the course of ongoing relationships.

APPENDIX A: CAUSAL CATEGORIES AND CLASSES

1. Circumstance/Environment
   a. Circumstance: temporary financial problems, lack of time, particular pressures (for example, I didn't spend the evening with partner because I had an exam the next day).
   b. State of environment: semi-permanent conditions, available opportunities, role and norm requirements (for example, I didn't look for work because occupational opportunities are limited at present).
2. People/Objects
   a. People: background, upbringing, present family or peer pressure, fault of other people (for example, I'm anti-social because people provoke me).
   b. Objects: something about an object is cause (for example, I can't drive her car because it is a piece of junk).
3. Actor's State
   a. Psychological state: frustration, anger, boredom, embarassment, anxiety, preoccupation, mood (for example, I didn't see my partner for a while because I needed to be alone).
   b. Physical state: fatigue, hunger, drunkeness, illness, nervousness (for example, I jump around from guy to guy at a party because I'm drunk and don't know what I'm doing).
4. Actor's Characteristics
   a. Inability: lack of coordination, poor ability, lack of knowledge or judgment, inability to handle consequences (for example, I don't ski because I don't know how to).
   b. Ability: (for example, I enjoy sports because I'm good at them).
   c. Characteristics: forgetfulness, irresponsibility, selfishness, insecurity, irritability, violence, immaturity, laziness, etc. (for example, my partner doesn't like parties because he is very shy).
   d. Habit: nervous habits, unconscious patterns of behavior (for example, I bite my nails because it's just a habit).
5. Actor's Preference/Belief
   a. Preference: personal criteria, likes or dislikes, tastes, priorities (for example, My partner doesn't watch TV because he prefers to read books).
   b. Belief: principles, religious beliefs, beliefs about people and the nature of the world (for example, I tend to be a women's libber because I've always believed in equal rights).
6. Actor's Concern
   a. Concern for partner: concern for partner's well-being, to please partner, to help partner (for example, I point out my partner's faults because I want to help her).
   b. Mutual benefit: to share pleasure, to benefit the relationship (for example, my partner took me to a basketball game because we would both have a great time).
   c. Concern for other people: helping others, to avoid hurting others (for example, I went home instead of staying with partner because I didn't want to leave my roommate alone).

7. Actor's Intention to Influence Partner
   a. To change partner's behavior: to get partner to conform to own tastes or behave more desirably, to get attention or affection from partner, to make partner jealous (for example, my partner complains about my ignoring her to make me go over to see her more often).
   b. To define the relationship: to set the ground rules for the relationship, define roles or dominance (for example, I date other people to make my partner realize that we don't own each other).
8. Actor's Negative Attitude toward the Partner
   a. Lack of concern for partner: being selfish, not caring about partner, thoughtless, lack of respect for partner (for example, my partner doesn't call very often because he doesn't care enough about me).
   b. Negative feelings toward partner: unsure of feelings toward partner, sick of partner, mad at partner, dissatisfied with partner (for example, I don't tell my partner I love her because I don't know if I really do love her).
   c. Insecurity about the relationship: fear of losing partner, uncertain about partner's commitment, jealousy, lack of trust in partner (for example, I was upset when my partner went to Berkeley because I was scared she would see her old boyfriend).
9. Partner is Responsible
   a. Partner's fault: partner's negligence or omissions, partner's instigation, revenge (for example, I left the house because my partner's nagging made me mad).
   b. Partner's influence: Direct: partner wants, asks, insists, expects. Indirect: to avoid partner's anger, to avoid offending partner (for example, she made it clear that I had to pay for our dates if I wanted to go out with her).
   c. Partner's characteristics: partner's state, inability characteristics, preferences and beliefs, habits (for example, I drove the car because my partner didn't feel well enough to drive).
   d. Partner's positive behavior: reciprocity, gratitude (for example, I gave my partner some flowers because she let my cousins stay with us for a week).
10. Activity is Desirable
    a. Activity is enjoyable: actor enjoys it, is entertained by it, makes actor feel good (for example, I go out with my buddies on the weekend because we have a good time).
    b. Activity has good direct consequences: activity is done to beautify self, be informed, develop skill, improve health, relieve tension (for example, I drink alcohol because it helps to loosen me up).
    c. Activity cuts costs: it cuts physical or psychological costs. Alternatives are worse, saves time, saves money, more convenient, easier (for example, I went out with someone else on New Year's Eve because otherwise I would have been at home twiddling my thumbs).
11. Activity is Undesirable (these are reasons for *not* doing something.)
    a. Activity is not enjoyable: not enjoyable, boring, not interesting (for example, I don't play monopoly with my partner because it is a long, dull game).
    b. Activity has bad direct consequences: bad for health, would cause guilt, discomfort, embarrasment, detracts from appearance (for example, I don't let my partner touch my face affectionately because it would make my skin break out).
12. Activity has Desirable Indirect Consequences
    a. Activity has good indirect consequences: projects image, shows off self, gives one social status, displaces hostility or aggression, strains unwanted friendship (for example, she eats health foods to project an organic image).

b. To elicit response from other people: to aggravate, impress, get praise from (for example, I talked to her to make her laugh).

13. Uncodeable

ACKNOWLEDGMENTS

The research reported here was conducted under a National Science Foundation Grant (GS-33069X) to Harold H. Kelley. We are grateful to Michael Passer and Bernard Weiner for their helpful comments on this paper. We are also deeply indebted to Penny Wood Fogel for performing the arduous task of check-coding the 1382 explanations.

REFERENCES

Broverman, I. K., Vogel, S. R., Borverman, D. M. Clarkson, F. E., & Rosenkrantz, P. S. Sex-role stereotypes: A current appraisal. *Journal of Social Issues,* 1972, **28,** 59–78.

de Charms, R. *Personal Causation.* New York: Academic Press, 1968.

Frieze, I. Studies in information processing and the attributional process in achievement-related contexts. Unpublished doctoral dissertation, University of California, Los Angeles, 1973.

Heider, F. *The psychology of interpersonal relations.* New York: John Wiley & Sons, 1958.

Jones, E. E., & Davis, K. E. From acts to dispositions: The attribution process in person perception. In L. Berkowitz, (Ed.), *Advances in experimental social psychology* (Vol. 2). New York: Academic Press, 1965.

Jones, E. E., & Nisbett, R. E. The actor and the observer: Divergent perceptions of the causes of behavior. In E. E. Jones, D. E. Kanouse, H. H. Kelley, R. E. Nisbett, S. Valins, & B. Weiner (Eds), *Attribution: Perceiving the causes of behavior.* Morristown, New Jersey: General Learning Press, 1972.

Jones, E. E., & Thibaut, J. W. Interaction goals as bases of inference in interpersonal perception. In R. Taguiri & L. Petrullo (Eds.), *Person perception and interpersonal behavior.* Stanford, California: Stanford University Press, 1958.

Kanouse, D. E. Language, labeling, and attribution. In E. E. Jones, D. E. Kanouse, H. H. Kelley, R. E. Nisbett, S. Valins, & B. Weiner (Eds.), *Attribution: Perceiving the causes of behavior.* Morristown, New Jersey: General Learning Press, 1972.

Kelley, H. H. Attribution theory in social psychology. In D. Levine (Ed.), *Nebraska Symposium on Motivation.* Lincoln: University of Nebraska Press, 1967.

Kruglanski, A. W. The endogenous–exogenous partition in attribution theory. *Psychological Review,* in press.

McArthur, L. A. The how and what of why: Some determinants and consequences of causal attribution. *Journal of Personality and Social Psychology,* 1972, **22,** 171–193.

Rosenbaum, R. M. A dimensional analysis of the perceived causes of success and failure. Unpublished doctoral dissertation, University of California, Los Angeles, 1972.

Lyman, S. M., & Scott, M. B. *A sociology of the absurd.* New York: Appleton–Century–Crofts, 1970.

Shaver, K. G. *An introduction to attribution processes.* Cambridge, Massachusetts: Winthrop, 1975.

Siegel, S. *Nonparametric statistics for the behavioral sciences.* New York: McGraw–Hill, 1956.

Smith, M. B. Allport, Murray, and Lewin on personality theory: Notes on a confrontation. *Journal of the History of the Behavioral Sciences,* 1971, **7,** 353–362.

Trope, Y. Inferential processes in the forced compliance situation: A Bayesian analysis. *Journal of Experimental Social Psychology,* 1974, **10,** 1–16.

Weiner, B., Frieze, I., Kukla, A., Reed, L., Rest, S., & Rosenbaum, R. M. Perceiving the causes of success and failure. In E. E. Jones, D. E. Kanouse, H. H. Kelley, R. E. Nisbett, S. Valins, & B. Weiner (Eds.), *Attribution: Perceiving the causes of behavior.* Morristown, New Jersey: General Learning Press, 1972.

Part IV

THEORETICAL INTEGRATIONS

The two chapters that appear in this final section are concerned with theoretical analysis and integration. In the first, "Correspondent Inferences and the Attribution Cube: A Comparative Reappraisal," Edward Jones and Daniel McGillis present a revision of correspondent inference theory which relates the theory more closely to Kelley's attributional formulations. They substitute the concept of "expected valence" for the concept of "assumed desirability"—a concept which was of central importance in the original statement of correspondent inference theory. The authors suggest that with this substitution the theory can deal with more extended behavior sequences. In this new conception, a correspondent inference constitutes a shift in expected valences so that a target person is seen as desiring certain consequences more or less than before the observed behavior.

In "Attribution Theory and Judgment under Uncertainty," Baruch Fischhoff compares and contrasts the views of a person's inferential ability which emerge from attribution theory and judgment and decision theory in cognitive psychology. While the former has been concerned mainly with how people explain events, the latter has been concerned principally with how people make predictive inferences about unknown events. Fischhoff points out that rather disparate characterizations of individuals as information processors often have evolved from these theoretical approaches. He proposes means by which meaningful convergence can be brought about between these two fundamentally complementary lines of research. Fischhoff suggests that we ought to take a more careful look at the notion of people as intuitive scientists and proposes that a more integrated attribution and judgment approach may prove to be useful in understanding both explanatory and predictive behavior.

17

Correspondent Inferences and the Attribution Cube: A Comparative Reappraisal

Edward E. Jones
Duke University

Daniel McGillis
Harvard University

A little more than 30 years ago, Heider (1944) planted the seed of an attribution theory in his classic paper on phenomenal causality. This contribution was admired as an extension of Gestalt principles to social perception, but with the notable exception of an experiment by Thibaut and Riecken (1955), the idea of using naive causal analysis as a basis for theoretical prediction bore little empirical fruit. Heider's 1958 *Psychology of Interpersonal Relations* transplanted the attribution seed into a richer soil by mixing Gestalt notions with cognitive functionalism, and the result was an accelerated interest in the processes and consequences of causal attributions for social behavior.

Influenced both by Heider and the empirical tradition of person perception research, Jones and Davis introduced correspondent inference theory in 1965. Their particular interest was in identifying the variables governing inferences that an act reflects an internal, personal disposition. Shortly thereafter, Kelley (1967) attempted to formulate another variant of attribution theory that would make more immediate contact with such central areas of social psychology as attitude change and social comparison. Not only did Kelley make effective use of Heider's and Festinger's (1954) ideas, but he explicitly extended the concern of attribution theory to self-attribution. In the process, Kelley showed the relevance and compatibility of Bem's "radical behaviorism" (1965, 1972) for the analysis of how one knows his own states and dispositions.

After allowing several years for the digestion of these theoretical proposals, Jones and Kelley joined several other colleagues (Kanouse, Nisbett, Valins, & Weiner) in coauthoring *Attribution: Perceiving the Causes of Behavior.* A highlight of this volume was the ubiquitous acknowledgement of the contributions of Schachter and his colleagues (Davison & Valins, 1969; Nisbett & Schachter, 1966; Schachter & Singer, 1962) to an attributional understanding of emotional experience. The book also illustrated the flexibility of attributional notions in their application to domains as diverse as language usage, reactions to success and failure, psychopathology, interpersonal trust, and psychotherapy.

The *Attribution* book was both a reflection and a stimulant of varied research efforts. Nelson and Hendrick's (1974) recent bibliographic survey shows that attribution oriented studies formed by far the largest category in the social psychology research literature of 1974. The growth of attribution research seems far from peaking and further from decline. A generation of young social psychologists is apparently comfortable with attributional concepts and their use in problem formulation and analysis. But the widespread usage of attributional language and style of approach is one thing, the growth of systematic deductive theory is another. In spite of the growing quantity of attribution research in the last decade, the growth of attribution *theory* has been somewhat stunted. With the thought that it might be an instructive exercise in the direction of theoretical development, we shall conduct a tenth anniversary reappraisal of correspondent inference theory and selective data of relevance. We shall then attempt to relate the theory of correspondent inference to Kelley's attributional formulation in the ultimate interest of solidifying the theoretical structure that will organize existing data and underpin future research.

CORRESPONDENT INFERENCE THEORY REVISITED

The main problem that Jones and Davis wanted to attack was the perception of another's personal attributes. When we see someone acting in a particular setting, how do we decide whether to attribute the cause of the action to the person or the setting? The point of departure was the recognition that people who are not constrained physically or socially will attempt to achieve desirable consequences by their behavior. Therefore, when the actor has behavioral freedom, we should be able to infer his intentions from the consequences or effects of his behavior. This should follow, at least, if the actor is believed to have the knowledge that the particular act will produce the consequences observed. Attributions of probable knowledge are facilitated by any evidence that the actor has the ability to achieve the consequences observed when he desires. After all, if a person cannot reliably achieve a set of effects, an observer is less likely to assume he deliberately intended to achieve them. The accurate attribution of intentions from behavioral effects, then, assumes knowledge, ability, and behavioral freedom. But the problem of attribution is immediately complicated by the fact that a given act invariably has multiple consequences. Not only that, but freedom to choose means there are options chosen and options foregone. Both the chosen and the rejected alternatives can, and most often do, have multiple effects. Some of the chosen and foregone effects may be the same; they are *common effects.* Any given action is uninformative with regard to such common effects since they do not provide a discriminating reason for the choice. Thus the only information about intentions is that contained in the *noncommon effects* of action. Something in the combination of noncommon effects chosen and foregone has guided the actor's behavior in the observed direction. As a first approximation, we might say that a dispositional inference is correspondent to the extent that an act and the disposition are similarly described by the inference (for example, "his domineering behavior reflects an underlying trait of dominance").

But correspondent inference theory is not just concerned with the attribution of intentions or other dispositions; it is concerned with information *gain* as a function of behavioral observation. In Bayesian terms (cf. Ajzen, 1971; Ajzen & Fishbein, 1975), correspondence reflects changes in the subjective probability of inferring a disposition given observed behavior. The perceiver may begin either with an expectancy based on normative considerations or one based on prior behavior of the same actor. Thus he may ask, what makes this actor differ from other actors? Or he may ask, what do I know about the actor now that I did not know about him before. A correspondent inference may be formally defined in the following terms: *Given an attribute–effect linkage which is offered to explain why an act occurred, correspondence refers to the degree of information gained regarding the probability or strength of the attribute.* Jones and Davis (1965) originally stated that "correspondence increases as the judged value of the attribute departs from the judge's conception of the average person's standing on that attribute [p. 234]" (for example, "he is more dominant than most people"). To this earlier formulation we would now like to add, "or the judge's prior conception of where the actor stood on that attribute" (for example, "he is more dominant than I thought he was").

When correspondence is defined as information gain, it is obvious that we need to consider more than just whether behavioral effects are noncommon. We also have to consider whether and to what extent they fit the judge's prior expectations concerning people in general or this particular actor. This determinant was originally described by Jones and Davis in terms of effects assumed to be desired in the culture. But the more generic case concerns the prior probability that the particular actor would desire a particular effect. This might be based on assumptions about cultural desirability, or it might be based on prior knowledge about the actor. Whichever may be the case, it seems useful to conceive of a valence attached to each effect having a possible range from −1 to +1. A maximally undesirable effect would have a valence of −1, indicating the highest probability that the individual concerned would seek to avoid that effect. A +1 valence, on the other hand, would signify the highest probability of effect desirability. From the point of view of information gain, pursuit of a low valence effect should lead to greater correspondence of inference, other things equal, than pursuit of a high valence effect.

Considering together the uniqueness and valence of the effects pursued, it now should be clear that correspondence of inference is greatest when the number of noncommon effects is low, and their valence is low or negative. The basic predictions of correspondent inference theory can be summarized, if oversimplified, in terms of the 2 X 2 design presented in Table 1. Correspondence is seen here as the joint inverse function of effect commonality and effect valence, as long as we keep in mind that the latter refers to the expected or assumed valence for the actor prior to his act. This relationship can be more simply described as a near tautology: the more distinctive reasons a person has for an action and the less surprising or unexpected these reasons are, the less informative the action is concerning the identifying attributes of the person. The table emphasizes that highly expected actions are trivial from an informational point of view. Also, inference ambiguity obviously increases when the number of potential determining effects increases.

With the few exceptions noted, this presentation is but a cryptic summary of the exposition presented at greater length in Jones and Davis (1965). The reader who is being exposed to these ideas for the first time may wish to retreat to that source. Jones and Davis (1965) also considered the operation of motivational sources as affecting correspondence of inference. There may be consequences for the perceiver as well as for the actor. Effects of action may be *hedonically relevant* in that the perceiver benefits from the action he has just observed or is disadvantaged by it. This could happen in the absence of any intention of the actor to help or hurt the perceiver. But the actor could also be *personalistic* in his provision of relevant effects, clearly intending benefit or harm for the perceiver. In spite of the potential importance of motivational effects on attribution, the present chapter will focus on the purely cognitive aspects of information processing. Readers interested in the effects of hedonic relevance

TABLE 1
Expected Valence and Effect Commonality as Determinants of Correspondence

Number of noncommon effects	Expected valence	
	High	Low
High	Trivial ambiguity	Intriguing ambiguity
Low	Trivial clarity	High correspondence

and personalism are referred to Jones and de Charms (1957) and Chaikin and Cooper (1973).

Sources of Expected Effect Valence

We would like to extend the theory from a conceptualization of initial contact with a stranger's behavior to a more general statement concerning information gain in person perception. One criticism of the earlier Jones and Davis formulation is that the analysis was concerned with the single behavioral episode, and there was no provision for inferences following more extended experience with a given target person. Indeed, their paper might better have been subtitled "From Acts to Intentions" than "From Acts to Dispositions." A disposition is inferred when an intention or related intentions persist or keep reappearing in different contexts.

The shift from assumed cultural desirability to assumed or expected valence makes it possible to treat correspondence of inference, or information gained, at any stage in a relationship. But is this extension legitimate, and can we treat prior probability inferences about the actor himself in the same way as inferences based on normative considerations? In order to explore the applicability of the theory to these two different cases, we introduce the distinction between category-based expectancies and target-based expectancies.

A category-based expectancy derives from the perceiver's knowledge that the target person is a member of a particular class, category, or reference group. In the case of extreme prior information poverty, the perceiver at least knows that the actor is a human being of a particular sex and approximate age. From this knowledge the perceiver doubtless can generate at least crude expectancies about probable attribute-effect linkages. Usually, of course, the perceiver has much more information to guide his expectancies, information that locates the actor in a sociocultural network. This might be information about occupation, group affiliation, country or state of origin, or social class and educational level. Whatever the particular pattern, the actor may be typically located as a member of overlapping categories, each contributing to a refinement of expectancies against which the actual behavior of the person will be judged. Feldman (1972) and Feldman and Hilterman (1975) have shown that such category combinations as race and occupation interact to produce different stereotypic impressions.

These category-based expectancies tend to be probabilistic: "since he is a Russian the chances are he favors state ownership," "women are apt to be more submissive than men," "he has narrow lapels, chances are 60-40 that he is a Republican." As these examples show, category expectancies are obviously held with varying degrees of probability. At one extrene, we are dealing with stereotypes where the probabilistic nature of the expectancy is simply ignored, and the individual actor is endowed with a characteristic believed to be modal in whatever "category" he represents. With extreme stereotyping, behavioral data may be entirely ignored or severely distorted to fit the expectancy (cf. Secord, Bevan, & Katz, 1956). At the other extreme, the perceiver may have the vaguest kind of expectancy which is quickly discarded in the face of behavioral data and assigned no weight whatsoever in the inference process. Brigham (1971) and Mann (1967) have shown that ethnic group stereotypes may or may not predict attributions toward individual ethnic group members on specific stereotyped traits, a reflection of this variation in the probabilistic nature of category-based expectancies.

Target-based expectancies are derived from prior information about the specific individual actor. The perceiver's task is to extrapolate from one set of judged dispositional attributes (traits, motives, attitudes) to a set of attributes relevant to the behavior observed. This extrapolation most likely follows the paths of the perceiver's "implicit personality theory." Target-based expectancies are formed with particular readiness in the attitude realm. From a single statement of support for George Wallace, we may readily infer a whole set of expected attitudes. Someone who opposes abortion is probably also in favor of restrictive laws governing massage parlors. Pro-busing advocates are more likely than those who oppose busing to favor the legalization of marijuana. The network that generates target-based expectancies is undoubtedly in part a product of the perceiver's experiences with dispositional contingencies in others and partly a product of logical implication.

Target-based expectancies are conceived of as probabilistic in the same way that category-based expectancies are. They undoubtedly range in strength from those which call for a direct generalization from prior behavior to those which must filter through an elaborate perceived dispositional structure.

Do these different kinds of expectancies operate differently in their influence on the attribution process? It first must be noted that category-based expectancies shade into target-based expectancies when the category is a voluntary membership or reference group. To know that the actor belongs to the ACLU is to have prior knowledge of his behavior. This should, then, really be classed as target-based information. But disregarding this kind of overlap, what evidence is there that this is a distinction of consequence?

The most well-developed line of research on correspondent inference theory is a series of studies dealing with attitude attribution. The typical paradigm presents a statement, an essay, or a speech authored by a target person under conditions of high or low environmental constraint. Crosscutting the constraint or choice variable is a variation in whether the statement reflects an expected or unexpected position. The subject's task is to estimate the true attitude of the target person. In terms of the theory, it is important to note that the constraint variation is a manipulation of noncommon effects; there are more reasons for making the statement in the no-choice than in the choice condition, since the no-choice case includes all the reasons of the choice case plus those associated with obedience to authority, avoidance of embarrassment, not appearing to misunderstand the instructions, and so on.

The cross-cutting variation of statement direction is a manipulation of expected valence. In the three experiments by Jones and Harris (1967), the expectancies were category based. For example, in one experiment a college student target person gave a pro-integration or pro-segregation speech. In some cases the target person was from the rural South and in other cases from the industrial North. Jones and Harris reasoned that a pro-segregation speech would be generally unexpected from a contemporary undergraduate, and even more unexpected from a northerner than from a southerner.

In a study by Jones, Worchel, Goethals, and Grumet (1971), on the other hand, expectancy was more directly manipulated by varying the related attitudes of the target person; the expectancy was thus target based. A pro- or anti-marijuana essay was attributed to a target person who had already endorsed a cluster of opinions concerning potential interference with an individual's rights to live his own life and control his own destiny. It was assumed, and the results support the assumption, that someone favoring personal freedom and autonomy would most likely favor the legalization of marijuana.

Regardless of whether the manipulated expectancy was category or target based, the results were basically the same in each of the various attitude attribution experiments. Correspondence was higher in the choice/unexpected than in the no-choice/unexpected conditions. That is, the attributed attitude was more in line with the behavior when noncommon effects were low and the expected valence of the position taken was low. In addition, in each of the attribution of attitude studies cited (and in several others described in Snyder & Jones, 1974), there was a main effect of speech direction even in the no-choice condition. Even the highly constrained target person was attributed an attitude in line with his statement, though much less strongly than his choice counterpart.

In a recent experiment, Jones and Berglas (in press) sought to compare directly subjects' responses to category-based versus target-based expectancies under different orders of receiving information. In all cases, the target person made a statement directly supporting a particular attitudinal position. The target person was also identified by either a category-based or a target-based expectancy. This expectancy was either strong or weak, as determined by pretesting with a similar subject population. The expectancy was either confirmed or disconfirmed by the statement. Examples of the stimulus materials are as follows:

1. Category-based expectancies (strong)
   a. Confirming: Actor is a female judge
   b. Disconforming: Actor is a North Carolina deputy sheriff

2. Target-based expectancies (strong)
   a. Confirming: Actor argues that homosexuals should not be discriminated against by employers
   b. Disconfirming: Actor argues that anything including violence must be used to prevent busing

Finally, there was an order variation. Half of the subjects were exposed to the expectancy first followed by the statement and then asked to attribute the target person's true attitude. The remaining half were given the statement first followed by the expectancy information.

The results showed that a strong order effect appeared. When the expectancy preceded the statement, it was essentially ignored and played no role in attitude attribution. When the expectancy followed the statement, however, the attribution was much less correspondent with the statement. What is of interest in the present context is the fact that this recency effect was almost identically strong with either category-based or target-based expectancies, and the strength of the expectancies did not play a significant role either.

The search for a contrast effect. The Jones and Berglas experiment was one of a number of studies searching for a contrast effect in attitude attribution. A contrast effect is defined by comparing an expectancy confirmation case to an expectancy disconfirmation case. There is contrast when the disconfirming behavior leads to a more extreme inference than the corfirming behavior. Let us suppose that we hear two people make similar statements favoring the unionization of municipal employees. The first person is a self-made financier, the second a postal clerk. We would speak of contrast if the financier were judged to be more liberal, more pro-union than the postal clerk. His behavior is contrasted with the prior expectancy, and the resulting inference moves further toward the liberal end of the scale than an inference based on the same behavior in the confirmation case.

It should be noted that correspondent inference theory would not necessarily make this strong a prediction. Although the theory does predict greater correspondence in the disconfirming case, this refers to information gain rather than the final extremity of the scale rating. Thus, relative to an initial rating of the financier on liberalism, the rating after his pro-union statement should show greater movement toward the liberal extreme than the postal clerk's rating after the same statement. The possibilities, including assimilation, the opposite of contrast, are spelled out in Fig. 1. Correspondence is measured by the changes calculable from the differences between Items 3 and 4 and between Items 2 and 5. In the contrast case, correspondence is obviously greater (+7) than the assimilation case (+3). But even the assimilation case is more correspondent than the confirmation case (+2).

As a sociopsychological phenomenon, a plausible case can be made for contrast. First, there is the simple possibility akin to psychophysical judgmental

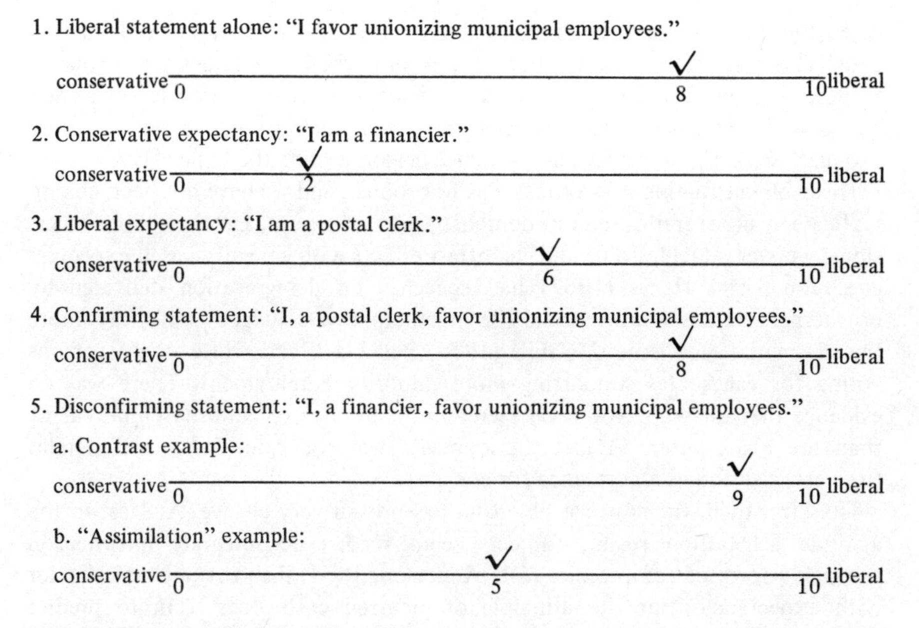

FIG. 1. Ratings of liberalism after different information about category-based expectancy and expressed opinions toward the unionization of municipal employees.

contrast: the disconfirming behavior is so perceptually salient because of its surprise value that its distance from the expectancy is exaggerated (cf. Sherif & Hovland, 1961). In addition, depending on the nature of the expectancy, various costs may be involved in taking a stand that is unexpected or unusual. The financier who favors unionization may suffer ridicule or disfavor within the paneled walls of his club. The white busing advocate from a small town in Alabama most likely has had to pay often and heavily for his maverick views. From the costs presumably incurred, one might infer greater intensity of belief for the southerner than for the young northerner who favors busing on general liberal ideological grounds. When expectancies are derived from reference group norms—and especially when these norms are supported by group sanctions—contrast effects might be more likely.

Following this reasoning, our initial hunch was that contrast is more likely after disconfirmation of category-based expectancies than of target-based expectancies. One is more likely to find a maverick in a group than to find a person with highly inconsistent attitudes or traits. However, there is at present absolutely no evidence to support this hunch. In fact, the only attitude attribution study that obtained anything like contrast involved *target*-based expectancies. Jones et al. (1971) found some marginal evidence that target persons writing essays on marijuana were judged to have attitudes more in line with their essays when the essays disconfirmed target-based expectancies than in the confirmation case. This only happened in those conditions where the target person had a choice to write an essay on either side. Thus an anti-autonomy target person choosing to write an essay favoring marijuana legalization was seen as more pro-marijuana than a pro-autonomy target person writing the same essay.

This isolated finding of contrast was not robust, and we have not been able to replicate it in several classroom demonstration followups. On three other occasions, we were unable to produce contrast effects with category-based expectancies. Jones and Harris (1967) had speeches on desegregation delivered by northerners versus southerners. The northerner was judged uniformly more liberal on the race issue. McGillis (1974) had black and white target persons voting for candidates supporting more minority employment. There was no evidence that the white voter was seen as more in favor of minority employment than the black voter. In fact the opposite was the case. Jones and Berglas (in press) also found no evidence for contrast.

Thus far, then, the contrast phantom has proved very elusive. At least in the attitude attribution realm, subjects seem readier to integrate information through a form of compromise (à la Anderson, 1974) than to contrast behavior with expectancy. But the principle of incurred costs does seem to predict contrast in other realms. In a persuasion study by Mills and Jellison (1967), the communicator who violated expectancy by delivering a speech against the interests of his audience was seen as more honest and sincere. He was also significantly more persuasive. Feldman (1972) found that black professionals were judged to be more "professional" than white professionals. This is consistent with the assumption that blacks have to incur more costs to achieve professional status. We are not yet in a position to identify the precise conditions favoring contrast, but it looks as though obvious perceived cost is an important ingredient.

It should be emphasized again that neither contrast nor category-target expectancy differences are basic correspondent inference theory predictions. In fact, it is comforting in a way that there appear to be no systematic differences between the effects of disconfirming category-based versus target-based expectancies. For the moment we can assume that correspondent inference theory holds regardless of the expectancy source. It is, however, premature to abandon the search for expectancy source differences. We hope others will join us in the search for such effects since the consequences of expectancy disconfirmation represent such a crucial problem in the study of person perception and stereotyping.

Some Problems with Correspondent Inference Theory

Boundary conditions for expectancy effects. The theory tells us that behavior which disconfirms prior expectancy is more likely to lead to a correspondent inference than is confirming behavior. It must be remembered, however, that inferences from effects back to intentions assume both knowledge and ability. There are undoubtedly cases where the expectancy is so strong and the disconfirmations so drastic that it is easier for the perceiver to question the knowledge assumption than for him to engage in cognitive gymnastics to achieve a correspondent inference. We therefore propose that as the discrepancy between expectancy and behavior grows, the perceiver will be increasingly skeptical concerning the knowledge assumption. And we would expect this skepticism to be positively accelerating, growing very slowly under low discrepancies and very rapidly as the discrepancy increases further. This skepticism factor is reminiscent of Osgood and Tannenbaum's (1955) credibility correction. They reasoned that a statement departing too far from expectation about the source will appear incredible and be discounted. A clear example of the skepticism factor was found by McGillis (1974) in his study of voter choice and attitude attribution. The procedure was to present a voter choosing between two candidates for the governorship of an unspecified state. The candidates in turn were alleged to favor passage of certain specified bills. The design varied both the number of noncommon effects and their prior probability. The two candidates either supported three identical bills and a different fourth (low noncommon effects) or one identical bill and three that were different (high noncommon effects). The voter was presented as black or white, and it was always the case that Candidate A favored a preventive detention bill whereas Candidate B favored a bill requiring preferential hiring of minorities by state agencies. This was the manipulation of expectancy or prior probability. It was assumed that subjects would expect the white voter to prefer A and the black voter B. This assumption was reinforced for the subject by some bogus poll data provided to him. A final variable was the inclusion or omission of further information concerning the degree of the voter's commitment to the chosen candidate. In the high-commitment treatment, it was alleged that the voter organized publicity drives and spoke at public rallies on behalf of his candidate. In the low-commitment treatment this information was omitted.

The major predictions were derived directly from correspondent inference theory. Attitude ratings on the issue relevant to voter choice were expected to be most extreme (most correspondent) in the conditions of low noncommon effects and low prior probability. For example, the white voter who chose the candidate favoring preferential minority hiring, with three other bills supported in common by the two candidates, was predicted to be, and was, seen as very pro-minorities. Also in line with prediction, commitment intensified this tendency.

In spite of the general support for the theory, certain reverse tendencies were discovered in the black voter conditions. Specifically, in the low noncommon effects condition, neutral attributions were made when the black voter supported the candidate favoring preventive detention and rejected the candidate favoring preferential hiring for minorities. But this only happened in the low-commitment condition. We suggest that this is probably an excellent example of the skepticism factor in operation. One way to handle the extreme disconfirmation of a black voting against his own interests is to assume that he was unaware of some of the positions endorsed by his chosen candidate. This would be especially likely in the low noncommon effects condition where there are no other differences between the issues supported by the candidates. However, this skepticism about voter awareness can hardly be maintained in the commitment condition, where it seems inconceivable that a voter would get deeply involved in campaigning for a candidate without knowing his position on issues of critical importance.

Another boundary condition concerns *differential* knowledge of the target person and his environment. If a perceiver has firm prior knowledge about the target person and the latter behaves in a highly unexpected way, the perceiver may attribute his behavior to the situation rather than change his conception of the person. If expectations about the situation are firmer than those about the actor, there should be a change in person attribution. Bell, Wicklund, Manko, and Larkin (in press) confirmed this hypothesis in a pair of experiments. In one, for example, they found that negative behavior from a positively regarded actor (an expectancy created by the experimenter) was attributed more to the situation than to the person. Bell et al. proposed that, in general, attributions will flow toward that aspect of the total person–environment complex about which the least is known. This may be true because it is harder to change a firm expectancy than one which is very vague and tentative.

Prior choice and effect magnitude. Jones and Davis were aware of an apparent absurdity that followed from a literal application of correspondent inference theory. There are occasions when it seems intuitively that much more information is gained when noncommon effects are numerous than when they are scarce. The example they invented pictured Miss Adams confronting the choice between medical school and law school at comparable universities and Miss Bagby trying to decide whether to enroll as a psychology graduate student at Duke or Colorado. Most people have the subjective experience of learning more from Adams' choice than from Bagby's. And yet, there are many more noncommon effects separating the law–medicine choice. As Jones and Davis (1965) point out, however, "the hidden factor in this comparison is . . . that considerably more information is contained in the datum that Bagby has already ruled out everything but psychology, than in the datum that Adams is still struggling with the choice between two basic professions, professions which differ from each other on many different dimensions [pp. 233–234]." In a sense, then, the effects common to the two psychology departments were noncommon effects when one backtracks to the decision to pursue a career in psychology.

This kind of consideration creates substantial problems at times when we try to apply correspondent inference theory to the analysis of attribution in the complex real world. Here choice is piled upon choice and presumably effects can shift from common to noncommon with a simple shift in temporal perspective.

Another way of approaching the Adams–Bagby comparison is to argue that some effects are more significant or important than others. A career choice is more important than a choice between two psychology departments where the choice is based, say, on a slight preference for mountain air. We view this as a defeatist alternative, however, because the problems of measuring effect importance are staggering. Hopefully, a retreat to this position can be avoided by equating importance to the number of positively valenced effects. In most cases effect magnitude can be seen to imply a quantity of discrete effects. Life versus death, for example, is a momentous decision for the potential suicide candidate because of the almost infinite number of effects of being alive.

Tying these considerations together, we might conclude that early choices in a behavior sequence are more likely than later choices to involve the exclusion of large numbers of noncommon effects. This may or may not be actuarially true. We can certainly think of cases where trivial or preliminary choices (sharpening the pencil) precede more important subsequent choices (choosing a topic to write about). Whether the choice is early or late in a behavior sequence, however, a more "important" choice implies that one learns less about more whereas with the "less important" choice he learns more about less. Thus, there seems to be a trade-off between the number of intentions about which one might make tentative inferences and the precision or confidence with which one can make them. This is inherent in the basic theoretical argument about noncommon effects. But we suspect that this is a point where the logic of the theory may be at odds with the cognitive-processing tendencies of the perceiver. At least it is an interesting question whether social perceivers overuse and are overaffected by ambiguous information. Thus, perceivers may feel better informed about the total personality of a target person whose behavioral choice involves many effects and yet not be able to rate him as confidently on any particular dimension. If a target person does something that has five effects in order to achieve one of them, the perceiver may conclude that he was disposed to achieve all five. Perceivers, in other words, may think they have learned more about a person from his choices than is warranted by the number of possible permutations and combinations that are involved in the chosen and rejected alternatives. They may have a low threshold for making solid inferences from behavioral data that are worthy only of the most tentative hypotheses.

This consideration is related to Jones and Nisbett's (1971) proposition that observers tend to attribute dispositions in circumstances where actors attribute their behavior to the situation. Jones and Nisbett argued that this was true both because actors typically know much more about their past behavior in similar situations and because actors and observers take a different perspective toward the information that is available. Now we are suggesting a further, more subtle reason for actor–observer differences. The actor is in a much better position than the observer to choose among effects achieved in explaining his intentions. In particular, if there are high-valence effects, the actor will be content to consider these as a sufficient explanation for his behavior. High-valence effects are usually effects which anyone would seek, that is, effects called for by the situation. The observer, on the other hand, may be more equally drawn to all effects and be motivated to include for consideration those that have low valence in the general population. Why might this be so? Because, whereas the actor is typically interested in seeing himself as rationally responsive to his changing environment, the observer has a special interest in discovering those dispositions that uniquely identify the actor. The actor's interest is in normalization; the observer's interest is in individualization.

We can summarize these speculations by suggesting that observers attach too much significance to each possible effect and become prematurely locked into inferential conclusions. If we add to this a further tendency for the observer to be drawn toward low-valence effects, we have one set of conditions that supports the actor–observer difference proposition. Whether or not one wishes to connect the present speculations with the earlier stated propositions, this is an intriguing area for further study because it suggests that observers are over-individuating in their inferences and overconfident about inferential validity. The subjective feeling that one has gained more information about a person in a high-effect than a low-effect situation may, in other words, be an illusion and support biased inferences. The high-effects case is analogous to an experiment in which the experimental and control groups differ from each other on many grounds. The trained experimenter is careful not to draw premature conclusions about the reason for any observed differences.

Effects chosen and effects foregone. Jones and Davis do not go into detail concerning the cognitive operations involved in identifying the common and noncommon effects of action. There is no question that the identification of effects is an arbitrary enterprise dependent on momentary perspective (see above). But at another stage in the operations specified by the theory, the perceiver must combine information on noncommon effects chosen and non-common effects foregone. Figure 2 presents an example of a target person choosing Act *X* over Act *Y*. The effects of each act and their attached valences are listed in the "choice circles." The target person's choice of *X* mildly disconfirms the perceiver's expectancy (indicated by the total valence). In general, then, this episode has the potential for a correspondent inference. But now the trouble begins. Effect b is common and can be ignored. This leaves not only the effects e and f of the choice alternatives; the foregone effects a, c and d must also be considered as reasons for the choice. Considering only effects f and c, the perceiver does not know whether the target person was trying to attain f, avoid c, or both. This, in concrete form, is the question of this section. Other

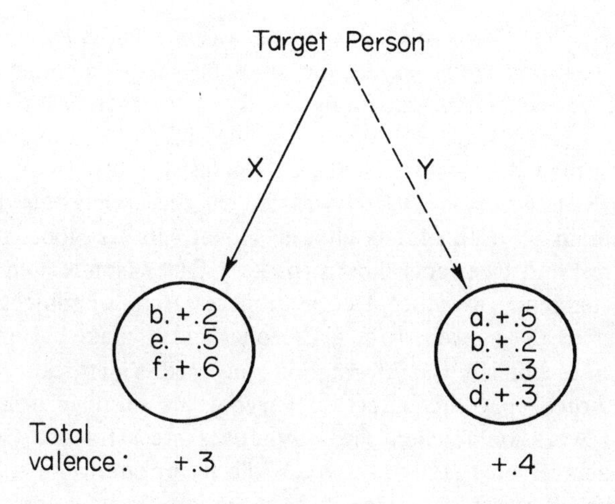

FIG. 2. A hypothetical choice of Act *X* over *Y*.

things being equal, do perceivers attach as much weight to foregone as to chosen alternatives?

Newtson (1974) has addressed a very similar question in an elegant experiment on the number of noncommon effects remaining and eliminated by a choice. Based on careful pretesting, Newtson selected three behavior alternatives, each having either one or two specified effects. Two target persons (Alex and Bob) were said to have chosen Alternative 3, "babysit for a professor." One effect of choosing this alternative was always listed as "in order to ingratiate himself with the professor." In the design, some subjects were given this as the only effect "remaining" after the choice. Other subjects were given this effect (ingratiation) plus "in order to get some extra studying done for all his courses." In the cases where Alex was said to have chosen to babysit for two reasons and Bob for one, Bob was seen as more ingratiating, in line with correspondent inference theory and the role of noncommon effect number.

In addition to varying the number of effects remaining, Newtson varied the number of effects eliminated by the choice to babysit. The alternatives were to go to the beach or to work in the library filling in for a friend. Each of these alternatives had one or two specified effects. In the case where Alex's choice to babysit eliminated four effects (two for each alternative) and Bob's choice eliminated only two, Alex was judged more ingratiating. Thus, the fewer the effects remaining and the more the effects foregone, the more correspondent the inference will be. The role of eliminating the effects follows directly from an information-theory analysis. It also follows from an "incurred cost" analysis that the more things Alex is willing to give up in order to babysit, the more ingratiating he must be.

The crucial part in Newtson's experiment was a clear-cut comparison of correspondence resulting from varying the number of effects eliminated versus varying the number of effects remaining. In the case where Alex's choice to babysit eliminates two effects with one remaining (E2, R1), he is judged to be more ingratiating than Bob, whose choice eliminates four effects but leaves two remaining (E4, R2). Newtson (1974) was led to the conclusion that "while persons can respond to both effects chosen and effects foregone, they apparently prefer to respond to effects chosen [p. 495]." Newtson related this to the well-established tendency in concept formation research for subjects to prefer positive instances of the concept to negative ones and to make better use of the former even when technically the information conveyed is identical.

It should be noted that this effect of varying the number of alternatives foregone should work only when the alternatives are attractive (better than nothing, with valences greater than zero). While it appears to be true that the more positive the alternatives foregone, the more significant is the meaning of the choice, it is also true that the more negative the alternatives foregone the more ambiguous is the meaning of the choice. In Steiner's (1970) term, perceived decision freedom is low to the extent that foregone alternatives are negatively valenced. When this negative valence clearly outweighs the positive valence of the chosen alternative, the behavior becomes "obvious," and there is no information gained.

What Kind of Theory?

Correspondent inference theory is essentially a rational baseline model. It does not summarize phenomenal experience; it presents a logical calculus in terms of which accurate inferences could be drawn by an alert perceiver weighing knowledge, ability, noncommon effects, and prior probability. But the role of the theory has been as much to identify attributional bias as to predict precisely the course of the social inference process. In a similar vein, Edwards (1968) has stated regarding decision theories that "they specify what an ideal decision-maker would do and thus invite comparison between performances of ideal and real decision-makers [p. 34]." The theory cannot be invalidated by experimental results any more than game theory can be invalidated by the choices of players in a prisoner's dilemma game. Of course, it may turn out that the theory is not very useful in stimulating research, and it might, of course, prove vulnerable to logical criticism as well.

Use of the theory as a rational baseline model can best be seen, perhaps, in the attitude attribution research outlined above. The main results of this line of research are consistent with correspondent inference theory in that attitudes in line with behavior are most strongly inferred when the actor has choice (low noncommon effects) and the behavior is low in prior probability (expected effect valence). But we have also noted that behavior under no-choice conditions leads to significant correspondence as well. A repeated finding has been that actors writing, for example, "pro" essays under no-choice instructions are seen as more pro than actors writing no choice "anti" essays, regardless of the topic involved and, within wide limits, regardless of the strength or persuasiveness of the essay.

Such a finding becomes interesting only after obvious sources of artifact are systematically ruled out. One might argue, for example, that the actors really have some choice in no-choice condition. They are not, after all, acting at gunpoint. They are going along with their instructor's examination question, the debating coach's assignment, the experimenter's request, and so on. Snyder and Jones (1974) showed, however, that no one in fact refuses these simple requests to compose a persuasive speech or essay even when the composition runs directly counter to his own attitude; and when the statements are exchanged so that actors become perceivers, they still assign attitudes in line with the constrained behavior. What this means, of course, is that perceivers are very inaccurate in making dispositional inferences about actors who are highly constrained. Instead of predicting an average or modal attitude, they are influenced by the direction of the opinion expressed. Correspondent inference theory provides a rational baseline in terms of which this inaccuracy can be assessed.

Similarly, the Newtson (1974) study presented above showed that perceivers make greater use of effects chosen than effects foregone. Jones and Davis were noncommital about the role of foregone effects in the inference process, but there seems to be no purely logical reason why chosen and foregone effects should not be equally weighted. The fact that empirically they are not becomes psychologically interesting.

As a final example, Snyder (1974) studied the attribution of anxious disposition (what clinicians sometimes refer to as trait anxiety) by having subjects observe a target person who allegedly was or was not going to be shocked in a few moments. Both correspondent inference theory and Kelley's (1971a) discounting principle would argue that more state anxiety would be perceived in the shock situation, but because the anxiety can be "explained" by the anticipated shock, trait anxiety would not be inferred. In fact, given a standard facial expression captured on videotape (as was the case in Snyder's study), the attribution of trait anxiety should be greater when no shock is involved. Snyder reasoned, on the other hand, that subjects would use cues about the shock setting to infer momentary anxiety in the target person and, when asked about anxious disposition, they would erroneously use the anxiety state inference to assign an anxiety trait. He found this to be true with his female subjects and, in a more recent replication (personal communication), the hypothesis was confirmed for males as well. Here is an instance, then, where correspondent inference theory was used as a kind of logical straw man to show how people make inappropriate use of situational data in the attribution of emotional dispositions.

In short, a theory can serve a variety of purposes. Correspondent inference theory has been useful both as a source of deductions that have been confirmed and as a model against which to identify and study attributional biases.

KELLEY'S ANOVA CUBE: A BRIEF SUMMARY

Kelley's (1967) theory has a different goal than that of correspondent inference theory. Kelley has stated the nature of this difference as follows:

> The observer's focus in the two [theories] is essentially at opposite ends of the person–environment polarity. In my earlier analyses . . . the person is concerned about the validity of an attribution regarding the environment. He applies the several criteria in an attempt to rule out person-based sources of "error" variance. In the problems specified by Jones and Davis the observer has exactly the opposite orientation. He is seeking for person-caused variance (that caused by the particular actor under scrutiny) and in doing so, he must rule out environmental or situation-determined causes of variations in effects [p. 209].

This divergence between the theoretical efforts is illustrated by the differences in dependent measures used in experiments to test them. In the attitude attribution experiments testing correspondent inference theory, specific attitude or trait attribution scales are used as dependent measures, and subjects are requested to indicate the degree to which the target person possesses the given attitude or trait. In research to test Kelley's theory, on the other hand, dependent measures reflect the theory's orientation towards allocating causal attributions to either the person or the environment. For example, in McArthur's (1972) experiment, dependent measures included "something about the person (for example, John) probably caused him to make response *X* (for example, laugh) to stimulus *Y* (for example, the comedian)" and "something about the stimulus probably caused the person to make response *X* to it." These measures provide global attributions regarding the perceived locus of causality for the observed behavior. Jones and Davis in their noncommon effect analysis provide a potential means for determining what the "something" is.

The four attributional criteria for decisions about causal allocation discussed by Kelley (1967) are (a) distinctiveness, (b) consensus, (c) consistency over time, and (d) consistency over modality. Kelley has discussed the use of these criteria both with respect to self- and other-attribution. In either case the perceiver wishes to determine whether his (or the other's) response to a stimulus is caused by properties inherent in the stimulus, by properties of the actor, or by circumstantial factors.

By distinctiveness, Kelley refers to whether the individual (self or target person) responds differentially to different entities (or stimuli). Thus, in cases of high distinctiveness, the individual would tend to respond uniquely to the given stimulus. The consensus variable involves whether or not the same response is produced by other people in the presence of the given stimulus. A high consensus would imply, in Kelley's terms, that the majority of individuals respond similarly in the presence of the stimulus. A low consensus could exist either if there were random responding or if there were a consensus differing from the target person's reaction. The consistency over time variable refers to whether the target person responds similarly to the stimulus whenever it is presented in similar circumstances, while the consistency over modality variable concerns whether the target person responds similarly to the stimulus regardless of the type of situation in which the stimulus is presented.

Kelley imbeds these variables in a factorial cube for presentational purposes and treats their influence on attributions by analogy to the analysis of variance. In this analogy, the distinctiveness variable is the numerator of the F ratio, standing for the between-conditions term in the analysis. The greater the difference in response to different entities, the more the response to one entity is determined by that entity. Consistency and consensus, on the other hand, are within-conditions "error" variance and belong in the denominator of the F ratio. These are the variables that inform about the stability and replicability of the effects being considered. The greater the inconsistency and lack of consensus, the larger the distinctiveness variation has to be to influence attribution significantly.

McArthur (1972) has manipulated the distinctiveness, consensus, and consistency variables factorially in an experiment presenting vignettes of accomplishment, emotion, opinion, and action. Her predictions were in line with those presented in Kelley (1967). Person attribution was predicted to be most frequent when a response was characterized by low distinctiveness, low consensus, and high consistency. Stimulus attribution was predicted to be maximized when a response was characterized by high distinctiveness, high consensus, and high consistency. Attribution to the specific circumstances in which the response occurred was predicted in cases of low consistency. McArthur's results supported the predictions based upon Kelley's theory for person attribution, stimulus attribution, and circumstance attribution.

In subsequent essays, Kelley (1971a, 1971b) has elaborated on the determinants of attribution, especially emphasizing the dichotomous possibilities of self versus other and person versus situation as causal loci. In his discussion of "Attribution in Social Interaction," Kelley (1971a) emphasized the covariation principle that underlies the effects of distinctiveness. Thus, an effect that occurs in the presence of one entity and not others is presumed to be caused by that entity.

In his consideration of various attributional "rules," Kelley (1971a) discussed the discounting effect and its obverse, the augmentation principle. The discounting effect holds that "the role of a given cause in producing a given effect is discounted if other plausible causes are also present [p. 8]." This can be mapped completely into the operation of noncommon effects as a variable in correspondent inference theory. In discussing the augmentation principle, Kelley (1971a) stated that: "if for a given effect, both a plausible inhibitory and a plausible facilitative cause are present, the role of the facilitative cause will be judged greater than if it alone were presented as a plausible cause of the effect [p. 12]." This principle can be restated in terms of expected negative valence or cost. Jones and Davis (1965) make the point that "any effects in the choice area which are not assumed to be negative will take on greater importance the more negative the remaining effects. Inferences concerning the intention to achieve desirable effects will increase in correspondence to the extent that costs are incurred, pain is endured and in general negative outcomes are involved [p. 227]."

While these are obvious points at which Kelley's analysis converges on that of correspondent inference theory, it is not as immediately evident how the analogical ANOVA cube relates to the noncommon effects, prior probability analysis of attribution. In the concluding section, we shall try to integrate the two theoretical statements with the eventual goal of stressing points of overlap and complementarity, as well as points of divergence where interesting empirical problems may lie.

CORRESPONDENCE AND THE CUBE: A COMPARISON

Category-Based Expectancies and the Consensus Variable

McArthur manipulated consensus in her vignettes simply by stating whether the target person's action was in agreement or disagreement with the majority. We have already discussed the attitude attribution research in which prior probabilities were manipulated by variations in category-based expectancies. It seems clear that these variables are operationally similar and that they play similar roles in the two theoretical analyses. Consensus, according to Kelley, has to do with veridicality, with the likelihood that behavior is caused by the situation or entity rather than the person. If everyone likes the movie, then the movie and not its viewers must be the prepotent causal factor. Prior probability variables in correspondent inference theory are treated in much the same way: behavior in line with expectation is not informative concerning the person. One only knows that he is like everyone else—by implication, that he places the same value and interpretation on the situation in which he finds himself.

There is no question that category-based expectancies play a significant role in attitude attribution research. We have already alluded to repeated experiments in which correspondence was greatest where consensus (prior probability deriving from category-based expectancy) was lowest. McArthur's results suggest, however, that consensus is a relatively weak determinant of internal versus external attribution in the fictitious settings with which she was concerned. Nisbett and Borgida (1975) have recently shown that there are remarkable insensitivities in the use of consensus data. Subjects appear unwilling to use data about the distribution of others' behavior to infer the responses of individual persons. McArthur suggested that consensus data may be more informative than distinctiveness when a person is evaluating his own behavior, but less informative when he is evaluating others'. The uninformativeness of consensus information in the perception of others may stem from the fact that the perceiver himself forms a limited consensus for evaluating the target person. In any event, the Nisbett and Borgida data certainly do not fit well with the attitude attribution data, and the reasons for this are not obvious. It may be that Nisbett and Borgida's subjects interpreted their task as one of not being fooled or misled by group data in dealing with individuals. This interpretation would be less likely in an attitude attribution study where the consensus manipulation is more implicit and poses a sociological problem for the subjects to solve.

Target-Based Expectancies and Distinctiveness

An obvious difference between correspondent inference theory and Kelley's cube is that Jones and Davis considered a given act in relation to other *potential* acts available to the target person whereas Kelley considered several already occurring acts and the information that could be abstracted concerning the comparability of responses across entities. There is obviously no point in trying to argue that these are equivalent cognitive operations. But if we shift again from the common effects part of the analysis to prior probability, there is some comparability between target-based expectancies and distinctiveness. McArthur's (1972) low-distinctiveness case involved information such as "John also laughs at almost every other comedian," whereas the obverse high-distinctiveness case stated "John does not laugh at almost any other comedian." This information was to be used in combination with the fact that John laughed at the specific comedian. As in the case of target-based expectancy manipulations (Jones et al., 1971), subjects were required to extrapolate from the target person's previous actions regarding similar stimuli to develop an expectancy regarding his response to the present stimulus. The low-distinctiveness case can be compared to the high prior probability case of Jones et al., and the high-distinctiveness case is comparable to the low prior probability manipulation. In all cases the subject's expectancy is based on his information regarding the target person's prior behavior toward similar stimuli. But here we confront a serious problem. In the Jones et al. study, the most correspondent inference is made when the person chooses to write an essay supporting marijuana legalization when she was expected to be against it, and vice versa. Distinctiveness thus leads to personal attribution in that case. The opposite is the case in McArthur's experiment. Here the person who laughs nondistinctively at all clowns is seen as most personally responsible for laughing at a particular clown. Why does the analogy between target-based expectancy and distinctiveness break down in this case?

The answer may lie in a consideration of the degree of similarity among the entities involved. In the Jones et al. (1971) study, subjects were confronted with a target person who made different responses to very similar attitude issues. In the McArthur study, the high-distinctiveness target person endorses one among all clowns, and clowns are known to differ greatly. Perhaps if the distinctive reaction were to one of several highly similar clowns, the cause would be located in the person. If, on the other hand, a very diverse group of opinions were endorsed by the same target person, we would make a personal attribution in line with McArthur's finding. We would say that the target person is a "yea sayer," or that he is response-acquiescent. This seems to be the analogy most consistent with the McArthur results.

Another way of reconciling the apparent conflict in predictions is to consider more carefully the situation in which a target person makes a statement toward Issue A which disconfirms the expectancy established by his previous statements toward related Issues B, C, and D. Perhaps an entity attribution to A (as Kelley and McArthur would predict) is the first step toward a correspondent inference that the target person really feels strongly about A. That is, the attributor must first see A as a distinctive issue before he can assign a special strength to the target person's attitude toward A in a direction opposite to his perceived attitudes toward B, C, and D. Otherwise, the target person must be seen as simply inconsistent.

Target-Based Expectancies and Consistency

This brings us rather naturally to our next comparison between target-based expectancy and the consistency variable. If it were stated that a specific behavior was performed consistently in the same or similar settings, this could be treated either as an example of target-based prior probability manipulation or as equivalent to Kelley's consistency over time and modality variable. McArthur (1972) manipulated the consistency variable in her experiment by stating (for a typical high-consistency manipulation) "in the past John has almost always laughed at the same comedian" or (for a low-consistency manipulation) "in the past John has almost never laughed at the same comedian." The comparable manipulation in Jones et al. was to inform subjects that, for example, the target person was against censorship, in favor of liquor by the drink, for student control of universities, for permissiveness in childrearing, *and* chose to write an antimarijuana speech.

A conflict of predictions again arises if we try to equate these two manipulations. As noted above Jones et al. predicted and found high correspondence (equivalent to high person attribution) in the unexpected behavior case. McArthur predicted (from Kelley) that person attribution would be facilitated by high consistency. We could avoid the conflict between these two predictions by emphasizing the differences between the respective manipulations of expectancy violation and inconsistency. They are not, of course, the same operations. But if we were to bring the opinion topics closer and closer to the marijuana issue (presenting, say, attitudes toward LSD, alcohol, and amphetamines), the manipulations begin to converge. Correspondent inference theory can handle the shift from strong person inferences after violated expectancies to weak person inferences after inconsistency by introducing the boundary condition of knowledge or the skepticism factor. When the expectancy violation reaches a particular threshold of extremity, the perceiver may be forced to decide that the target person was unaware of what he endorsed or at least of the consequences and implications of his endorsement.

But we know from the McArthur study that the typical inference from inconsistency is "something about the particular circumstances," and there is no specific provision for this inference in correspondent inference theory. Since the theory is explicitly concerned with attributions of personal dispositions, it does not distinguish between stimulus and circumstance. Both would be treated together as category-based valence effects. In effect, subjects' use of the "circumstance" dependent variable is nothing more than a confession of ignorance. Inconsistency cannot be caused either by the entity or the person, since both remain constant as the actor's response changes. Therefore, the perceiver has failed to identify the relevant entities, and he can only retreat to a residual category in assigning causation.

Presumably, a more complete analysis of the setting in terms of the full range of action effects would eliminate the difference between stimulus and circumstance. Since the person has behaved differently in commerce with the same entity, we can say that his behavior has had different effects at two different times. Then the problem is to identify what is different about the total context of action at Time 1 and Time 2. This amounts to analysis of potential but foregone action alternatives. We suggest, then, that a careful noncommon effects analysis may rule out the need for circumstance attributions. A circumstance refers to features of a situation that suggests various response options. If these, often very subtle, features can be made prominent by the perceiver in his analysis, then he may not need to resort to a circumstance attribution.

TOWARD AN INTEGRATED FRAMEWORK FOR ATTRIBUTIONAL PROCESSES

It is clear from the preceding discussion that correspondent inference theory and the Kelley ANOVA cube make a number of similar or at least complementary distinctions. Both theoretical statements approach a lay version of experimental design and analysis. The experiment and the attribution episode each involves an attempt to isolate distinctive causes for observed behavior. Each proceeds by a combination of comparison and control procedures. The experiment manipulates potential causal variables and compares the effects observed with those effects present in the absence of these variables. More complicated experimental designs are often generated by the need to eliminate the confounding of potential causes. The attributor makes use of experimental logic whenever possible, but he usually must cope with the natural confounding of potential causes that occurs in the real world. The kinds of attributional analyses we have described in this chapter are similar to the kinds of "quasi-experimental" designs that Campbell and Stanley (1963) talk about.

Although they have much in common, correspondent inference theory and the ANOVA cube are divergent in important respects. We have noted that correspondent inference theory focuses on the attribution of identifiable personal dispositions and addresses the case in which the attributor is trying to establish the cause for a single behavior episode. While some prior probability manipulations have provided information regarding the earlier behavior of the target person, the theory does not explicitly incorporate the impact of information regarding the target person's past behavior. This factor is systematically dealt with by Kelley, whose theory is concerned with judgments based upon behavioral data gathered over multiple behavior episodes. We have already noted the importance of consistency information in the attribution process. Kelley (1967) has stated, "it has been postulated . . . that physical reality takes precedence over social reality information . . . the implication is that the consistency criteria may be more important to the individual than the consensus criterion [p. 207]." In her analysis of the proportion of the variance accounted for by the various independent variables in her experiment, McArthur (1972) observed that consistency information accounted for more of the total variance than consensus information. This finding argues for the importance of including the consistency variable in an integrated theory of attributional processes.

Kelley's theory is also limited. This point is illustrated by the dependent measures used to test the theory. Kelley's theory was designed to provide global attributions either to the person, the stimulus, or the circumstances in which the behavior occurred. Thus, dependent measures result in attributions that "something" about the person, stimulus, or environment caused the behavior. The noncommon effects analysis of correspondent inference theory provides a mechanism for determining what specifically the "something" is. That is, specific causal factors are focused upon in the noncommon effects analysis, particularly specific causal factors residing in the person. Perhaps, each theory would gain analytic power by incorporating variables of the other. The following integrated framework is offered as an attempt in this direction.

The Conceptual Variables of an Integrated Attributional Analysis

The following terms are proposed as concepts in an integrated framework of attribution theory. Each is described in terms of an observer trying to understand the behavior of another, but the same terms can also be used by an actor in a self-attribution process.

I. Prior probability variables (expected effect valences)
    A. Category-based expectancies (*consensus*). Based on inferences derived from the target person's membership in a social category such as sex, age, occupation, or ethnic origin. May be further differentiated into:
        Type 1: Stereotypic. Category membership suggests a modal behavior expectancy or the presence of one categorizing feature (obesity) suggests other correlated features (jolliness).
        Type 2: Normative. There is a modal behavior expectancy created and maintained by social sanctions against deviance. Thus, behavior departing from a normative expectancy incurs personal cost.
    B. Target-based expectancies. Based on inferences drawn from knowledge about the behavior of the specific target person at other times. May be further differentiated into:
        Type 1: Replicative. Expectancy derived from previous observation of behavior in identical situation. Helps to validate ability and knowledge assumptions (*consistency over time*).
        Type 2: Conceptual replicative. Expectancy derived from previous observation of consistent behavior toward the same entity in conceptually similar but descriptively different settings (*consistency over modality*).
        Type 3: Structural. Expectancy derived from observing correlated responses to other entities seen as related through attitude structures or implicit personality theory (*distinctiveness*).

II. Noncommon effect variables. Derived from an analysis of the observed consequences of actions undertaken and plausible action alternatives foregone. Each effect carries a valence denoting its desirability or prior probability as a purpose of action (see I above).

III. Knowledge and ability variables. Evaluation of the target person's awareness that his action was going to produce the resulting effects. As the overall prior probability of the observed action (relative to alternative actions) declines, skepticism concerning knowledge increases in a positively accelerating fashion. Ability may be separately inferred from expressive behavior or other indicators of the skill apparently required, but both ability and knowledge are validated by consistency over time.

The Action Sequence

In this final attempt at theoretical integration, we can imagine an attributional "flow chart" in which the above variables are embedded. It is important to emphasize that perceivers do not make attributions from every act they observe.

And when attributions are drawn, many of the steps in the sequence depicted in Fig. 3 may be bypassed or occur at different points in the order of events. The sequence is not expected to bear much of a relationship to the phenomenology of a naive perceiver. Many attributional routines and subroutines run themselves off as overlearned inference paths. Also, it is usually the case that the attributor must work with very incomplete data, and Kelley (1971b) has made some ingenious suggestions about the "schemata" people use in such cases. The depicted sequence is a somewhat idealized version of how dispositions are inferred from acts. It includes the variables that might be consciously reviewed and analyzed in a setting where the perceiver has unlimited time and where it is important that inferences be made as precisely and validly as possible.

With these caveats in mind, then, let us review the various steps in the sequence and note how they fit into the previous discussions of correspondent inference theory and the Kelley ANOVA cube. First of all, of course, there is the observation of a target person acting in commerce with his environment. We can imagine the environment as consisting of multiple entities at any point in time. The perceiver's first cognitive problem is to resolve the situation into a figural entity and a background context. This context includes a sense of the setting as well as a residual or interactive category known as circumstances. The target person is assumed to direct his action toward the figural entity, though contextual factors may be recognized as influential in determining the nature (direction, intensity, etc.) of the inference.

This first step in the process is no doubt a crucial one, and the choice of figural entity is quite probably determined by the purposes of the perceiver as well as by objective, Gestalt properties of the situation. Jones and Thibaut's (1958) analysis of inferential sets is quite relevant here, as are papers by Zadny and Gerard (1974) and by Jeffery and Mischel (unpublished).

The next step in the depicted sequence is a noncommon effects analysis. At this point our idealized perceiver considers the act in its context of other responses to the situation that might reasonably have occurred. In particular, the analysis concerns consequences brought about by the act and consequences foregone. It is out of this pattern of actual and potential effects that the perceiver weaves a preliminary picture of what the target person was trying to accomplish in the situation.

Although valences are shown as attached to each effect, their value is determined in the next stage of prior probability analysis. The perceiver relates his observations and effect analyses to expectations derived both from social placement (recall and estimates of how others would have acted) and his recall of prior acts by this target person in the same or similar situations. Expectations derived from social placement may or may not involve normative sanctions. Such sanctions would appear as negative valences in cases where the individual's behavior violates reference group norms. The generation of category-based and target-based expectations not only determines the valences to be attached to

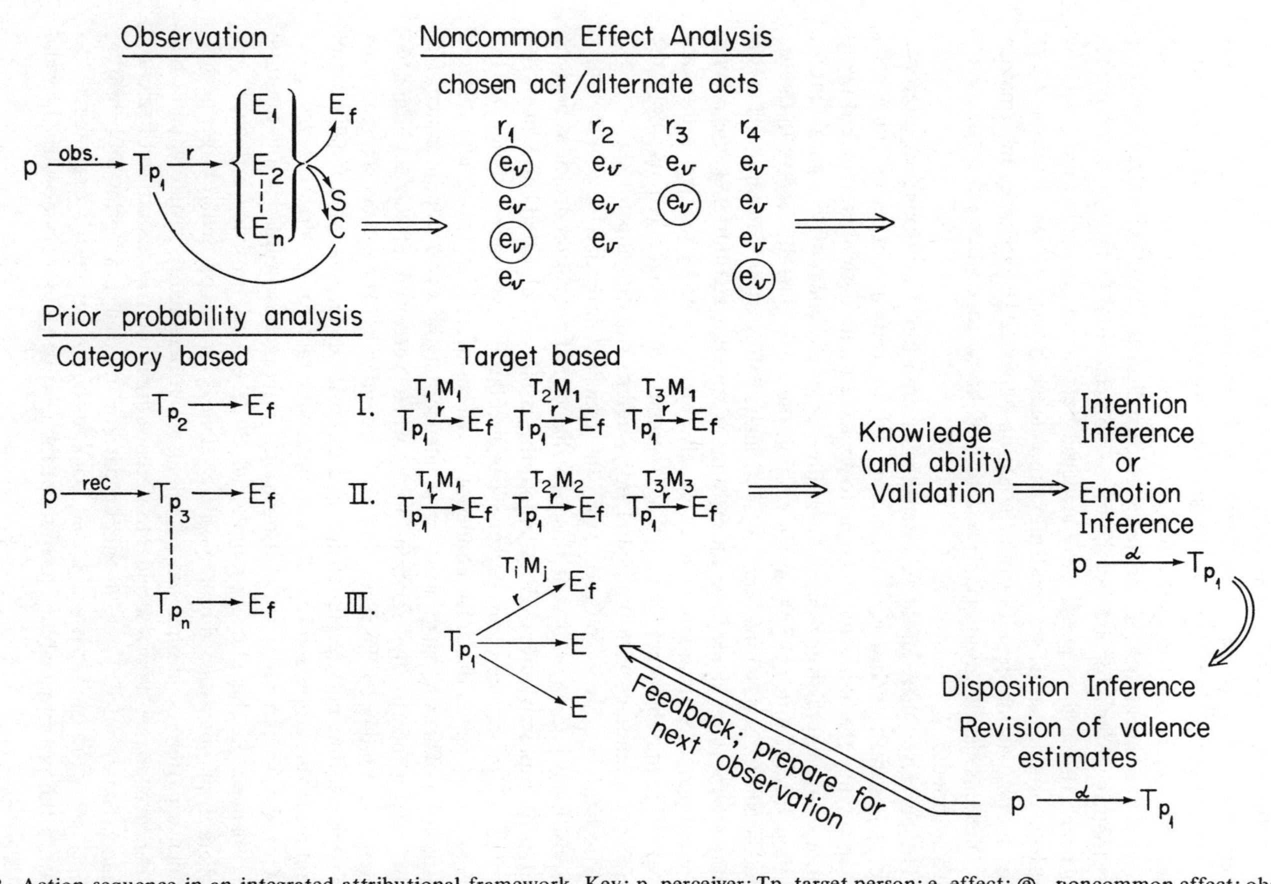

FIG. 3. Action sequence in an integrated attributional framework. Key: p, perceiver; Tp, target person; e, effect; ⓔ, noncommon effect; obs. observe; rec, recall; α, attribution; T, time; M, modality; E, entity; E_f, figural entity; S, setting; c, circumstance; v, valence.

effects, but the information shaping the expectancies is also used to validate the assumption of knowledge and ability. In order to know what to make of his effect analysis, the perceiver must consider the evidence regarding intentionality. Did the target person know that his actions would have the effects obtaining? And the closely related questions, does he have the power, ability, control, to produce the act at will? Consistency information is valuable here since effects persistently striven for are not likely to be accidental or unanticipated consequences. Also, effects persistently achieved betoken ability rather than luck. As we have noted, skepticism concerning knowledge grows at an accelerating rate as the observed action departs from the action expected. Consistency information becomes increasingly important as a test of knowledge when expectancies are disconfirmed.

The perceiver then infers an intention or emotion. Correspondent inference theory has dealt exclusively with intentional, instrumental behavior. It has been assumed that the individual is acting to achieve certain desired effects. But many actions are expressive *re*actions to the mere presence of a figural entity. Although McArthur (1972) has shown that there are some differences in the use of consistency, consensus, and distinctiveness information for emotions versus actions (emotions yielded significantly more stimulus attributions), the process of emotional attribution may be tentatively encompassed by the depicted framework. This may be done by including expressive emotional reactions among the effects of commerce with the figural entity and its situational context.

The next step is to proceed from the transient intention or emotion to a dispositional inference. It is at this step that causal allocation decisions become critically important. In ANOVA jargon, the variance is allocated between entity, setting, circumstance, and person. The greater the variance assigned to the person, the greater the correspondence between inferred intention and inferred disposition. This suggests that correspondence may be a two-step inference. The first step concerns the relationship between perceived effects and inferred intent: he dominated the group and really intended to. The second step concerns the relationship between intention and disposition: he was intentionally dominant in this situation and is really a dominant character in general.

One way to conceive of the dispositional inference is to imagine the resetting of valences. In the above example, all those effects involving influence over others should be assigned more positive valences for this particular target person. Valence pattern is another way of talking about personality structure, and when a perceiver makes a correspondent inference, he is inferring a pattern of valences that is somewhat different from the pattern inferred as a "prior probability." This newly inferred pattern is fed back into the perceiver's expectancies to govern the prior probability analysis of the next behavior episode with the same target person.

Once again it should be stressed that there is considerable arbitrariness in any specification of a sequence from observation to dispositional inference. Hopefully, however, by concretizing the steps in this way, we can show more explicitly the relationships between correspondence inference theory and Kelley's cube. In spite of the apparent complexity of the flow chart, it is, of course, merely a series of sign posts that point to further complexities. The problem of effect identification and analysis remains as nettlesome as ever, and perhaps the inclusion of emotion creates new difficulties that at the moment are unforeseen. The relationship between act and expected act, and whether the resultant inference is a product of contrast, assimilation, or information integration, is an intriguing problem about which much more needs to be known. The links between intention or emotion and disposition have not been systematically explored, and Snyder's (1974) data (discussed earlier) suggest that a distinction between emotion and intention may be very crucial in considering this link. Finally, there is the intriguing problem of the impact of the inference drawn about the target person on category-based expectancies. When our stereotypes are disconfirmed by behavior, under what circumstances do they shatter or disappear, and when do they become even more rigid ("the exception proves the rule")? Hopefully, the integrated framework will suggest where these and other meaningful research questions may be lurking.

SUMMARY

In the present chapter, we have been primarily concerned with modifying correspondent inference theory to bring it more in line with Kelley's theoretical statements on attribution and with experimental data generated in the past decade. A major change has been the substitution of the concept of expected valences for the more restricted notion of assumed desirability. The valence concept permits reference to target-based expectancies and thus opens up the possibility of dealing with more extended behavior sequences. A correspondent inference constitutes a shift in expected valences so that the target person is seen as desiring certain consequences more or less than before the behavioral observation. Certain problems with the theory remain: locating the threshold of skepticism concerning the actor's knowledge of his behavioral effects, dealing with the question of prior choice and effect magnitude, and determining the relative importance of effects chosen and effects foregone. By means of an idealized attributional flow chart, we have attempted to encompass the various subprocesses of person perception. The would-be attributor appraises the effects of the observed act and of plausible alternative acts, considers the effects in terms of his prior expectancies of people in general and the actor in particular, validates the knowledge-of-effects assumption, and makes inferences about intentions and ultimately more stable dispositions. These dispositional attributions become expectancies influencing subsequent attributional inferences when more behavior by the same actor is observed. The integrated framework proposed seems to accomodate existing data and helps to identify the points at which intriguing research problems remain.

ACKNOWLEDGMENTS

The work reported in this chapter was facilitated by a grant from the National Science Foundation (No. GS-1114X).

REFERENCES

Ajzen, I. Attribution of dispositions to an actor: Effects of perceived decision freedom and behavioral utilities. *Journal of Personality and Social Psychology,* 1971, **18,** 144–156.

Ajzen, I., & Fishbein, M. A Bayesian analysis of attribution processes. *Psychological Bulletin,* 1975, **82,** 261–277.

Anderson, N. H. Cognitive algebra: Integration theory applied to social attribution. In L. Berkowitz (Ed.), *Advances in experimental social psychology* (Vol. 7). New York: Academic Press, 1974.

Bell, L., Wicklund, R. A., Manko, G., & Larkin, C. When unexpected behavior is attributed to the environment. *Journal of Research in Personality,* in press.

Bem, D. J. An experimental analysis of self-persuasion. *Journal of Experimental Social Psychology,* 1965, **1,** 199–218.

Bem, D. J. Self-perception theory. In L. Berkowitz (Ed.), *Advances in experimental social psychology* (Vol. 6). New York: Academic Press, 1972.

Brigham, J. C. Racial stereotypes, attitudes, and evaluations of and behavioral intentions toward Negroes and Whites. *Sociometry,* 1971, **34,** 360–380.

Campbell, D. T., & Stanley, J. C. *Experimental and quasi-experimental designs for research.* Chicago: Rand–McNally, 1963.

Chaikin, A. L., & Cooper, J. Evaluation as a function of correspondence and hedonic relevance. *Journal of Experimental Social Psychology,* 1973, **9,** 257–264.

Davison, G. C., & Valins, S. Maintenance of self-attributed behavior change. *Journal of Personality and Social Psychology,* 1969, **11,** 25–33.

Edwards, W. Psychological aspects of decision making. In D. Sills (Ed.), *International encyclopedia of the social sciences.* New York: MacMillan, 1968.

Feldman, J. M. Stimulus characteristics and subject prejudice as determinants of stereotype attribution. *Journal of Personality and Social Psychology,* 1972, **21,** 333–340.

Feldman, J. M., & Hilterman, R. J. Stereotype attribution revisited: The role of stimulus characteristics, racial attitude, and cognitive differentiation. *Journal of Personality and Social Psychology,* 1975, **31,** 1177–1188.

Festinger, L. A theory of social comparison processes. *Human Relations,* 1954, **7,** 117–140.

Heider, F. Social perception and phenomenal causality. *Psychological Review,* 1944, **51,** 358–374.

Heider, F. *The psychology of interpersonal relations.* New York: John Wiley & Sons, 1958.

Jeffrey, K., & Mischel, W. Effects of purpose on the organization and recall of information in person perception. Unpublished manuscript, Stanford University.

Jones, E. E., & Berglas, S. A recency effect in attitude attribution. *Journal of Personality,* in press.

Jones, E. E., & Davis, K. E. From acts to dispositions: The attribution process in person perception. In L. Berkowitz (Ed.), *Advances in experimental social psychology* (Vol. 2). New York: Academic Press, 1965.

Jones, E. E., & de Charms, R. Changes in social perception as a function of the personal relevance of behavior. *Sociometry,* 1957, **20,** 75–85.

Jones, E. E., & Harris, V. A. The attribution of attitudes. *Journal of Experimental Social Psychology,* 1967, **3,** 1–24.

Jones, E. E., Kanouse, D. E., Kelley, H. H., Nisbett, R. E., Valins, S., & Weiner, B. *Attribution: Perceiving the causes of behavior.* Morristown, New Jersey: General Learning, 1972.

Jones, E. E., & Nisbett, R. E. *The actor and the observer: Divergent perceptions of the causes of behavior.* Morristown, New Jersey: General Learning, 1971.

Jones, E. E., & Thibaut, J. W. Interaction goals as bases of inference in interpersonal perception. In R. Tagiuri & L. Petrullo (Eds.), *Person perception and interpersonal behavior.* Stanford, California: Stanford University Press, 1958.

Jones, E. E., Worchel, S., Goethals, G. R., & Grumet, J. F. Prior expectancy and behavioral extremity as determinants of attitude attribution. *Journal of Experimental Social Psychology,* 1971, **7,** 59–80.

Kelley, H. H. Attribution theory in social psychology. In D. Levine (Ed.), *Nebraska Symposium on Motivation.* Lincoln: University of Nebraska Press, 1967.

Kelley, H. H. *Attribution in social interaction.* Morristown, New Jersey: General Learning Press, 1971. (a)

Kelley, H. H. *Causal schemata and the attribution process.* Morristown, New Jersey: General Learning Press, 1971. (b)

Mann, J. W. Inconsistent thinking about group and individual. *Journal of Social Psychology,* 1967, **71,** 235–245.

McArthur, L. The how and what of why: Some determinants and consequences of causal attribution. *Journal of Personality and Social Psychology,* 1972, **22,** 171–193.

McGillis, D. A correspondent inference theory analysis of attitude attribution. Unpublished doctoral dissertation, Duke University, 1974.

Mills, J., & Jellison, J. M. Effect on opinion change of how desirable the communication is to the audience the communicator addressed. *Journal of Personality and Social Psychology,* 1967, **6,** 98–101.

Nelson, C. A., & Hendrick, C. Bibliography of journal articles in social psychology. Mimeo, Kent State University, 1974.

Newtson, D. Dispositional inference from effects of actions: Effects chosen and effects foregone. *Journal of Experimental Social Psychology,* 1974, **10,** 489–496.

Nisbett, R. E., & Borgida, E. Consensus information and the psychology of prediction. *Journal of Personality and Social Psychology,* 1975, **32,** 932–943.

Nisbett, R. E., & Schachter, S. Cognitive manipulation of pain. *Journal of Experimental Social Psychology,* 1966, **2,** 227–236.

Osgood, C. E., & Tannenbaum, P. H. The principle of congruity in the prediction of attitude change. *Psychological Review,* 1955, **62,** 42–55.

Schachter, S., & Singer, J. E. Cognitive, social and psychological determinants of emotional states. *Psychological Review,* 1962, **69,** 379–399.

Secord, P. F., Bevan, W., & Katz, B. The Negro stereotype and perceptual accentuation. *Journal of Abnormal and Social Psychology,* 1956, **53,** 78–83.

Sherif, M., & Hovland, C. I. *Social judgment: Assimilation and contrast effects in communication and attitude change.* New Haven: Yale University Press, 1961.

Snyder, M. The field engulfing behavior: An investigation of attributing emotional states and dispositions. Unpublished doctoral dissertation, Duke University, 1974.

Snyder, M., & Jones, E. E. Attitude attribution when behavior is constrained. *Journal of Experimental Social Psychology,* 1974, **10**, 585–600.

Steiner, I. Perceived freedom. In L. Berkowitz (Ed.), *Advances in experimental social psychology* (Vol. 5). New York: Academic Press, 1970.

Thibaut, J. W., & Riecken, H. R. Some determinants and consequences of the perception of social causality. *Journal of Personality,* 1955, **24**, 113–133.

Zadny, J., & Gerard, H. B. Attribution intentions and informational selectivity. *Journal of Experimental Social Psychology,* 1974, **10**, 34–52.

18

Attribution Theory and Judgment under Uncertainty

Baruch Fischhoff

Oregon Research Institute

Two of the most active areas of research on inferential behavior are the approaches generally known as "judgment and decision-making under conditions of uncertainty" (here, "judgment") and "attribution theory." The former deals primarily with predictive inferences about unknown events–typically set in the future. The latter deals with how people attribute causes to or explain events which have already transpired.

Formally, the two areas differ in their respective subject matter: prediction and explanation. More striking, however, is the difference in the picture of men and women which emerges from them. Attribution researchers find people to be effective processors of information who organize their world in a systematic manner prone to relatively few biases. Judgment researchers reveal people to be quite inept at all but the simplest inferential tasks–and sometimes even at them–muddling through a world that seems to let them get through life by gratuitously allowing for a lot of error.

To illustrate this contrast, in a central article in attribution theory, Kelley (1973) compares man to an intuitive scientist; in a central article in judgment research, Slovic and Lichtenstein (1971) seriously question the notion of man as an intuitive statistician. Whereas it has been recommended (Kelley, 1972b, p. 171) that in the future attribution researchers make greater and more explicit assumptions about people's causal sophistication, judgment researchers have often gone in quite the opposite direction, looking for ever more biases in people's judgments and for ways in which fallible people can be wholly or partially removed from their own decision-making processes.

Why do these divergent images emerge from research in these two areas? One possible explanation is that people are excellent explainers, but poor predictors.

What empirical evidence there is, however, seems to indicate that people may be even worse at explanation than prediction. Indeed, this evidence, albeit collected in the judgment tradition, suggests that people's problems with prediction are due in part to the poor quality of their explanations (Fischhoff, 1974, 1975).

The explanation of this contrast will be sought in the paradigmatic properties of the two research areas. Both judgment and attribution research have many of the characteristics of full-blown research paradigms: pet problems, unquestioned assumptions, recognized centers of research activity. In contrast with Kuhn's (1962) conception of the paradigm, however, the two areas are not completely incommensurable. As we shall see, many of the incapacities which each approach has built into its activists are due more to investigators' fairly arbitrary conventions and interests than to their allegiance to metatheoretical assumptions. Thus, contrasting the two has, hopefully, a good chance of generating some light as well as heat.

Ideally, such a discussion should begin with either definitive statements of judgment research and attribution theory or at least a summary of relevant research. As the former is presently unavailable–owing to the youth and diversity of the two areas–and the second beyond the scope of this paper, I begin instead with brief descriptions of several studies which seem to typify work done in each area ("Characteristic Research"). In the second section, "Mutually Relevant Studies," results emerging in each area which are directly relevant to the other are considered. Then I will discuss several paradigmatic assumptions that emerge from analysis of the studies cited, assumptions that seem to be the source of the divergence of judgment and attribution research. Finally, in the conclusion, directions for future research and cross-fertilization are suggested.

CHARACTERISTIC RESEARCH

Judgment

Three focal topics of judgment researchers have been the ways in which people (a) make subjective probability estimates, (b) sequentially update such estimates upon receipt of additional information, and (c) simultaneously combine probabilistic information from multiple sources. In each area subjects' performance has been compared with normative criteria of judgmental adequacy and found lacking. With many probabilistic tasks, people appear neither to produce the responses demanded by the appropriate normative models nor to process information in ways indicated by the models.

Probability estimates. Perhaps the simplest task involving probabilistic inference is that used in probability-learning experiments. In a typical study of this type, subjects might be asked to predict the color of each of a series of marbles drawn from an urn containing an unspecified mixture of red and blue marbles. In reviewing these studies, Vlek (1970; also Luce & Suppes, 1965; Peterson & Beach, 1967) noted that although subjects are able to estimate accurately the proportion of marbles of each type, they do not use this information effectively. Rather than consistently predicting the more frequent type of marble, a strategy which would maximize their accuracy, subjects typically predict the less frequent color in a substantial proportion of trials. Provision of payoffs weakens but does not eliminate the bias (for example, Messick & Rapoport, 1965). The patterns of subjects' predictions appear to reflect complex, idiosyncratic theories about sequential dependencies in random series, theories which have no basis in the mathematical theory of binomial processes. "Gamblers' fallacy" is a related example of people's misconceptions of how random sequences should look.

When frequentistic data, such as the proportion of red and blue marbles, are lacking, it is difficult, if not impossible, to assess the veracity of any individual probability estimate. If someone assigns a probability of .80 to there being a Democrat in the White House in 1984, there is no way now, nor will there be a way in 1985, to tell how good that estimate was. It is, however, possible to evaluate the validity of a set of probability estimates. The measure of their validity is their degree of calibration. For perfectly calibrated judges, XX% of the events to which they assign .XX probability of occurrence will, in fact, occur. In empirical tests, calibration is typically quite remiss. Fischhoff and Beyth (1975), for example, found underestimation of low probabilities and overestimation of high ones. Such predictions meet too many big surprises, very unlikely events that occur and very likely ones that do not. They indicate overconfidence, with people making more extreme predictions than their knowledge justifies. Further evidence of miscalibration and other forms of miscalibration are reviewed by Lichtenstein, Fischhoff, and Phillips (in press).

When asked to quantify their confidence about knowing the correct answer to general knowledge questions (that is, to estimate the probability that their chosen answer is correct), subjects typically overestimate how much they know, For example, Alpert and Raiffa (1968) had subjects set upper and lower limits for possible values of quantities like the population of Outer Mongolia, so that there was only a 2% chance that the true answer fell outside of the limits. Across problems, some 40% of the true values fell outside of the confidence intervals. People also have been found to exaggerate their ability to predict entities like horse races (Scott, 1968), the stock market (Fama, 1965), and natural hazards (Kates, 1962).

Perhaps even more disturbing is the fact that people's probability estimates frequently violate the most basic laws of probability theory. For example, Kahneman and Tversky (1973), Hammerton (1973), and Lyon and Slovic (1975) have found that when people are called upon to combine base-rate information with evidence regarding a specific case, they consistently ignore the base-rate information, even when the "individuating evidence" has negligible validity. Years ago, Meehl and Rosen (1955) found a similar problem to affect the developers of psychometric tests.

Tversky and Kahneman (1971, 1974) found that people are largely oblivious to questions of sample size, exaggerating the stability of results obtained from small samples and failing to see the increased stability to be found in larger samples. In their earlier article (Tversky & Kahneman, 1971), this bias was called "belief in the law of small numbers."

In a number of studies, Wyer (1974; Wyer & Goldberg, 1970) has found that people consistently overestimate the likelihood of the conjunction of two events. Slovic, Fischhoff, and Lichtenstein (1976) have shown that the judged probability of compound events may actually be larger than the probability of the constituent events [that is, at times $P(A \cap B) > P(A)$ or $P(B)$]. The probabilities assigned to an exhaustive set of mutally exclusive events have frequently been found not to sum to 100, another violation of internal consistency in people's intuitive judgments (see Peterson & Beach, 1967, p. 36). Summarizing his work on the interrelations between people's probabilistic beliefs, McGuire (1968) concluded that they do not have the sort of internal coherence demanded by the laws of probability.

Opinion revision. One crucial aspect of functioning in a probabilistic environment is being able to update properly one's beliefs about that environment upon receipt of additional information. The consensual normative model for opinion revision is Bayes' theorem, implications of which are explicated in Edwards, Lindman, and Savage (1963), Phillips (1973), and Slovic and Lichtenstein (1971).

People's intuitive adherence to Bayesian inference has been investigated in an extensive research program. Whereas initial work appeared to indicate that people were generally quite sensitive to the parameters of the Bayesian model (for example, Peterson & Beach, 1967), more recent reviewers have been considerably more pessimistic. Slovic and Lichtenstein (1971) conclude,

> . . . the intuitive statistician appears to be quite confused by the conceptual demands of probabilistic inference tasks. He seems capable of little more than revising his response in the right direct upon receipt of a new item of information (and the inertia effect is evidence that he is not always successful in doing even this). After that, the success he obtains may be purely a matter of coincidence–a fortuitous interaction between the optimal strategy and whatever simple rule he arrives at in his groping attempts to ease cognitive strain and to pull a number "out of the air" [p. 714].

The simple rules that people appear to use (for example, Dale, 1968; Kahneman & Tversky, 1972; Lichtenstein & Feeney, 1968; Pitz, Downing, & Reinhold, 1967) not only fail to produce accurate Bayesian estimates, but they also have no analog in the formal model. Thus, the model fails both to predict subjects' responses and to capture the essential determinants of their judgment processes.

Information integration. One crucial skill for anyone living in an uncertain world is the ability to combine information from a variety of sources into a single diagnostic or prognostic judgment. Such tasks constitute the life work of stock analysts, investigative radiologists, and major league scouts; they confront all of us daily in enterprises as diverse as forming impressions and deciding when to cross the street.

Perhaps the best known work on how well people integrate information from multiple sources is the research on clinical psychologists' judgmental processes summarized and inspired by Meehl's (1954) *Clinical versus Statistical Prediction.* These studies, reviewed more recently by Goldberg (1968, 1970), indicate that rather simple actuarial formulae typically can be constructed to perform at a level of validity no lower than that of the clinical expert.

This disturbing finding produced a great deal of research into why clinical judges did no better–research that in turn produced even more disturbing findings. For example, the accuracy of judges' inferences appears to be unrelated to either the amount of information in their possession or their level of professional training and experience (see Goldberg, 1968, and references therein).

Other researchers indicate that the "rules" of intuitive information integration are fundamentally inconsistent with normative principles of optimal information utilization. Kahneman and Tversky (1973) found that when making inferences on the basis of a given set of cues, people are more confident when they believe that those cues are redundant than when they believe the cues to be independent. Normatively, in such a situation, redundant cues carry less information and, thus, justify less confident judgments. Tversky (1969) found that subjects' preferences between multiattribute alternatives are sometimes intransitive. In a numerical prediction task requiring the utilization of only two cues, Lichtenstein, Earle, and Slovic (1975) discovered that subjects used a nonnormative averaging heuristic. Slovic and MacPhillamy (1974) found that when called upon to choose between multiattribute alternatives, people are unreasonably influenced by commensurable dimensions, those that can be readily compared across possible choices. An example might be choosing the cheapest of several alternative summer vacation plans, not because cost is of utmost importance, but because it provides the one dimension on which all possibilities can be unambiguously characterized and compared.

Although recent work (for example, Dawes & Corrigan, 1974) suggests that the superiority of actuarial predictions may say more about the power of the models than about the impotence of their human competition, these other disturbing results remain on the record. (A related approach, but one in which researchers have typically not concerned themselves with questions of optimality, is the theory of information integration advanced by Anderson and his associates. Insightful introductions may be found in Anderson [1970, 1971, 1973].)

Interpretation. One central notion in explaining these biases is that of "cognitive strain." In many tasks judges are confronted with more information than they are able to process. To cope with this overload, they develop ad hoc algorithms, or heuristics, for information processing. In this view, people are seen as computers that have the right programs but frequently cannot execute them properly because their central processor is too small.

Some of these biases (for example, the insensitivity to sample size or the preference for redundant information), illustrate more serious deficiencies. Here, people's probabilistic judgments are not only biased or incomplete, but fundamentally wrong. Returning to the computer analogy, it appears that people lack the correct programs for many important judgmental tasks. Even more disturbing is the fact that neither decisions made by experts nor decisions with grave social consequences are immune to these biases (Goldberg, 1968; Slovic, 1972; Slovic et al. 1976; Slovic, Kunreuther, & White, 1974; Tversky & Kahneman, 1971).

Attribution

The central concern of attribution research is people's intuitive perceptions of causality, specifically, their attributions of reasons for the occurrence of behavioral events. Although attribution research is quite diverse, much of it can be traced to work of Heider (1958), as developed and operationalized by Jones and Davis (1965), Kelley (1967, 1972a, 1972b, 1973), Weiner (1974), and others. As several recent reviews in addition to the present volume are readily available (Ajzen & Fishbein, 1975; Miller & Ross, 1975; Shaver, 1975), this section will be even more abbreviated than the previous one.

Kelley's analyses provide perhaps the most general framework for studying attributional processes.[1] In them he distinguishes between situations in which historical data regarding the behavior in question are available and situations in which they are not. In the former situations he hypothesizes three characteristics of the behavior to be explained that will govern an observer's attributions. They are the behavioral act's consistency over time (does the actor always respond that way?), its distinctiveness (is it elicited by other stimulus situations as well?), and its degree of consensus (do other actors respond similarly?). These historical data are seen as being organized in a three-dimensional matrix from which attributions are derived in keeping with J. S. Mill's method of difference. Kelley translates this method into the "covariation principle," according to which acts are attributed to possible causes with which they covary. From this hypothesis,

[1] Ajzen and Fishbein's (1975) attempt to conceptualize attribution tasks in Bayesian terms would, of course, provide a common framework for much judgment and attribution work—were it successful. There appear, however, to be some difficulties with their interpretations of both Bayes' Theorem and of the empirical evidence regarding intuitive Bayesian inference, leaving the unifying theory as yet unfound (Fischhoff & Lichtenstein, 1975).

he derives many interesting predictions which have been most thoroughly examined by MacArthur (1972).

MacArthur systematically varied consensus, distinctiveness, and consistency information pertaining to a given behavioral act (for example, John laughs at the comedian). Subjects were told, "Your task is to decide on the basis of the information given, what probably caused the event to occur. You will be asked to choose between four alternative causes . . . the cause which you think is most probable." The alternatives were (a) something about the *person* (for example, John), (b) something about the *stimulus* (for example, the comedian), (c) something about the particular *circumstances,* and (d) some combination of a, b, and c. Two main findings were (a) that each of the three sorts of information affected attributions to some degree and (b) that there was a preponderance of person attributions. This latter finding has drawn considerable attention from those concerned with possible discrepancies in attributions for oneself and others (for example, Jones & Nisbett, 1972; Weiner & Seirad, 1975).

When historical data are lacking, Kelley (1972b, 1973) sees people relying on what he calls causal schemata. In his words (Kelley, 1972b), "Given information about a certain effect and two or more possible causes, the individual tends to assimilate it to a specific assumed analysis of variance pattern and from that to make a causal attribution [p. 152]." These schemata might be thought of as general types of laws of behavior. For example, one principle of behavior which many people appear to accept is that in order to succeed on a difficult task, one must both be capable and try hard (Kun & Weiner, 1973). This principle is an example of a multiple necessary causal schema, the sort of rule which, according to Kelley (1972b), is invoked to account for unusual occurrences (see also Enzel, Hansen, & Lowe, 1975) or events of great magnitude (Cunningham & Kelley, 1975).

In trying to account for the way in which people infer underlying dispositions from observation of behavioral acts, Jones and Davis (1965) hypothesized that the observer does the following: (a) identifies the choice options facing the actor, (b) lists the actor's possible reasons for selecting each act, (c) eliminates those reasons that could have motivated the selection of acts other than the one chosen, and (d) assesses the importance for the actor of each of the remaining reasons. The basis of this last assessment is the perceived importance of each reason (the desirability of the anticipated effect) for members of the various reference groups to which the actor belongs. If the observer can identify a reason that could only motivate selection of the chosen act and that is not highly valued by others, then he or she will attribute the act to that reason and the underlying personal disposition that it represents. "We can be certain that a politician who advocates achieving cuts in government spending by lowering social security payments to an audience of senior citizens *really* means what he says" (Shaver, 1975, p. 48).

The attribution paradigm has been used to advance understanding in substantive areas such as achievement motivation (Weiner, 1974), therapeutic effectiveness (Bowers, 1975), order effects in impression formation (Jones & Goethals, 1972) and sex role stereotyping (Feldman–Summers & Kiesler, 1974). Much of this work has focused on questions of social perception, presumably because attribution theory has grown within the context of social psychology. In many ways those using the paradigm seem to have encouraged and legitimized asking a variety of new and illuminating questions about behavior and have also somewhat restructured the role of the social psychologist which Kelley (1973) believes "is not to confound common-sense, but rather to analyze, refine, and enlarge it [p. 172]."

Contrast

The similarities between judgment and attribution research must be apparent from even these brief reviews. Both study how people interpret, organize, and use multivariate information in an uncertain environment. Both are largely phenomenological in their theorizing, attempting to understand in common-sense terms the ways in which people think about their world. There are even some vague similarities between the theories developed in each. For example, Kahneman and Tversky's (1972) representativeness heuristic, which leads a judge to view a possible event as likely if it embodies or "represents" the main features of the situation or person creating it, seems related to Jones and Davis' (1965) notion that a high degree of correspondence of inference, "the extent that the act and the underlying characteristic or attribute are similarly described [p. 223]," is necessary for confident dispositional attributions.

The image of people's information-processing ability that emerges from these two areas is, however, strikingly different. In judgment research, people seem to do so poorly that cataloguing and shoring up their inadquacies has become the focal topic of research. In attribution research, they either are found to do quite well or the question of adequacy never comes up.

These generalizations are not without exceptions. Peterson and Beach (1967) did identify a number of tasks, primarily those requiring intuitive estimates of univariate descriptive statistics, that people perform quite well. A continuing goal of current research has been to find out what tasks people perform well, what tasks they might perform more adequately with training, and what tasks are best taken out of their hands entirely and allocated to machines and actuarial formulae. Students of biases have also indirectly acknowledged people's inferential abilities by the great ingenuity they have shown in trying to elicit clearly biased behavior.

Similarly, there has been study of "attributional biases." In particular, investigators have looked at ways in which attributors distort incoming data to better serve their own ego-defensive functions or their sense of control over their world (for example, Cialdini, Braver, & Lewis, 1974; Kelley, 1967; 1972a; Luginbuhl, Crow, & Kahan, 1975). The finding that behavioral acts are often attributed internally (to the actor) by observers but externally (to stimulus circumstances) by the actor himself has also been discussed as reflecting a bias toward inferring personal traits where there are none—a bias that is shared by psychologists and laypeople alike (for example, Jones & Nisbett, 1972; Mischel, 1968; Ross, in press). Kanouse (1972) has presented some evidence indicating that people are prone to primacy effects (relying on the first sufficient explanation that comes to mind) in their attributions. Walster (1966, 1967) and others (for example, Vidmar & Crinklaw, 1974) have looked at people's defensive attribution of responsibility for accidents and their tendency to exaggerate the predictability of accidents that threaten them—and, thus, their ability to avoid the danger.

However, not only is the study of biases somewhat the exception in attribution work, but the robustness of even these biases has recently been seriously questioned by attribution researchers (Ajzen & Fishbein, 1975; Miller & Ross, 1975; Ross, Bierbrauer, & Polly, 1974; Taylor, 1975). In addition, it will be noted that in these examples it is not the attributional information processing that is being questioned but the information that the attributor uses. Attributional biases are essentially proper conclusions drawn from improper premises. The impropriety of the premises arises either from inefficient information gathering or hedonic distortion of what is happening, not from difficulties in handling or combining information. Judging from the literature, attribution researchers appear to assume that (a) people use the attributional techniques which they (the psychologists) hypothesize, (b) they use them properly, and (c) these techniques provide adequate guides for making attributions. Returning to the computer analogy, the naive attributor is seen as having both the proper programs and the capacity to execute them.

In Kelley (1972b), for example, this viewpoint emerges not only from his presentation of the schemata concept but also in his discussions of its limitations (pp. 171–173)—which are essentially ways in which the theory underestimates people's attributional sophistication. So much faith is placed in people's inferential abilities that attributional theories are often produced by first discerning (often with great ingenuity) the data to which people will attend in a given situation and then formally working through the conclusions which may be properly deduced from them.

MUTUALLY RELEVANT STUDIES

Although attribution and judgment researchers have generally gone on their separate courses with few glances to the side, there are a number of studies of mutual interest. Several, such as the work in Bayesian opinion revision and on the inconsistency in personal belief systems have been presented above. Others appear below. Most of these, it seems, suggest ways in which judgmental biases may intrude on attributional tasks.

Use of Base-Rate Information

Perhaps the most direct and creative integration of judgment and attribution work is a recent study by Nisbett and Borgida (1975). After reviewing studies testing Kelley's covariance model, they concluded that although there is much evidence showing people's sensitivity to consistency and distinctiveness information, there is little evidence of a similar sensitivity to consensus information. They noted that this failure to use information about what most people do in a particular situation is directly analogous to Kahneman and Tversky's (1973) finding that people ignore base-rate information in favor of individuating infirmation about the case at hand. In an ingenious study, they replicated this finding using behavioral base-rate information, providing a judgmental reason for the failure of Kelley's model in this respect.

It is unclear how these results can be reconciled with the heavy reliance on base-rate or reference group data in making dispositional attributions postulated by Jones and Davis (1965). One possibility is that the norms of the reference group are considered only insofar as they are embedded in the description of the actor. They then become some of the actor's characteristics which should be "represented" (in Kahneman & Tversky's, 1972, sense) in any behavioral act. In that case, they would be considered more for their associational value than their informational worth.

Nisbett and Borgida (1975) suggest that "perhaps in fact, it is only when we have rather well-rehearsed schemata for dealing with certain types of abstract, data-summary information that it is used in a fashion that the scientist would describe as rational [p. 4]." What evidence there is of subjects' responding to consensus information (Frieze & Weiner, 1971; Weiner & Kukla, 1970) comes, indeed, from attributions for success and failure, situations for which "well-rehearsed schemata" do seem to be available (Weiner, 1974).

Defensive Attribution and Hindsight

A reverse confluence of research efforts emerged in some of my own work (Fischhoff, 1974, 1975). As mentioned above, Walster (1967) found that when confronted with news of an unfortunate accident, people tend to exaggerate in retrospect its predictability. I found this result to be a special case of a more general phenomenon. In general, events which are reported to have occurred are seen in hindsight as having appeared more likely (and thus predictable) in foresight than they actually did appear. I argued that by exaggerating the predictability of the past, people underestimate what they have to learn from it.

A corollary of this bias is that in hindsight we find it very difficult to reconstruct the uncertainties which faced other decision makers in the past. In second-guessing others, we typically overestimate the clarity with which they foresaw what was going to happen (Fischhoff, 1974, 1975, Experiment 3).

These results suggest a further source of bias in dispositional attributions produced by users of Jones and Davis' (1965) algorithm.

Perceived Correlation

According to Kelley's (1972b, 1973) ANOVA conception, people organize behavioral information as they receive it into a data matrix whose three dimensions are entities, time, and persons. When called upon to make attributions, they base them upon the stored covariation information. Any inaccuracies in perceived covariation could, of course, lead to erroneous attributions. Just such a discrepancy was found by Chapman and Chapman (1967, 1969; also Golding & Rorer, 1972) who showed that clinical psychologists and clinically naive undergraduates perceive correlations which are purely illusory between patients' symptoms and their responses to diagnostic tests. Kelley (1973, p. 119) attributed these misperceptions to conflict between covariation information and causal preconceptions (schemata). In slightly later work, Tversky and Kahneman (1973) showed that such illusory correlations may be due to the differential memorability or availability of various symptom–response pairs. Smedslund (1963), who had nurses judge disease–symptom correlations, and Ward and Jenkins (1965; also Jenkins & Ward, 1965), using more artificial tasks, found that when looking at 2 × 2 cooccurrence tables (for example, [disease, no disease] × [symptom, no sympton]) people base their perceptions of causality solely on the number of cases in which both the disease and the symptom are present. Miller and Ross (1975) have capitalized on some of these results to provide a nonmotivational explanation of findings that have been interpreted as reflecting a bias toward distorting information to facilitate making self-serving attributions.

Perceptions of Randomness

Before attributing a cause to an event, an observer must decide whether it was caused at all, or whether, in the light of the information at his or her disposal, it should be treated as a random event. Certainly causal attributions for random events are worthless. There is a good deal of judgmental evidence showing that people have a very poor conception of randomness. In particular, they do not recognize it when they see it and offer deterministic explanations of random phenomena (for example, Kahneman & Tversky, 1972, pp. 434–437). Gamblers' fallacy in the interpretation of random binary series is one well-known example (for example, Jarvik, 1951; Tune, 1964). Less well known is the corollary of the law of small numbers by which scientists rarely attribute deviations of results from expectations to sampling variation, because they are always able to find causal explanations for such discrepancies (Tversky & Kahneman, 1971). Another is the persistence and often destructiveness with which people provide causal explanations for regression toward the mean phenomena (Kahneman & Tversky, 1973). An example involving high stakes is Londoners' causal explanations for the pattern of German bombing during World War II, explanations which frequently guided their decisions about where to live and when to seek shelter. Upon later examination, it was found that the clustering of bomb hits closely approximated a Poisson (random) distribution (Feller, 1968, p. 160). Burton and Kates (1964) and Kates (1962) provided further costly examples in people's responses to natural hazards.

Cognitive Control

Making proper attributions requires some fairly sophisticated and complicated use of the knowledge accumulated in covariation matrices and causal schemata. There is a good deal of evidence showing that people are poorly equipped for this sort of conditional, multivariate thinking. Hammond and Summers (1972) have shown that cognitive control, or ability to apply knowledge, may lag well behind the acquisition of that knowledge. They also argue that everyday learning experiences are typically not structured to develop cognitive control.

A related problem is people's poor insight into the information integration policies that they are following. Goldberg (1968, 1970) and others have found that in situations in which clinical judges believe that they are performing complicated multivariate judgments, their information-processing policies can be effectively captured by simple linear models, utilizing a relatively small number of variables (for example, Hoffman, Slovic, & Rorer, 1968). Slovic, Fleissner, and Bauman (1972), studying the judgmental policies of stockbrokers, found a substantial negative correlation between years of experience as a broker and accuracy of self-insight.

One common type of error found in such studies (see Slovic & Lichtenstein, 1971, pp. 683–684) is a tendency for judges to overestimate the importance they place on minor cues and underestimate their reliance on a few major variables. All of these results indicate that when introspecting about their own judgmental processes, people tend to exaggerate their information-processing sophistication (see also Michael, 1968; Shepard, 1964).

Field Studies

Although an adequate review is beyond the scope of this paper, sociologists have also identified a number of rich attributional phenomena, primarily pathologies of explanation. Garfinkel's (for example, 1964, 1966) ethnomethodological works is one source; labeling theory (for example, Prus, 1975; Schur, 1971), a second; the cataloguing of accounts (Scott & Lyman, 1968) and techniques of neutralization (Rogers & Buffalo, 1974; Sykes & Matza, 1967), a third; and observational studies of gambling behavior (Oldman, 1974; Scott, 1968), a fourth.

Further evidence of the differences between the way social scientists and laypeople think can be found in O'Leary, Coplin, Shapiro, and Dean's (1974) study of the explanatory protocols used by U.S. Department of State foreign affairs analysts. They found that whereas academic international relations researchers tended to use small numbers of continuous variables interlinked by simultaneous linear relationships, applied analysts relied on multivariate, explanatory models using discrete variables with nonlinear, time-lagged relationships between them. O'Leary et al. observed that, "the kinds of relationships found in the majority of [State Department] analyses represent such complexity that no single quantitative work in the social sciences could even begin to test their validity [p. 228]."

PARADIGMATIC ASSUMPTIONS

In the previous section I attempted to show the natural interface between judgment and attribution research. In the present section some possible reasons for the preponderance of judgment results suggesting sources of bias in attributional tasks will be considered. If attribution and judgment research are seen as tapping the same basic information-processing facility, the divergence in results is better sought in the minds doing the research than the minds being researched.

Probabilistic versus Deterministic Processes

The inferential processes hypothesized by both judgment and attribution researchers represent highly deterministic ways of relating to one's environment. The judgment subject is seen as looking for patterns in random sequences, ignoring probabilistic base-rate information in favor of individuating information, and using relatively few cues from a multitude of potentially valid ones. Probabilistic considerations are almost totally absent in the use of Tversky and Kahneman's heuristics. By like token, the naive attributor seems typically to be viewed as a puzzle-solver, who by process of elimination whittles down a set of possible alternative hypotheses.

However, although researchers in both fields agree on the basically nonprobabilistic nature of people's inference, they disagree on the propriety of that nature. A fundamental notion in judgment research is Brunswik's (1952; also Hammond, 1966) "probabilistic functionalism," the idea that the role of psychologists is to study the adaptive interaction between an organism and its uncertain environment. Insofar as that environment is probabilistic, considering the information at the organism's disposal, deterministic rules of inference are at best approximations. A judgment rule that allows no reflection of probabilistic phenomena known to be operating is by its nature suspect.

Similar suspicion seems lacking in attribution research. Certainly there must be situations in which this is an adequate policy, in which the underlying process generating the behavioral data to be explained has no major probabilistic components, and in which deterministic reasoning will suffice. If the situations studied in attribution research fall into this category, it would not be surprising that attribution subjects perform more adequately than judgment subjects, who are typically confounded by the counterintuitive nature of the probabilistic processes about which they are called upon to make inferences.

Aside from helping to make people look good, reliance on deterministic tasks (or deterministic aspects of probabilistic tasks) seems to have masked some questions which attribution theorists themselves might find extremely interesting. For example: to what extent are events perceived to be explicable or attributable; in analysis of variance terms, how much of their variance is viewed as explainable? How well do people believe that they have succeeded in explaining events when they have given the best available explanation? If people were asked to estimate the percentage of variance explained by each of several causes, how flat or peaked would their distributions of causal responsibility be in different situations? How many causes would be assigned at least some responsibility? Are there tasks or individuals with predispositions toward unicausal attributions?

Some attribution researchers, particularly those concerned with perceived causes of success and failure, have elicited attributions to the category of "luck." Presumably, any chance factors impinging upon a success–failure outcome do constitute either good or bad luck–depending upon how things turn out. Yet it is not clear how these results may be generalized to situations in which success–failure is not the primary criterion characterizing outcomes. Nor is it clear whether chance and luck are indeed synonymous even in success–failure situations. It appears, for example, that "luck" is a person attribution, whereas "chance" is a property of the environment. Nor is it clear how either term relates to those causes or forces which the attributor believes that he could understand with some additional information, although for the moment they appear inexplicable and random. A further example of the potentially interesting issues missed by avoidance of probabilistic considerations is Kukla's (1972) attributional theory of performance, which attempts to integrate attribution theory and expectancy theory. After formulating and describing the subjective (probability) expected utility model of expectancy theory, he eliminates all probabilities for the sake of simplification, leaving a U or expectancyless model of expectancy. One might also wonder how MacArthur's subjects would have produced attributions from stimuli like "X can fool some of the people some of the time."

Stimulus and Response Modes

In the typical judgment study, stimuli are designed to be as complete and unambiguous as possible. The desire to provide subjects with data that cannot be misconstrued and that are sufficient to make the required judgments has, indeed, often resulted in extremely artificial experimental tasks.

Subjects are, however, given little help to knowing how to encode these data properly or relate them to their own previous experience. They are seldom told either what the formal analog of the experimental task is or which of their previous experiences are at all relevant to handling it. The response mode in most judgment tasks (for example, giving a single subjective probability) provides them no additional hint as to what data are relevant.

Tversky and Kahneman (1974) have claimed that probabilistic inference is difficult to learn because the proper way to characterize tasks is often unintuitive or even counterintuitive, and because life experiences are not organized or juxtaposed to reveal their common underlying statistical properties. They write:

> A person could conceivably learn whether his judgments are externally calibrated by keeping a tally of the proportion of events that actually occur among those to which he assigns the same probability. However, it is not natural to group events by their judged probability. In the absence of such grouping, it is impossible for an individual to discover, for example, that only 50 percent of the predictions to which he has assigned a probability of .9 or higher actually came true [p. 1130].

If this is the case, then failure to help subjects with encoding may be a source of their downfall in judgment experiments as well as in life.

By contrast, whereas attribution researchers often put forth great effort to guarantee the verisimilitude (nonartificiality) of their experimental tasks, critical details of these "slices of life" are typically either presented ambiguously or left entirely unstated. The attribution subject is then asked to infer his or her own values. For example, the observer of a tutorial session might be asked about the person being tutored, "Was he really trying hard to succeed?" At times, as in some studies relevant to attributional therapy, the stimulus information presented to subjects may be deliberately distorted, leading subjects to reach erroneous attributions by reasoning correctly from false premises (for example, Loftis & Ross, 1974; Storms & Nisbett, 1970).

Attributional subjects are, however, given considerable assistance in structuring this stimulus data. A typical example is MacArthur's (1972) study in which subjects were explicitly told all the dimensions of stimulus characterization relevant to their response. Additional hints of how to look at and organize data can frequently be found in the response format used. For example, asking instructors in a tutoring study to rate the importance of four instructor factors and four student factors in determining their students' success or failure (Ross et al., 1974) must help them know how to structure the data and what the experimenters are looking for. Although the degree of such structuring varies from study to study, it appears to be consistently greater than the degree found in most judgment studies. It must be wondered whether this increase in structure is not enough to get people on the road to reasonable inference. As noted above, the moderate structuring in Bayesian inference experiments appears to create demand characteristics which usually push people to revise their probabilities in the correct direction even when their intuitive inferential processes are almost unrelated to the proper ones.

Aside from subtly directing subjects' efforts (and possibly improving their performance), the exclusive use of structured, forced-choice response modes seems to have obscured many questions that would be of interest to attribution researchers. How do people organize data for attribution when undirected? Does this subjective organization resemble Kelley's matrices, and if so, what are its dimensions? If asked to fill in such a matrix with data points, could people do it appropriately? If presented with such a matrix in terms of data points rather than empirical generalizations of the type used by MacArthur, would people make the proper generalizations, or would they have difficulties as suggested by the illusory correlation results? If policy-capturing techniques (Goldberg, 1968; Hammond, Hursch, & Todd, 1964) were applied to sets of attributional judgments, how much insight would people have into their inferred policies? If asked to explain an event, when do people naturally produce explanations of "why" rather than of "how" or "what"? When do people feel that events need no explanation at all or that no explanation is adequate? When do people even consider alternative explanations? Does it matter for whom an explanation is being prepared? Frieze's work (cited in Weiner, 1974) is a first step toward answering these questions. Garland, Hardy, and Stephenson's (in press) finding that when subjects were allowed to ask for more information, (with stimuli similar to MacArthur's, 1972) 77% of their requests were not codable into categories of consistency, consensus, or distinctiveness suggests that the answers may be very interesting.

Judgment researchers, too, with a few notable exceptions (Kleinmuntz, 1968; Payne, in press) have relied on forced choice or probability estimation responses. Elicitation of decision-making protocols or predictive scenarios would probably reveal hitherto neglected aspects of judgment. To the judgment researchers' credit, though, is the fact that they have experimented systematically with the effects of using different response modes to elicit the probability estimates central to their enterprise (Slovic & Lichtenstein, 1971, p. 698ff). Comparable response mode research has been scarce in attribution work.

Focal Issues

Many inferential tasks have two aspects: (a) evaluating the meaning and relevance of the stimuli presented and (b) combining the information derived from these stimuli into a judgment. Both of these operations allow for psychological inputs of considerable interest, as the subject decides what's happening and how to put it all together to make sense out of it. Since it is difficult to study two sorts of interrelated psychological phenomena simultaneously, most researchers (with notable exception of Anderson and his colleagues) have made simplifying assumptions or experimental manipulations regarding either data "valuation" or "integration" (to use Anderson's, 1973, terms).

Most judgment researchers appear to have resolved this dilemma by attempting to eliminate the need for valuation. As suggested in the previous section, stimuli are usually unambiguous and complete, so that the researcher can assume that the subject takes them at face value and adds nothing to them. In addition to the Bayesian inference experiments, excellent examples of this strategy are the lens model studies of subjective judgment (for example Hammond et al., 1964) in which the relevant dimensions of the stimulus object are defined and its value on each dimension expressed by a simple number. Most experiments are designed to guarantee that subjects have no previous relevant experiences to bring to the task and have no emotional involvement (other than desire to utilize the information supplied optimally) which might lead them to distort the data.

As noted by Anderson (1973), much attribution work appears to be concentrated on questions of valuation. Jones and Nisbett (1972), for example, have argued that differences in perspective lead to differences in the way in which observers and actors valuate behavior. Lerner's work (for example, 1970) on the attributional consequences of people's "belief in a just world" has shown how generally accepted beliefs can distort people's valuation of data used in their attributions. Research into the attribution of causes for success and failure may be seen as studies of subjects' naive laws about how the world works, revealing whether or not they believe that to succeed on hard tasks one must invest both ability and effort. In order to make confident inferences about how people valuate events from the ways in which they respond to them, the researcher must assume rather simple, straightforward and manageable integration processes. Such an assumption is reasonable when the integration called for is quite simple, either because of the nature of the problem or the help extended to the subject by the experimenter. Most attribution studies fit this requirement. By tapping subjects' previous experiences, attribution studies may afford their subjects an additional advantage. In life, even when people do not possess the programs necessary to solve inferential problems, they may still arrive at a proper solution through trial and error. Judgment research has been directed specifically at helping people solve urgent problems (such as dealing with nuclear power) for which we lack both the proper cognitive programs and the time and resources to learn by trial and error (Slovic et al., 1976).

Simplification of either the valuation or the integration operation is, of course, a valid research strategy. The price it carries is some loss of generality. The generalization of current attribution theories to complex events with probabilistic underlying processes is not obvious. Nor is it clear how existing theories of judgment can incorporate phenomena like wishful thinking or the effects of time and social pressure.

Although a complete understanding of how specific individual decisions are made is impossible without relating to these issues, it is worth noting what judgment researchers hope to gain by their strategy. The goal is an understanding of how well people perform judgmental tasks under the best of conditions. Performance there establishes an asymptotic level beneath which people will perform when confronted with motivational pressures, misleading information, and the like. Establishing this asymptote appears to precede logically encumbering people with these factors.

Use of Normative Theories

Obviously, identifying biased behavior requires a clear conception of behavioral adequacy. Generally speaking, such a conception is central to judgment research, tangential to attribution work. It is, then, little wonder that one area has discovered biases and the other has not.

In part, this difference is due to the basic theoretical or philosophical orientation of the two fields. Attribution theory is a fundamentally phenomenological approach, to understand what people do when they do what seems right to them. In one context, Kelley (1972a) remains unconvinced of the apparent irrationality of a number of seeming "attribution biases"; in another context (1972a) he suggests that even when time pressure prevents a complete covariance analysis, "the lay attributor [proceeds] in a reasonable and unbiased manner [p. 2]." Although his covariation principle, which people are believed to follow intuitively, is derived from a normative theory of behavior, Mills' "Law of Difference," the optimality of subjects' adherence to it is nowhere seriously questioned.

Judgment researchers have been concerned from the start with the question of subjective optimality: how well people are able to maximize their attainment of subjective criteria. Researchers view cognitive limitations as a major obstacle in such attainment. Moreover, they see suboptimality as a problem with serious social and personal consequences. Slovic, Fischhoff, and Lichtenstein (1976, in press) state:

> The regulation of risk poses serious dilemmas for society. Policy makers are being asked, with increasing frequency, to "weigh the benefits against the risks" when making decisions about social and technological programs. These individuals often have highly sophisticated methods at their disposal for gathering information about problems or constructing technological solutions. When it comes to making decisions, however, they typically fall back upon the technique which has been relied upon since antiquity–intuition. The quality of their intuitions sets an upper limit on the quality of the entire decision-making process and, perhaps, the quality of our lives. There is an urgent need to link the study of man's judgmental and decision-making capabilities to the making of decisions that affect the health and safety of the public [p. 1].

In addition, they have sought to improve decision-making in areas as diverse as admission to graduate school (Dawes, 1971), investigative radiology (Slovic, Rorer & Hoffman, 1971), experimental design (Tversky & Kahneman, 1971), and adjustment to natural hazards (Slovic et al., 1974).

Ironically, it would appear as though the more humanistic, phenomenological approach (attribution) shows greater respect for people's intuitive capacities at the price of being able to do relatively little to help them. The ostensibly less humanistic, more mathematically inclined field of judgment may produce useful tools for helping people manage their decisions and learn from experience.

Recent judgment work by Tversky and Kahneman (1974) and others has actually gone in a more phenomenological direction, attempting to understand in nonmechanistic terms, the judgmental heuristics, or rules of thumb, that people use in trying to interpret their world and predict the future. Their focus, however, has been on seeing how these phenomenological rules may lead people astray. A similar compromise would seem possible in the attribution context. Carroll and Payne's (1976) study of the parole decision process is oriented in just this direction.

To some extent, however, the relative predominance of normative theories in judgment research seems to reflect extrapsychological realities. Perhaps the foremost of these is the differential accessibility of normative theories for predictive and attributional behavior.

The judgment researcher interested in exploring the descriptive validity of a normative theory of decision-making or probabilistic reasoning has no difficulty in finding and learning well-developed, easily operationalized, and widely agreed upon examples. The subjective expected utility model (Edwards & Tversky, 1967; Feather, 1959; Raiffa, 1968) and the familiar postulates of the probability calculus are but two examples.

By way of contrast, there is no generally agreed upon criterion for explanatory adequacy. There is, instead, a continuing discussion among philosophers of science about what constitutes the proper normative theory. In this debate even the intuitively obvious criterion that a good explanation is one that increases predictive ability is not without its critics. Furthermore, the debate itself is buried in philosophical literature which few psychologists are either familiar with or trained to interpret. Nor have many serious attempts been made to bring to the general public useful statements of the current state of the art, showing how it currently appears that one might best go about making attributions.

In view of this confused state of affairs, it may be tempting to ignore the work of philosophers, at least until they get their normative standards in order. This would, however, be a mistake, for even without producing the "ultimate truth" about explanation, the philosophers have identified logical subtasks that are common to many modes of explanation and attribution and for which behavioral adequacy can be readily defined. If people have difficulty performing any of these subtasks, their ability to meet the demands of any plausible normative theory is suspect. For example, attribution of almost any sort requires simple syllogistic reasoning, testing behavioral laws, and recognizing counterexamples. Wason and Johnson-Laird (1972) have found that subjective inference on tasks like these is often erroneous.

Philosophers can also contribute insight into how to characterize inferential tasks properly. Often there are subtle distinctions between types of tasks that might be casually passed over in planning experiments or developing theories but that may have the greatest consequences in the modes of inference they elicit. Shope (1967), for example, has analyzed in depth the conditions in which it makes sense to speak of the (one) cause of an event rather than the causes, a distinction of obvious importance to attribution work.

The incredible confusion in the attribution of responsibility literature (see Fishbein & Ajzen, 1973; Vidmar & Crinklaw, 1974) due to failure to define the term "responsibility" both precisely and consistently might serve as an illustrative example of how psychologists' vagueness about their basic concepts can strip their work of its value (see also Rozeboom, 1972, 1974). Since conceptual analysis requires training that many psychologists lack, it seems appropriate to exploit the groundwork already laid by others. Without such broad and detailed conceptualizations it may be difficult for attribution research ever to go beyond the substantive areas in which it has made great headway to date and evolve into a general psychology of explanation.

Among the many other topics that should interest attribution theorists and that have received detailed philosophical analysis are: How may causes be weighted (Martin, 1972)? How is an "event," the object of an explanation, defined (Pachter, 1974)? When and how may motives be imputed to someone (MacIver, 1940; Paulson, 1972)? How should dispositional attributions be interpreted (Rozeboom, 1973)? What is the logical form of causal statements (Davidson, 1967)? Insightful comments on these questions, and many others, may be found in Hempel (1965). A well-developed multivariate normative model of causal attribution that is quantitative and probabilistic and which has a considerable following is path analysis (for example, Alwin & Tessler, 1974; Blalock, 1971; Lewis-Beck, 1974). It certainly seems possible to study how good people are as intuitive path analysts. Some general thoughts on the use of normative models in the study of cognition may be found in Barclay, Beach, and Braithwaite (1971) and Little (1972).

It seems appropriate to mention in this context an issue which arose quite early in judgment research, apparently because of the use of quantitative models, and which attribution researchers may find it insightful to work through. It is what Hoffman (1960) called the paramorphic representation of clinical judgment. Kelley (1972b) stated that "We do indeed wish and need to know the terms in which the lay attributor thinks about causal problems [p. 171]." Regarding the way to obtain this knowledge, he wrote, "It seems unlikely that an 'as if' model that has little correspondence to the attributor's actual modes of information processing will succeed in anticipating and summarizing all the important details of his activities [p. 171]." Hoffman's analysis showed, however, that very different models of information processing may be reflected in identical input–output relationship and may have indistinguishable formal characterizations.

CONCLUSION

If the above characterization is generally correct, probably the most intriguing question to emerge from it is, "Just how good are people as intuitive information processors?" The answer, like the answer to most intriguing questions, appears to be "It depends." I have tried to give some idea of what it depends on, particularly discussing those features of attributional research that appear to encourage or highlight good performance and those features of judgmental research that appear to discourage or obscure it. A more definitive answer requires both the collection of more directly comparable results, and some conceptual reconciliation between the areas. Whereas I believe that such integration is possible and have attempted to stress the basic commensurability of these two paradigms, some thorny—and not uninteresting—issues remain to be resolved.

The Role of Error Analysis

One underlying question is whether one learns more about behavior by asking "What do people do?" or "What do they do wrong?" Attribution and judgment researchers seem to end up asking the first and second questions, respectively. Whereas a general answer to this question is beyond the scope of the present paper, it is worth noting that a similar conflict has faced investigators in at least one other area of verbal behavior, the applied linguistics of second language learning. There, the respective approaches are called "contrastive" and "error" analysis. As described by Hammarberg (1974), the error analyst attempts to understand the types, frequency, and causes of linguistic errors, as well as the degree of disturbance that they cause and how they can be ameliorated. The perceived advantages of this approach to understanding and improving language use arise from the assumptions that errors (a) are evidence of speakers' basic linguistic strategies, (b) reveal how far speakers have progressed in language acquisition and how stable their performance is at that level, and (c) can be used to instruct speakers about their own inadequacies. For those concerned with demand characteristics that might lead subjects to respond unnaturally in order to impress the experimenter, error analysis has an additional advantage. There appears to be no reason why subjects would deliberately respond incorrectly unless they really believed that they were right. The primary difficulties with error analysis are that it is often difficult to define unambiguously what is an error and what is not and that to help someone, it is frequently crucial to know what people are capable of doing correctly.

A related analogy is Hexter's (1971) comparison between explaining historical events and playing the field in baseball. In both pursuits, most "chances" (calls to explain an event or catch a ball) are quite routine, often having habitual, preprogrammed acceptable responses.

> Fielding easy chances calls for a very complex set of motions not vastly different from what it takes to field the hard ones, but no one becomes a big leaguer because he can catch an easy grounder and make the easy throw to first. A big leaguer may even make more errors than an amateur, but that is because he gets within reach of balls that others would not even get near. In deciding who is fit to stay in the big leagues, the question is 'Can he field the hard ones?' [p. 51].

If the analogy drawn above between the respective positions of error analysis and judgment research is indeed valid, detailed examination of the specific issues in the two areas may well prove illuminating.

Mainstreaming

To some extent, the actual interrelation of the aspects of inferential behavior revealed by judgment and attribution researchers will only be understood when work in these two areas is properly coupled with what is going on in the rest of psychology. Although the concept of "bounded rationality" so central to judgment work arose out of developments in cognitive psychology in the 1950s (for example, Bruner, Goodnow, & Austin, 1956; Miller, 1956; Simon, 1957), little contact has been made since then even with such closely related fields as the study of nonprobabilistic information processing. With attribution research the situation is somewhat better (Kanouse, 1972; Kelley, 1973; Weiner, 1974; Weiner & Kun, 1976), yet the field is still autonomous in many respects.

Just as the phenomena described here cannot be fully understood without consideration of their underlying cognitive mechanisms, some of them should provide stimulating inputs for general work in cognition. The hindsight results (Fischhoff, 1975), for example, indicate one way in which semantic memory is reorganized to accommodate new information. The paucity of results and theories showing exactly how Tversky and Kahneman's (1973) availability mechanism might work suggests a need for further research into the process of constrained associates production.

Even more exciting might be exploration of the developmental implications of this work. For example, no theory of cognitive development appears to relate fully to the notion of judgmental biases and heuristics as presented here. Many conceptualizations of cognitive development are primarily concerned with how children acquire the skills that will make them fully functioning adults. Consideration of judgment work might lead us to look at how children acquire the heuristics that lead them to be substantially biased adult information processors, why neither age nor experience appears to eradicate these biases (Goldberg, 1968), and what we might do to educate people to do probabilistic, multivariate thinking (Michael, 1968, 1973).

Prediction and Explanation

One obvious prerequisite to integrating work in judgment and attribution is to understand the formal and psychological relationships between prediction and explanation. Common sense appears to hold that the two are highly interrelated, both because of perceived similarities in the underlying processes and because increased prowess in one is seen as conferring increased prowess in the other. When we manage to explain the past, we fell that we have increased our ability to predict the future. The main perceived difference seems to be that we can adequately explain more things than we can predict because we know more in hindsight than in foresight.

Both our dominant philosophy of science, which holds that scientific prediction and explanation are formally identical (for example, Hempel, 1965), and those attribution theorists who have related to the question (for example, Kelley, 1972a; Weiner, Frieze, Kukla, Reed, Rest, & Rosenbaum, 1972, p. 96) appear to subscribe to this view.

In studies of hindsight (Fischhoff, 1974, 1975; Fischhoff & Beyth, 1975) however, it is indicated that people process information about the past in a way that systematically reduces its perceived suprisingness. We argued that only when confronted with surprises do we feel any need to change our way of looking at and responding to events, that is, any need to learn. Thus, the very feeling that we have explained, or made sense out of, an event may be the best guarantee that we are not learning anything from it that will improve our predictive efficacy.

One way of interpreting this result would be in terms of the ego defensive bias noted by Kelley (1972a) which reflects our need to feel that the world is controllable. Such a need is certainly served by exaggerating how much we know about it. Here, as with other forms of denial, the long-range acquisition of coping skills is sacrificed for the short-range illusion of coping ability.

Yet before "resorting" to such a motivational explanation, it is worth considering whether prediction and explanation are formally part of the same process. One argument to the contrary is offered by Hintikka (1968) who distinguished between local and global theorizing. He described these, respectively, as "on the one hand, a case in which we are predominantly interested in a particular body of observations *e* which we want to explain by means of a suitable hypothesis *h*, and on the other hand, a case in which we have no particular interest in our evidence *e* but rather want to use it as a stepping stone to some general theory *h*, which is designed to apply to other matters, too, besides *e* [p. 321]." In the present context, *e* may be likened to a reported event and *h* to the set of data and laws from which that event is to be inferred.

Hintikka (1968) further notes, "Regarding local theorizing, we want to choose the explanatory hypothesis *h* such that it is maximally informative concerning the subject matter with which *e* deals. Since we know the truth of *e* already, we are not interested in the substantive information that *h* carries concerning the truth of *e*. What we want to do is to find *h* such that the truth of *e* is not unexpected, given *h* [p. 321]." Hintikka then shows that this leads to the choice of *h* according to the maximum likelihood principle, "a weapon of explanation rather than of generalization." [p. 321] The extent to which the maximum likelihood principle is also a weapon of generalization seems to depend upon the regularity of the universe from which *e* has been drawn and to which *h* may be applied, i.e., to the extent to which "whatever observations we make concerning a part of it can be carried over intact so as to apply to others [pp. 322–323]."

With the unique events considered in most explanations, this irregularity is likely to be substantial. Often our explanations will be so good in the specific case that generalizability is sacrificed. An analogous case can be seen in the regression equation which is "overfit" to a set of data (for example, by inclusion of too many predictor variables). The price paid for closeness of fit is loss of predictive validity–shrinkage (see also Stover, 1967).

Additional biases in explanation will doubtless be forthcoming, particularly in future research in which open-ended questions like "When do people explain events?" "For whom do they prepare their explanations?" and "What are their subjective criteria for explanatory adequacy?" are asked. In discussing "social psychology's rational man," Abelson (1974) has compiled a partial list of reasons why people may hold beliefs other than for the sake of rationality. Most of these reasons are concerned with systems-maintenance, ways of keeping oneself going in a difficult and unpredictable world. They include: as protection against anxiety, as a way to organize vague feelings, and as a means of providing a sense of identity. A similar list of reasons for explaining events other than to increase one's predictive abilities may one day be forthcoming. On that day, we may also be able to help those who are interested in increasing the positive transfer between prediction and explanation.

A Possible Reconceptualization

Kelley (1973, p. 112) has likened man to an intuitive scientist. In doing so, he has defined scientist by projection, adopting the interpretation commonly accepted among experimental psychologists, that of the intuitive analyzer of variance. Yet, there certainly are other reconstructions of the scientific process (for example, Kaplan, 1964; Lakatos, 1970), and one might ask whether another conceptualization might be more appropriate.

Probably the most insightful discussions of how and why people do and should explain the past events may be found in the ruminations of historians over the state and nature of their craft (for example, Beard, 1935; Carr, 1961; Commager, 1965; Hexter, 1961; Marwick, 1970; Plumb, 1969). Many of these analysts (for example, Dray, 1962; Gallie, 1964; Hexter, 1971; Passmore, 1962; Scriven, 1959; Walsh, 1967) have focused on how historical explanation differs in form and purpose from the notion of scientific explanation as proposed in the covering law model advanced by Hempel (1965) and others. Typically, they argue that historians explain for much the same reasons and in much the same way as ordinary people do. They also discuss the particular training needed to produce effectively explanations of this type. Some of their analyses offer rich inputs to understanding the nature and purpose of "explanation in every-day life, science, and history" (Passmore, 1962, p. 105).

Before abandoning, as suggested by the judgment results, or embracing as suggested by attribution theory, the notion of people as intuitive scientists, we should ask what sort of scientists they are or attempt to be. We might get a good deal of mileage out of thinking of ourselves as intuitive historians and attempting to produce an integrated psychology of predictive and explanatory behavior that accommodates the historians' observations, the philosophers' formalizations, and the psychologists' and the sociologists' theories and empirical findings.

ACKNOWLEDGMENTS

Support for this paper was provided by the Advances Research Projects Agency of the Department of Defense (ARPA Order No. 2449) and was monitored by Office of Naval Research under Contract No. N00014-73-C-0438 (NR 197-026). I am indebted to Tom Climo, Lita Furby, Sol Fulero, Bernie Goitein, Sarah Lichtenstein, Paul Slovic, and Bernard Weiner for comments that greatly improved the clarity of this paper.

REFERENCES

Abelson, R. Social psychology's rational man. In G. W. Mortmer & S. I. Benn (Eds.), *The concept of rationality in the social sciences.* London: Routledge & Kegan Paul, 1974.

Ajzen, I., & Fishbein, M. A Bayesian analysis of attribution processes. *Psychological Bulletin,* 1975, **82,** 261–277.

Alpert, M., & Raiffa, H. A progress report on the training of probability assessors. Unpublished manuscript, Harvard University, 1968.

Alwin, D. F., & Tessler, R. C. Causal models, unobserved variables, and experimental data. *American Journal of Sociology,* 1974, **80,** 58–86.

Anderson, N. H. Functional measurement and psychophysical judgment. *Psychological Review,* 1970, 77, 153–170.

Anderson, N. H. Integration theory and attitude change. *Psychological Review,* 1971, **78,** 171–206.

Anderson, N. H. Cognitive algebra: Integration theory as applied to social attribution. In L. Berkowitz (Ed.), *Advances in experimental social psychology* (Vol. 7). New York: Academic Press, 1973.

Barclay, S., Beach, L. R., & Braithwaite, W. P. Normative models in the study of cognition. *Organizational Behavior and Human Performance,* 1971, **6,** 387–413.

Beard, C. A. The case for historical relativism: The noble dream. *American Historical Review,* 1935, **40,** 74–87.

Blalock, H. M. *Causal models in the social sciences.* Chicago: Aldine–Atherton, 1971.

Bowers, K. S. The psychology of subtle control: An attributional analysis of behavioral persistance. *Canadian Journal of Behavioral Science,* 1975, **7,** 78–95.

Bruner, J. S., Goodnow, J. J., & Austin, G. A. *A study of thinking.* New York: John Wiley & Sons, 1956.

Brunswick, E. *The conceptual framework of psychology.* Chicago: University of Chicago Press, 1952.

Burton, U., & Kates, R. W. The perception of natural hazards in resource management. *Natural Resources Journal,* 1964, **3,** 412–441.

Carr, C. E. *What is history?* Hammondsworth, England: Penguin, 1961.

Carroll, J. S., & Payne, J. W. The psychology of the parole decision process. In J. S. Carroll & J. W. Payne (Eds.), *Eleventh Carnegie symposium on cognition.* Hillsdale, New Jersey: Lawrence Erlbaum Associates, 1976.

Chapman, L. J., & Chapman, J. P. Genesis of popular but erroneous psychodiagnostic observations. *Journal of Abnormal Psychology,* 1967, **72,** 193–204.

Chapman, L. J., & Chapman, J. P. Illusory correlation as an obstacle to the use of valid psychodiagnostic signs. *Journal of Abnormal Psychology,* 1969, **74,** 271–280.

Cialdini, R. B., Braver, S. L., & Lewis, S. K. Attributional bias and the easily persuaded other. *Journal of Personality and Social Psychology,* 1974, **30,** 631–637.

Cohen, J., & Christensen, I. *Information and choice.* Edinburgh: Oliver & Boyd, 1970.

Commager, H. S. *The nature and study of history.* Columbus, Ohio: C. E. Merrill, 1965.

Cunningham, J. D., & Kelley, H. H. Causal attributions for personal events of varying magnitude. *Journal of Personality,* 1975, **43,** 74–93.

Dale, H. C. A. Weighing evidence: An attempt to assess the efficiency of the human operator. *Ergonomics,* 1968, **11,** 215–230.

Davidson, D. Causal relations. *Journal of Philosophy,* 1967, **64,** 691–703.

Dawes, R. M. A case study of graduate admissions. *American Psychologist,* 1971, **26,** 180–188.

Dawes, R. M., & Corrigan, B. Linear models in decision making. *Psychological Bulletin,* 1974, **81,** 95–106.

Dray, W. The historian's problem of selection. In E. Nagel, P. Suppes, & A Tarski (Eds.), *Logic, methodology, and the philosophy of science,* Stanford, California: Stanford University Press, 1962.

Edwards, W., Lindman, H., & Savage, L. J. Bayesian statistical inference for psychologists. *Psychological Review,* 1963, **70,** 193– 242.

Edwards, W., & Tversky, A. *Decision making.* Baltimore: Penguin, 1967.

Enzel, M. E., Hansen, R. A., & Lowe, C. E. Causal attributions in the mixed motive game. *Journal of Personality and Social Psychology,* 1975, **31,** 50–54.

Fama, E. F. Random walks in stock market prices. *Financial Analysts Journal,* 1965, **21,** 55–60.

Feather, N. T. Subjective probability and decision under uncertainty. *Psychological Review,* 1959, **66,** 150–163.

Feldman–Summers, S., & Kiesler, S. B. Those who are number two try harder. *Journal of Personality and Social Psychology,* 1974, **30,** 846–855.

Feller, W. *An introduction to probability theory and its applications* (Vol. 1). New York: John Wiley & Sons, 1968.

Fischhoff, B. Hindsight: Thinking backward? *Oregon Research Institute Research Monograph,* 1974, **14**(1). (Reprinted in *Psychology Today,* 1975, **3,** 4.)

Fischhoff, B. Hindsight ≠ Foresight: The effect of outcome knowledge on judgment under uncertainty. *Journal of Experimental Psychology: Human Perception and Performance,* 1975, **1,** 288–299.

Fischhoff, B., & Beyth, R. "I knew it would happen"–remembered probabilities of once-future things. *Organizational Behavior and Human Performance,* 1975, **13,** 1–16.

Fischhoff, B., & Lichtenstein, S. Don't blame this on Reverend Bayes. *Oregon Research Institute Research Bulletin,* 1975, **15**(5).

Fishbein, M., & Ajzen, I. Attribution of responsibility: A theoretical note. *Journal of Experimental Social Psychology,* 1973, **9,** 148–153.

Frieze, I., & Weiner, B. Cue utilization and attributional judgments for success and failure. *Journal of Personality,* 1971, **39,** 591–606.

Gallie, W. B. *Philosophy and the historical understanding.* London: Chatto & Windus, 1964.

Garfinkel, H. Studies of the routine grounds of everyday activities. *Social Problems,* 1964, **11,** 225–250.

Garfinkel, H. *Studies in ethnomethodology.* Englewood Cliffs, New Jersey: Prentice–Hall, 1966.

Garland, H., Hardy, A., & Stephenson, L. Information search as affected by attribution type and response category. *Personality and Social Psychology Bulletin,* in press.

Goldberg, L. R. Simple models or simple processes? Some research on clinical judgments. *American Psychologist,* 1968, **23,** 483–496.

Goldberg, L. R. Man vs. model of man: A rationale, plus some evidence, for a method of improving on clinical inferences. *Psychological Bulletin,* 1970, **73,** 422–432.

Golding, S. L., & Rorer, L. G. "Illusory correlation" and the learning of clinical judgment. *Journal of Abnormal Psychology,* 1972, **80,** 249–260.

Hammarberg, B. The unsufficiency of error analysis. *International Review of Applied Linguistics in Language Teaching,* 1974, **12,** 185–192.

Hammerton, M. A case of radical probability estimation. *Journal of Experimental Psychology,* 1973, **101,** 252–254.

Hammond, K. R. Probabilistic functionalism: Egon Brunswik's integration of the history, theory and method of psychology. In K. R. Hammond (Ed.), *The psychology of Egon Brunswik.* New York: Holt, Rinehart, & Winston, 1966.

Hammond, K. R., Hursch, C. J., & Todd, F. J. Analyzing the components of clinical inference. *Psychological Review,* 1964, **71,** 438–456.

Hammond, K. R., & Summers, D. A. Cognitive control. *Psychological Review,* 1972, **79,** 58–67.

Hempel, C. G. *Aspects of scientific explanation.* New York: Free Press, 1965.

Hexter, J. *Reappraisals in history.* Evanston, Illinois: Northwestern University Press, 1961.

Hexter, J. *The history primer.* New York: Basic Books, 1971.

Hintikka, J. The varieties of information and scientific explanation. In N. van Rootselaar & R. Stael (Eds.), *Logic, methodology, and the philosophy of science.* Amsterdam: North-Holland, 1968.

Hoffman, P. J. The paramorphic representation of clinical judgment. *Psychological Bulletin,* 1960, 57, 116–131.

Hoffman, P. J., Solvic, P., & Rorer, L. G. An analysis–of–variance model for the assessment of configural cue utilization in clinical judgment. *Psychological Bulletin,* 1968, **69,** 338–349.

Jarvik, M. E. Probability learning and negative recency effect in the serial anticipation of alternative symbols. *Journal of Experimental Psychology,* 1951, **41,** 291–297.

Jenkins, H. M., & Ward, W. C. Judgment of contingency between responses and outcomes. *Psychological Monographs,* 1965, **79,** (1, Whole No. 594).

Jones, E. E., & Davis, K. E. From acts to dispositions: The attribution process in person perception. In L. Berkowitz (Ed.), *Advances in experimental social psychology* (Vol. 2). New York: Academic Press, 1965.

Jones, E. E., & Goethals, G. R. Order effects in impression formation: Attribution context and the nature of the entity. In E. E. Jones, D. E. Kanouse, H. H. Kelley, R. E. Nisbett, S. Valins, & B. Weiner (Eds.), *Attribution: Perceiving the causes of behavior.* Morristown, New Jersey: General Learning Press, 1972.

Jones, E. E., & Nisbett, R. E. The actor and the observer: Divergent perceptions of the causes of behavior. In E. E. Jones, D. E. Kanouse, H. H. Kelley, R. E. Nisbett, S. Valins, & B. Weiner (Eds.), *Attribution: Perceiving the causes of behavior.* Morristown, New Jersey: General Learning Press, 1972.

Kahneman, D., & Tversky, A. Subjective probability: A judgment of representativeness. *Cognitive Psychology,* 1972, **3,** 430–454.

Kahneman, D., & Tversky. A. On the psychology of prediction. *Psychological Review,* 1973, **80,** 237–251.

Kanouse, D. E. Language, labeling, and attribution. In E. E. Jones, D. E. Kanouse, H. H. Kelley, R. E. Nisbett, S. Valins, & B. Weiner (Eds.), *Attribution: Perceiving the causes of behavior.* Morristown, New Jersey: General Learning Press, 1972.

Kaplan, A. *The conduct of inquiry.* San Francisco: Chandler, 1964.

Kates, R. W. *Hazard and choice perception in flood plain management.* Chicago: University of Chicago, Department of Geography, Research Paper No. 78, 1962.

Kelley, H. H. Attribution theory in social psychology. In D. Levines (Ed.), *Nebraska Symposium on Motivation.* Lincoln: University of Nebraska Press, 1967.

Kelley, H. H. Attribution in social interaction. In E. E. Jones, D. E. Kanouse, H. H. Kelley, R. E. Nisbett, S. Valins, & B. Weiner (Eds.), *Attribution: Perceiving the causes of behavior.* Morristown, New Jersey: General Learning Press, 1972. (a)

Kelley, H. H. Causal schemata and the attribution process. In E. E. Jones, D. E. Kanouse, H. H. Kelley, R. E. Nisbett, S. Valins, & B. Weiner (Eds.), *Attribution: Perceiving the causes of behavior.* Morristown, New Jersey: General Learning Press, 1972. (b)

Kelley, H. H. The processes of causal attribution. *American Psychologist,* 1973, **28,** 107–128.

Kleinmuntz, B. The processing of clinical information by man and machine. In B. Kleinmuntz (Ed.), *Formal representation of human judgment.* New York: John Wiley & Sons, 1968.

Kuhn, T. *The structure of scientific revolution.* Chicago: University of Chicago Press, 1962.

Kukla, A. Foundations of an attributional theory of performance. *Psychological Review,* 1972, 79, 454–470.

Kun, A., & Weiner, B. Necessary versus sufficient causal schemata for success and failure. *Journal of Research in Personality,* 1973, 7, 197–207.

Lakatos, I. Falsification and scientific research programmes. In I. Lakatos & A. Musgrove (Eds.), *Criticism and the growth of scientific knowledge.* Cambridge: Cambridge University Press, 1970.

Lerner, M. J. The desire for justice and reactions to victims. In J. Macauley & L. Berkowitz (Eds.), *Altruism and helping behavior.* New York: Academic Press, 1970.

Lewis-Beck, M. S. Determining the importance of an independent variable: A path analytic solution. *Social Science Research,* 1974, **3,** 95–107.

Lichtenstein, S., & Feeney, G. J. The importance of the data-generating model in probability estimation. *Organizational Behavior and Human Performance,* 1968, **3,** 62–67.

Lichtenstein, S., Fischhoff, B., & Phillips, L. D. Calibration of probabilities: The state of the art. In H. Jungerman & G. de Zeeuw (Eds.), *Proceedings of the fifth conference on subjective probability, utility and decision-making, Darmstadt, 1975.* In press.

Lichtenstein, S., Earle, T., & Slovic, P. Cue utilization in a numerical prediction task. *Journal of Experimental Psychology: Human Perception and Performance,* 1975, **104,** 77–85.

Little, B. R. Psychological man as scientist, humanist and specialist. *Journal of Experimental Research in Personality,* 1972, **6,** 95–118.

Loftis, J., & Ross, L. Retrospective misattribution of a conditioned emotional response. *Journal of Personality and Social Psychology,* 1974, **30,** 683–687.

Luginbuhl, J. E. R., Crow, D. H., & Kahan, J. P. Causal attribution for success and failure. *Journal of Personality and Social Psychology,* 1975, **31,** 86–93.

Luce, R. D., & Suppes, P. Preference, utility and subjective probability. In R. D. Luce, R. R. Bush, & E. H. Galanter (Eds.), *Handbook of mathematical psychology* (Vol. 3). New York: John Wiley & Sons, 1965.

Lyon, D., & Slovic, P. Dominance of accuracy information and neglect of base rates in probability estimation. *Acta Psychologica,* 1976, in press.

MacArthur, L. A. The how and what of why. *Journal of Personality and Social Psychology,* 1972, **22,** 171–193.

McGuire, W. J. Theory of the structure of human thought. In R. P. Abelson, E. Aronson, W. S. McGuire, T. M. Newcomb, M. S. Rosenberg, & P. H. Tannenbaum (Eds.), *Theories of cognitive consistency: A sourcebook.* Chicago: Rand McNally, 1968.

MacIver, R. M. The imputation of motives. *American Journal of Sociology,* 1940, **46,** 1–12.

Martin, R. On weighing causes. *American Philosophical Quarterly,* 1972, **9,** 21–29.

Marwick, A. *The nature of history.* London: Macmillan, 1970.

Meehl, P. E. *Clinical versus statistical prediction.* Minneapolis: University of Minnesota Press, 1954.

Meehl, P. E., & Rosen, A. Antecedent probability and the efficiency of psychometric signs, patterns or cutting scores. *Psychological Bulletin,* 1955, **52,** 194–216.

Messick, D. M., & Rapoport, A. A comparison of two pay-off functions in multiple choice decision behavior. *Journal of Experimental Psychology,* 1965, **69,** 75–83.

Michael, D. N. *The unprepared society.* New York: Basic Books, 1968.

Michael, D. N. *Learning to plan and planning to learn.* San Francisco: Jossey–Bass, 1973.

Miller, D. T., & Ross, M. Self-serving biases in the attribution of causality: Fact or fiction? *Psychological Bulletin,* 1975, **82,** 213–225.

Miller, G. A. The magical number seven, plus or minus two: Some limits on our capacity for processing information. *Psychological Review,* 1956, **63,** 81–92.

Mischel, W. *Personality and assessment.* New York: John Wiley & Sons, 1968.

Nisbett, R. E., & Borgida, E. Attribution and the psychology of prediction. *Journal of Personality and Social Psychology,* 1975, **32,** 932–945.

Oldman, D. Chance and skill: A study of roulette. *Sociology,* 1974, **8,** 407–426.

O'Leary, M. K., Coplin, W. D., Shapiro, H. B., & Dean, D. The quest for relevance. *International Studies Quarterly,* 1974, **18,** 211–237.

Pachter, H. M. Defining an event: Prolegomenon to any future philosophy of history. *Social Research,* 1974, **44,** 439–466.

Passmore, J. Explanation in everyday life, in science, and history. *History and Theory,* 1962, **2,** 105–123.

Paulson, S. L. Two types of motive explanation. *American Philosophical Quarterly,* 1972, **9,** 193–199.

Payne, J. W. Task complexity and contingent processing in decision making: An information search and protocol analysis. *Organizational Behavior and Human Performance,* in press.

Peterson, C. R., & Beach, L. R. Man as an intuitive statistician. *Psychological Bulletin,* 1967, **68,** 29–46.

Phillips, L. D. *Bayesian statistics for social scientists.* London: Nelson, 1973.

Pitz, G. F., Downing, L., & Reinhold, H. Sequential effects in the revision of subjective probabilities. *Canadian Journal of Psychology,* 1967, **21,** 381–393.

Plumb, J. H. *The death of the past.* London: Macmillan, 1969.

Prus, R. C. Labeling theory: A reconceptualization and a propositional statement on typing. *Sociological Focus,* 1975, **8,** 79–96.

Raiffa, H. *Decision analysis.* Reading, Massachusetts: Addison–Wesley, 1968.

Rogers, J. W., & Buffalo, M. D. Neutralization techniques. *Pacific Sociological Review,* 1974, **17,** 313–333.

Ross, L. Distortion in the social perception process: The production and perseverance of attributional biases and errors. In L. Berkowitz (Ed.), *Advances in social psychology,* in press.

Ross, L., Bierbrauer, G., & Polly, S. Attribution of educational outcome by professional and non-professional instructors. *Journal of Personality and Social Psychology,* 1974, **29,** 609–618.

Rozeboom, W. W. Problems in the psychology of knowledge. In J. R. Royce & W. W. Rozeboom (Eds.), *The psychology of knowing.* New York: Gordon & Breach, 1972.

Rozeboom, W. W. Dispositions revisited. *Philosophy of Science,* 1973, **40,** 59–74.

Rozeboom, W. W. Metathink. Paper presented at 82nd Convention of the American Psychological Association, New Orleans, 1974.

Schur, E. M. *Labeling deviant behavior: Its sociological implications.* New York: Harper & Row, 1971.

Scott, M. B. *The racing game.* Chicago: Aldine, 1968.

Scott, M. B., & Lyman, S. Accounts. *American Sociological Review,* 1968, **33,** 46–62.

Scriven, M. Definitions, explanations and theories. *Minnesota Studies in the Philosophy of Science,* 1959, **2,** 99–195.

Shaver, K. G. *An introduction to attribution processes.* Cambridge, Massachusetts: Winthrop, 1975.

Shepard, R. N. On subjectively optimum selection among multiattribute alternatives. In M. W. Shelly, II., & G. L. Bryan (Eds.), *Human judgments and optimality.* New York: John Wiley & Sons, 1964.

Shope, R. K. Explanations in terms of "the cause." *Journal of Philosophy,* 1967, **64,** 312–318.

Simon, H. *Models of man: Social and rational.* New York: John Wiley & Sons, 1957.

Slovic, P. From Shakespeare to Simon: Speculations–and some evidence–about man's ability to process information. *Oregon Research Institute Research Monograph,* 1972, **12**(1).

Slovic, P., Fischhoff, B., & Lichtenstein, S. Cognitive processes and societal risk-taking. In J. S. Carroll & J. W. Payne (Eds.), *Eleventh Carnegie Symposium on Cognition,* Hillsdale, New Jersey: Lawrence Erlbaum Associates, 1976, in press.

Slovic, P., Fleissner, D., & Bauman, W. S. Analyzing the use of information in investment decision making: A methodological proposal. *Journal of Business,* 1972, **45,** 283–301.

Slovic, P., Kunreuther, H., & White, G. F. Decision processes, rationality and adjustment to natural hazards. In G. F. White (Ed.), *Natural hazards, local, national and global.* New York: Oxford University Press, 1974.

Slovic, P., & Lichtenstein, S. Comparison of Bayesian and regression approaches to the study of human information processing in judgment. *Organizational Behavior and Human Performance,* 1971, **6,** 649–744.

Slovic, P., & MacPhillamy, D. J. Dimensional commensurability and cue utilization in comparative judgment. *Organizational Behavior and Human Performance,* 1974, **11,** 172–194.

Slovic, P., Rorer, L. G., & Hoffman, P. J. Analyzing the use of diagnostic signs. *Investigative Radiology,* 1971, **6,** 18–26.

Smedslund, J. The concept of correlation in adults. *Scandiavian Journal of Psychology,* 1963, **4,** 165–173.

Storms, M. D., & Nisbett, R. E. Insomnia and the attribution process. *Journal of Personality and Social Psychology,* 1970, **16,** 319–328.

Stover, R. D. *The nature of historical understanding.* Chapel Hill: University of North Carolina Press, 1967.

Sykes, G. M., & Matza, D. Techniques of neutralization: A theory of delinquency. *American Sociological Review,* 1957, **26,** 664–670.

Taylor, S. E. On inferring one's attitudes from one's behavior: Some limiting conditions. *Journal of Personality and Social Psychology,* 1975, **31,** 126–131.

Tune, G. S. Response preferences. A review of some relevant literature. *Psychological Bulletin,* 1964, **61,** 286–302.

Tversky, A. Intransitivity of preferences. *Psychological Review,* 1969, **76,** 31–48.

Tversky, A., & Kahneman, D. The belief in the "law of small numbers." *Psychological Bulletin,* 1971, **76,** 105–110.

Tversky, A., & Kahneman, D. Availability: A heuristic for judging frequency and probability. *Cognitive Psychology,* 1973, **5,** 207–232.

Tversky, A., & Kahneman, D. Judgment under uncertainty: Heuristics and biases. *Science,* 1974, **185** 1124–1131.

Vidmar, N., & Crinklaw, L. D. Attributing responsibility for an accident: A methodological and conceptual critique. *Canadian Journal of Behavior Science,* 1974, **6,** 112–130.

Vlek, C. A. J. Multiple probability learning. In A. F. Sander (Ed.), *Attention and performance* (Vol. III). Amsterdam: North–Holland, 1970.

Walsh, W. H. *Philosophy of history: An introduction* (rev. ed.). New York: Harper & Row, 1967.

Walster, E. Assignment of responsibility for an accident. *Journal of Personality and Social Psychology,* 1966, **3,** 73–79.

Walster, E. Second-guessing important events. *Human Relations,* 1967, **20,** 239–250.

Ward, W. C., & Jenkins, H. M. The display of information and the judgment of contingency. *Canadian Journal of Psychology,* 1965, **19,** 231–241.

Wason, P. C., & Johnson–Laird, P. N. *Psychology of reasoning: Structure and content.* London: Batsford, 1972.

Weiner, B. *Achievement motivation and attribution theory.* Morristown, New Jersey: General Learning Press, 1974.

Weiner, B., Freize, I., Kukla, A., Reed, L., Rest, S., & Rosenbaum, R. M. Perceiving the causes of success and failure. In E. E. Jones, D. E. Kanouse, H. H. Kelley, R. E. Nisbett, S. Valins, & B. Weiner (Eds.), *Attribution: Perceiving the causes of behavior.* Morristown, New Jersey: General Learning Press, 1972.

Weiner, B., & Kukla, A. An attributional analysis of achievement motivation. *Journal of Personality and Social Psychology,* 1970, **15,** 1–20.

Weiner, B., & Kun, A. The development of causal attributions and the growth of achievement and social motivation. In S. Feldman & D. Bush (Eds.), *Cognitive development and social development*, Hillsdale, New Jersey: Lawrence Erlbaum Associates, 1976.

Weiner, B., & Seirad, J. Misattribution for failure and enhancement of achievement strivings. *Journal of Personality and Social Psychology*, 1975, **31**, 415–421.

Wyer, R. S. *Cognitive organization and change: An information processing approach.* Potomac, Maryland: Lawrence Erlbaum Associates, 1974.

Wyer, R. S., & Goldberg, L. R. A probabilistic analysis of the relationship between beliefs and attitudes. *Psychological Review*, 1970, **77**, 100–120.

Author Index

Subject Index